NEW DEVELOPMENTS IN DIETARY FIBER
Physiological, Physicochemical, and Analytical Aspects

ADVANCES IN EXPERIMENTAL MEDICINE AND BIOLOGY

Recent Volumes in this Series

A Continuation Order Plan is available for this series. A continuation order will bring delivery of each new volume immediately upon publication. Volumes are billed only upon actual shipment. For further information please contact the publisher.

NEW DEVELOPMENTS IN DIETARY FIBER

Physiological, Physicochemical, and Analytical Aspects

Edited by

Ivan Furda

General Mills, Inc.
Minneapolis, Minnesota

and

Charles J. Brine

FMC Corporation
Princeton, New Jersey

PLENUM PRESS • NEW YORK AND LONDON

Library of Congress Cataloging in Publication Data

New developments in dietary fiber: physiological, physicochemical, and analytical aspects / edited by Ivan Furda and Charles J. Brine.
    p.   cm. — (Advances in experimental medicine and biology; v. 270)
    Proceedings of a symposium held Apr. 9–14, 1989 in Dallas, Tex. in conjunction with the 197th American Chemical Society National Meeting.
    Includes bibliographical references.

    1. Plant Fibers — Physiological effect — Congresses. 2. Fiber in human nutrition — Congresses. 3. Plant fibers — Analysis — Congresses. [1. Dietary Fiber — congresses.] I. Furda, Ivan, date. II. Brine, Charles J. III. American Chemical Society. Meeting (197th: 1989: Dallas, Tex.) IV. Series.
    [DNLM: W1 AD559 v. 270 / WB 427 N532 1989]
    RM237.6.N48   1990
    612.3'96 — dc20
    DNLM/DLC   90-7171
    for Library of Congress   CIP

ISBN-13:978-1-4684-5786-5   e-ISBN-13:978-1-4684-5784-1
DOI: 10.1007/978-1-4684-5784-1

Softcover reprint of the hardcover 1st edition 1990

Proceedings of the Agricultural and Food Chemistry
Division Symposium on Dietary Fiber-New Developments:
Physiological Effects and Physicochemical Properties,
held April 9–14, 1989, in Dallas, Texas, in conjunction with the
197th American Chemical Society National Meeting

© 1990 Plenum Press, New York
A Division of Plenum Publishing Corporation
233 Spring Street, New York, N.Y. 10013

Ivan Furda dedicates this book to
Jana, Peter, and Tom
Magdalena and Martin

Charles J. Brine dedicates this book to
Irene and Charles

PREFACE

It has been acknowledged that the physiological effects of dietary fiber are an exceedingly complex matter which requires a multidisciplinary research effort. The increased scientific involvement of the medical community, nutritionists, chemists and physicists is not only warranted but it has become mandatory. This is because we are entering a more advanced research phase in which the observed in vivo effects should not be only recorded, but they should be systematically correlated with the physicochemical and analytical properties of the individual dietary fibers.

The Division of Agricultural and Food Chemistry of the American Chemical Society has recognized this for some time, and has asked us to organize another International Symposium, similar to one in 1982, which would address the latest developments in this field. We decided to ask a cross section of leading experts from industrial and academic research institutions to assess the state of the art in dietary fiber, namely in the areas of the physiological effects, physicochemical attributes, and in existing and proposed analytical methods. We also felt that chemistry and physical chemistry should play a greater role in fiber research to complement and better explain the existing in vivo data. There is a large volume of animal and human physiological and nutritional data available. Unfortunately, the generated information is frequently confusing. One reason is that this research is not conducted with well characterized compounds, but rather with loosely defined complex mixtures or entities.

Thus, the need for better and more comprehensive characterization of the investigated fibers seems inevitable. This is particularly required to reduce the existing controversies and conflicting findings. We feel that an approach which will focus systematically on the physicochemical and analytical attributes of selected fibers will not only reduce the number of existing controversies but eventually it will enable us to correlate the molecular structure of individual fibers with observed in vivo efficacy. This could lead ultimately to the designing of new and improved fibers possessing man-tailored physiological effects.

These proceedings are organized in two major sections - physiological and physicochemical effects, and analysis. While most of the chapters focus on new scientific findings, some chapters deal with more provocative topics. These include critical evaluations and limitations of existing analytical methods, theories explaining hypolipidemic effects of specific fibers, or they describe unique undigestible oligosaccharides which seem to behave physiologically as dietary fiber. It is our hope that these proceedings will become a useful tool, not only for the broad spectrum of scientists, but also for students studying these and related disciplines. They may also attract other individuals and institutions who are interested in this relatively new and exciting field.

Ivan Furda
Minneapolis, Minnesota

Charles J. Brine
Princeton, New Jersey

ACKNOWLEDGEMENTS

The authors acknowledge the Division of Agricultural and Food Chemistry of the American Chemical Society for sponsoring this symposium at the National meeting in Dallas.

Financial contributions and support from General Mills, Inc., FMC Corporation, RJR Nabisco, Inc., Protein Technologies International, Farma International, Meiji Seika Kaisha, Ltd., The Quaker Oats Company, The Procter & Gamble Company, Kellogg Company, Ross Laboratories, American Crystal Sugar Company, General Foods Corp., D. D.Williamson & Co., Inc., and Kraft Inc., are sincerely acknowledged.

We appreciate the technical assistance of Mrs. Diane K. Ekstedt whose dedication to the production of this book is greatly appreciated.

CONTENTS

PHYSIOLOGICAL EFFECTS AND PHYSICOCHEMICAL PROPERTIES

ANALYSIS

DIETARY FIBER: A GLANCE INTO THE FUTURE

David Kritchevsky

The Wistar Institute of Anatomy and Biology
3601 Spruce Street
Philadelphia, PA 19104

The idea of physiological benefits deriving from dietary fiber is not new. Burkitt (1986) has reviewed the genesis of what we now call the dietary fiber hypothesis going back to Hippocrates and citing workers in the 19th and early 20th centuries who contributed to the progress in the field. However, until the middle 1950's most studies of or allusions to fiber were concerned primarily with its laxative properties. Surgeon-Captain T. L. Cleave (1956) pointed out the overall metabolic role(s) of diets low or high in complex carbohydrates. His work influenced the later observations of Burkitt and Trowell whose writings, together with those of A.R.P. Walker laid the groundwork for the modern phase of this developing field. The popular fiber era in the United States was ushered in by the paper by Burkitt, Walker and Painter (1974) which pointed out differences in health conditions between the United States and Africa and showed where fiber intake might impinge upon these differences. Whereas a number of investigators had been concerned with effects of dietary fiber before 1974, the Burkitt, Walker and Painter paper brought dietary fiber to the attention of lay and professional audiences.

This three day symposium is evidence of the current broad interest in fiber. I have been asked to peer into my clouded crystal ball and suggest directions for the future. Some of these will have already been touched upon in this state-of-the-art meeting.

The field of dietary fiber research has grown so rapidly that orderly progression has been by-passed and we must now go back to flesh out the outlines which research has provided.

Analytical work, which owes much to the pioneering efforts of Southgate (1969a,b) and van Soest and McQueen (1973) has progressed to the point where generalized methods are available (Prosky et al., 1984, 1985; Asp, 1986; Southgate 1986; Englyst and Cummings, 1986). There are still questions related to fibers in specific foods but these are being addressed. Many of the early problems were addressed in the multi-author book edited by James and Theander (1981).

New Developments in Dietary Fiber
Edited by I. Furda and C. J. Brine
Plenum Press, New York, 1990

One problem which requires resolution may be stated simply as: does fiber in a food matrix behave metabolically the same way it does in a pure form? More comparisons of pure and indigenous fiber are needed. This is particularly important since so much of our research base is predicated on the use of purified (pectin, guar gum) or partially identifiable (wheat bran) fibers whereas epidemiological data, on which much of the fiber hypothesis is based, comes from comparison of diets and dietary components of populations. A review of the physiological effects of fiber (Pilch, 1987) pointed out research needs for assessment of dietary fiber intake which included evaluation of equivalency of fiber in foods. This review also mentioned the need for more data on fiber intake in groups of different age, sex and ethnic origin.

It is now becoming apparent that starch which resists intestinal digestion (a simple acronym is needed desperately) may provide physiological effects which mimic or augment those of fiber (Jenkins et al., 1986) and thus better knowledge of fiber *in situ* is required.

The effects of fiber on digestive physiology and morphology have been studied but more accurate delineation would be helpful. Schneeman (1982) and Schneeman and Gallagher (1986) reviewed much of the data on effects of purified fibers on pancreatic enzyme activity in rats. Studies in other species are needed. Uniform methodology for study of transit time would be welcome (Wiggins, 1983). The jejunal villi of adults in Western societies are finger-like and regular whereas those of vegetarians or subjects in developing countries are broad and leaf shaped (Chacko et al., 1969; Cook et al., 1969; Owen and Brandburg, 1977). Cassidy (1981, 1986) and Tasman-Jones (1982) and their colleagues have demonstrated how fiber intake can affect the intestinal morphology of rats. Not all fibers give precisely similar results, would it be possible to find some structure-function relationship?

The products of fiber digestion are gases (CO_2, CH_4, O_2, H_2) and short chain fatty acids. Fleming and Calloway (1983a, b) have reviewed the data and methodologies but more data vis-a-vis specific fibers would be useful. The production and functions of short chain fatty acids (SCFA) has been reviewed by Cummings and Branch (1986). This is an important and rapidly expanding area of research. The principal SCFA are acetate, propionate and butyrate; which are absorbed from the colonic lumen (McNeil et al., 1978). These can all be cleared, to some extent, by the liver (Cummings and Branch, 1986). Acetate is metabolized in the peripheral tissues (Skutches et al., 1979). Propionate can be a significant precursor of glucose in remnants (Bergman et al., 1966) but little is known of its metabolism in other species. Chen et al. (1984) have reported that dietary propionate is hypocholesterolemic in rats. Butyrate is metabolized primarily by the epithelial cells of the colon (Roediger, 1982). Hellendoorn (1978) suggested that many of the observed gastrointestinal effects of dietary fiber might be due to the production and metabolic effects of SCFA. Hoverstad and Bjorneklett (1984) found that levels of SCFA in feces were correlated significantly with mean fecal weight and transit time. In addition to the major SCFA they also found isobutyrate, siovalerate, valerate and caproate. The function and fate of SCFA merits study, especially in humans.

The ecology of the colon has been studied by Savage (1983) and Salyers (1983, 1986) among others. The effects of fibers, starches and their metabolic products on the numbers and types of intestinal flora is difficult to study but may provide clues to differences in intraspecies effects of fiber.

Another aspect of dietary fiber which requires further attention is its effects on xenobiotics. Ershoff (1960, 1972) showed that various fibers could reduce the toxicity of surface active agents or cyclamates observed when those xenobiotics were added to a fiber-free diet. On the other hand are data which suggest that dietary fiber may alter the bacterial metabolism of xenobiotics and thus affect their toxicity (Wise et al., 1983; deBethizy and Goldstein, 1985).

The foregoing has been a brief overview of some areas of fiber research which bear further examination. No effort has been made to discuss effects of fiber on disease states which have been reviewed and discussed in a number of recent publications (Trowell, Burkitt and Heaton, 1985; Vahouny and Kritchevsky, 1986; Spiller, 1986; Dreher, 1987; Pilch, 1987; Kritchevsky, 1988).

Great advances in our knowledge of fiber, its chemistry, analysis and metabolic functions, have been made in the last decade. This has been due, in part, to talented investigators working together in an atmosphere of collegiality and cooperation. These qualities and attitudes bode well for the future advancement of this discipline.

ACKNOWLEDGEMENT

Supported, in part, by a Research Career Award (HL00734) from the National Institutes of Health and by funds from the Commonwealth of Pennsylvania.

REFERENCES

Asp, N. G., 1986, Enzymatic gravimetric methods, in: "Handbook of Dietary Fiber in Human Nutrition," G. A. Spiller, ed., CRC Press, Boca Raton.

Bergman, E. N., Roe, W. E., and Kon, K., 1966, Quantitative aspects of propionate metabolism and gluconeogenesis in sheep, Am. J. Physiol., 211:793.

Burkitt, D., 1986, Foreword, in: "Dietary Fiber: Basic and Clinical Aspects," G. V. Vahouny and D. Kritchevsky, eds., Plenum Press, New York.

Burkitt, D. P., Walker, A. R. P., and Painter, N. S., 1974, Dietary fiber and disease, J. Am. Med. Assoc., 229:1068.

Cassidy, M. M., Fitzpatrick, L. R., and Vahouny, G. V., 1986, The effect of fiber in the postweaning diet on nutritional and intestinal morphological indices in the rat, in: "Dietary Fiber: Basic and Clinical Apsects," G. V. Vahouny and D. Kritchevsky, eds., Plenum Press, New York.

Cassidy, M. M., Lightfoot, F. G., Grau, L., Story, J. A., Kritchevsky, D., and Vahouny, G. V., 1981, Effect of chronic intake of dietary fibers on the ultrastructural topography of the rat jejunum and colon, Am. J. Clin. Nutr., 34:218.

Chacko, C. J. G., Paulson, K. A., Mathon, V. I., and Bahu, S. J., 1969, The villus architecture of the small intestine in the tropics. A necropsy study, J. Pathol., 98:146.

Chen, W. L., Anderson, J. W., and Jennings, D., 1984, Propionate may mediate the hypocholesterolemic effects of certain soluble plant fibers in cholesterol-fed rats, Proc. Soc. Exp. Biol. Med., 175:215.

Cook, G. C., Kajcibi, S. K., and Lu, F. D., 1969, Jejunal morphology of the African in Uganda, J. Pathol., 98:157.

Cummings, J. H., and Branch, W. J., 1986, Fermentation and the production of short-chain fatty acids in the human large intestine, in: "Dietary Fiber: Basic and Clinical Aspects," G. V. Vahouny and D. Kritchevsky, eds., Plenum Press, New York.

deBethizy, J. D., and Goldstein, R. S., 1985, The influence of fermentable dietary fiber on the disposition and toxicity or xenobiotics, in: "Xenobiotic Metabolism: Nutritional Effects," J. W. Finley, and D. E. Schwass, eds., Am. Chem. Soc., Washington.

Dreher, M. L., 1987, "Handbook of Dietary Fiber. An Applied Approach," Marcel Dekker, New York.

Englyst, H. N., and Cummings, J. H., 1986, Measurement of dietary fiber as nonstarch polysaccharides, in: "Dietary Fiber: Basic and Clinical Aspects," G. V. Vahouny and D. Kritchevsky, eds., Plenum Press, New York.

Ershoff, B. H., 1960, Beneficial effects of alfalfa meal and other bulk-containing or bulk-forming materials on the toxicity of non-ionic surface active agents in the rat, J. Nutr., 70:484.

Ershoff, B. H., 1972, Comparative effects of a purified diet and stock ration on sodium cyclamate toxicity in rats, Proc. Soc. Exp. Biol. Med., 112:362.

Fleming, S. E., and Calloway, D. H., 1983a, Determination of intestinal gas excretion, in: "Dietary Fiber," G. G. Birch and K. J. Parker, eds., Applied Science Publishers, London.

Fleming, S. E., and Calloway, D. H., 1983b, Food and intestinal gas, Roy. Soc. New Zealand Bulletin, 20:157.

Hellendoorn, E. W., 1978, Fermentation as the principal cause of the physiological activity of indigestible food residue, in: "Topics in Dietary Fiber Research," G. A. Spiller, ed., Plenum Press, New York.

Hoverstad, T., and Bjornklett, A., 1984, Short chain fatty acids and bowel functions in man, Scand. J. Gastroenterol 19:1059.

James, W. P. T., and Theander, O., 1981, "The Analysis of Dietary Fiber in Food," Marcel Dekker, New York.

Jenkins, D. J. A., Jenkins, A. L., Wolever, T. M. S., Rao, A. V., and Thompson, L. V., 1986, Fiber and starchy foods: gut function and implications in disease, Am. J. Gastroenterol., 81:920.

Kritchevsky, D., 1988, Dietary fiber, Ann. Rev. Nutr., 8:301.

McNeil, N. I., Cummings, J. H., and James, W. D. T., 1978, Short chain fatty acid absorption by the human large intestine, Gut, 19:819.

Owen, R. L., and Brandborg, L. L., 1977, Jejunal morphologic consequences of vegetarian diet in humans, Gastroenterology, 72:A88.

Pilch, S. M. ed., 1987, "Review of Physiological Effects and Health Consequences of Dietary Fiber," FASEB, Bethesda.

Prosky, L., Asp, N. G., Furda, I., DeVries, J., Schweizer, T. F., and Harland, B., 1984, Determination of total dietary fiber in foods, food products and total diets: interlaboratory study, J. Assoc. Off. Agric. Chem., 67:1044.

Prosky, L., Asp, N. G., Furda, I., DeVries, J., Schweizer, T. F., and Harland, B., 1985, The determination of total dietary fiber in foods and food products: collaborative study, J. Assoc. Off. Agric. Chem., 68:677.

Roediger, W. E. W., 1982, Utilization of nutrients by isolated epithelial cells of the rat colon, Gastroenterology, 83:424.

Salyers, A. A., 1983, Enzymes involved in degradation of unabsorbed polysaccharides by bacteria of the large bowel, Roy. Soc. New Zealand Bulletin, 20:135.

Salyers, A. A., 1986, Diet and the colonic environment: measuring the response of human colonic bacteria to changes in the hosts' diet, *in*: "Dietary Fiber Basic and Clinical Aspects." G. V. Vahouny and D. Kritchevsky, eds., Plenum Press, New York.

Savage, D. C., 1983, Effects of food and fibre on the intestinal luminal environment, Roy. Soc. New Zealand Bulletin, 20:125.

Schneeman, B. O., 1982, Pancreatic and digestive function, *in*: "Dietary Fiber in Health and Disease," G. V. Vahouny and D. Kritchevsky, eds., Plenum Press, New York.

Schneeman, B. O., and Gallaher, D., 1986, Effects of dietary fiber on digestive enzymes, *in*: "Handbook of Dietary Fiber in Human Nutrition," G. A. Spiller, ed., CRC Press, Boca Raton.

Skutches, C. L., Holroyde, C. P., Meyers, R. N., Paul, P., and Reicherd, G. A., 1979, Plasma acetate turnover and oxidation. J. Clin. Invest., 64:708.

Southgate, D. A. T., 1969a, Determination of carbohydrates in foods. I. Available carbohydrate. J. Sci. Food. Agric., 20:326.

Southgate, D. A. T., 1969b, Determination of carbohydrates in foods. II. Unavailable carbohydrate. J. Sci. Food Agric., 20:331.

Southgate, D. A. T., 1986, The Southgate method of dietary fiber analysis in: "Handbook of Dietary Fiber in Human Nutrition," G. A. Spiller, ed., CRC Press, Boca Raton.

Spiller, G. A., ed., 1986, "Handbook of Dietary Fiber in Human Nutrition," CRC Press, Boca Raton.

Tasman-Jones, C. T., Owen, R. L., and Jones, A. L., 1982, Semipurified dietary fiber and small bowel morphology in rats, Dig. Dis. Sci., 27:519.

Trowell, H., Burkitt, D., and Heaton, K., eds., 1985, "Dietary Fibre, Fibre-depleted Foods and Disease," Academic Press, London.

Vahouny, G. V., and Kritchevsky, E., eds., 1986, "Dietary Fiber: Basic and Clinical Aspects, Plenum Press, New York.

Van Soest, P. J., and McQueen, R. W., 1973, The chemistry and estimation of fibre, Proc. Nutr. Soc., 32:123.

Wiggins, H. S., 1983, Gastroenterological functions of dietary fibre, *in*: "Dietary Fibre," G. G. Birch, and K. J. Parker, eds., Applied Science Publishers, London.

Wise, A., Rowland, I. R., and Mallett, A. K., 1983, The influence of dietary fibre on xenobiotic metabolism of gut bacteria, *in*: "Dietary Fibre," G. G. Birch and K. J. Parker, eds., Applied Science Publishers, London.

EFFECT OF SOLUBLE FIBERS ON PLASMA LIPIDS, GLUCOSE TOLERANCE AND MINERAL BALANCE

K.M. Behall

Carbohydrate Nutrition Laboratory
Beltsville Human Nutrition Research Center
Agricultural Research Service
U.S. Department of Agriculture
Beltsville, Maryland 20705

ABSTRACT

Fibers are broadly classified as soluble and insoluble based on their physical or analytical properties. Two human studies have been carried out in this laboratory utilizing soluble gums. The first compared a low fiber diet to the diet with an average of 19.5 g of added fiber per day from cellulose, an insoluble fiber, or carboxymethylcellulose gum, karaya gum or locust bean gum, all soluble fibers. Plasma cholesterol levels but not triglycerides were significantly lower when the soluble gums were consumed for 4 weeks each. Glucose and insulin response curves after a standard glucose tolerance test were not significantly different between the 5 diets. Adding refined fibers to the basal diet did not significantly affect apparent mineral balance of calcium, magnesium, manganese, iron, copper or zinc, with the exception of a a negative manganese balance after carboxymethylcellulose. The second study added an average of 31.7 g of guar gum per day to the diets of non-insulin dependent diabetic individuals for 6 months. Lipid levels observed at the beginning of the study were not reduced in either group, placebo or guar gum supplemented. Consumption of guar gum significantly reduced the C-peptide but not the glucose response curve. The number of insulin receptors increased while affinity remained the same. Apparent mineral balance was not affected by the consumption of guar gum for 6 months. The combined results of these studies indicates that soluble refined gums may have therapuetic value in reducing cholesterol and improving glucose metabolism without adversely affecting most mineral balances.

INTRODUCTION

High fiber diets have been suggested as beneficial for a number of conditions including diverticulitis, hyperlipidemia and non-insulin dependent diabetes (1-8). Research has shown that some fiber sources are hypocholesterolemic including oat bran (9), rolled oats (10), and refined gums including, locust bean gum (11), pectin (12,13) and guar (14). The use of these fiber sources has been suggested for lowering total blood cholesterol in individuals with hypercholesterolemia one of the risk factors of atherosclerotic disease (3,12). High fiber diets have been fed to patients with non-insulin dependent diabetes as a means of lowering fasting blood glucose, and improving the blood glucose control.

Two approaches have been used to study the effects of fiber: 1) replace low fiber foods with higher fiber foods (1,7,8,15-18) and 2) add refined fibers to the diet (4-6,19). Many foods thought of as 'high fiber foods', such as wheat bran, contain primarily fibers insoluble in water. These fibers are important in the bowel for laxation, however, addition of some insoluble fiber sources have been implicated in decreased mineral retention (1,20,21). Many of the highly refined or isolated fibers are soluble in water and form gels. It is the soluble fibers that have shown promise in the treatment of hypercholesterolemia or hyperglycemia since they appear to slow the absorption of nutrients from ingested foods. The refined fiber most extensively studied is guar gum (1,8,13-15,17) which has been suggested as a useful therapeutic agent either alone or when combined with a high carbohydrate intake.

The effect of soluble fibers on mineral balance has not been investigated to the same extent that the insoluble fibers have. A few studies have reported the effects of consuming refined fibers, primarily cellulose, hemicellulose and pectin, on mineral retention (1,20-25) and mineral bioavailability based on serum levels has been reported to a limited extent (26). The research approach in this laboratory has been one to associate structure of fibers with their metabolic effects on blood, fecal and urinary metabolites. We investigated the metabolic effects of consuming refined fibers on plasma glucose, insulin, lipids and apparent mineral balance when the fibers were consumed as part of a diet.

METHODS STUDY I

The first study investigated the effect of four refined fiber sources that contained a range of sugar subunits and side-chain structures. Twelve men completed the 20 week study. The study design was approved by the Human Studies Committees of the US Department of Agriculture, University of Maryland, and Georgetown University. Written informed consent was obtained from the subjects. A basal diet with a 4-day rotating menu containing a relatively low fiber content (6.22 g neutral detergent fiber (27) per 2550 kcal), was fed throughout the study. Four refined fibers, cellulose, carboxymethylcellulose gum (CMC), locust bean gum (LBG), and karaya gum (KG), were added to the basal diet in the form of baked muffins or fruit juice gel for 4 weeks each at the level of 7.5 g of refined fiber per 1000 kcal. Refined fiber intake ranged from 19.1 to 27.0 g per day depending on the calculated caloric intake (28) which ranged from 2550 to 3600 kcal/day. The five diets, basal and basal with each fiber added, were fed in a randomized rotation pattern with each diet proceeded and followed by every other diet. The basal diet provided approximately 14.5% of the calories from protein, 35.0% from fat, and 50.4% from carbohydrate (28). Zinc and magnesium were supplemented to meet the Recommended Dietary Allowance for men (29) based on a prestudy calculation of the diets to be fed. All meals were prepared and weighed in the Beltsville Human Nutrition Research Center's Diet Kitchen Facility. Weekday breakfast and dinner were consumed at the diet facility while lunch and weekend meals were packed by the staff for work or home consumption. Subjects were instructed to maintain their activity at a relatively constant level throughout the study and to record their major activities on a daily worksheet. Caloric intake was adjusted by increasing or decreasing all foods when the subjects had gained or lost 2 Kg to try to maintain their initial weight throughout the study. Additional information on dietary protocol has been previously published (30-32).

Fasting blood was drawn and blood pressure was recorded weekly. A glucose tolerance test (1 g glucose/kg body weight) was given at the end of each period with blood samples drawn before and 1/2, 1, 2 and 3 hours after the glucose load. Duplicate food samples were collected the last 8 days of each period. All urine and fecal samples were collected from each subject to match the 8 day food composites. Fecal collections were marked on the first and last day of collection with 50 mg of Brilliant Blue dye. Serum glucose, cholesterol and triglycerides were determined enzymatically with prepared kits for the Centrifichem automated system (30,31). Plasma lipoprotein cholesterol fractions were determined by the Lipid Research Clinic at Johns Hopkins Hospital, Baltimore, MD. Insulin was determined using radioimmunoassay reagent kits from Amersham Co., Arlington Heights, IL. (31). Food and fecal composites were treated by a dry ash and then a wet ash technique before mineral content was read by atomic absorption spectrophotometry with the appropriate hollow cathode lamp (32). Data was analyzed statistically using SAS general linear model procedure (33). Least square means were used because some periods were unbalanced.

RESULTS STUDY I

Serum cholesterol and triglyceride levels determined prior to the controlled diets averaged 221 and 127 mg/dl, respectively (30). Mean cholesterol was 196 + 4 mg/dl after all subjects consumed the basal diet and 204 + 4 mg/dl after the diet with cellulose added. The mean cholesterol levels after feeding KG, LBG and CMC gums were significantly lower than after the basal or cellulose diets, averaging 177 + 4, 169 + 4 and 164 + 4 mg/dl, respectively. Serum cholesterol was 10 to 16% lower after the gums than after the basal diet. Changes in low density lipoprotein cholesterol (LDL) levels followed the same pattern as was seen with total cholesterol. LDL levels after the cellulose diet were significantly higher (142 mg/dl) and the levels after CMC (107 mg/dl), KG (118 mg/dl) and LBG (109 mg/dl) were significantly lower than those observed after the basal diet (131 mg/dl). No significant difference between the diets was observed in the very low density lipoprotein (VLDL) or high density lipoprotein (HDL) cholesterol levels. The HDL to LDL + VLDL cholesterol ratio was higher when the gums were fed (0.32 or more) compared to either the basal diet or cellulose (both 0.28). The change in the ratio reflects the change in LDL cholesterol observed. Average serum triglyceride levels ranged from 110 + 8 mg/dl after the diet with CMC to 136 + 8 mg/dl after cellulose. The differences seen between the diets were not statistically significant. Fasting glucose and insulin levels tended to be lower after 4 weeks on the added fibers than with the basal diet, however; these findings were not statistically significant. Mean serum glucose and insulin response curves to the oral glucose tolerance test were not significantly different between the diets. The order that the fibers were fed also had no effect on the response curves observed at the end of each period.

Variation in mineral intake, fecal and urinary excretion occurred between the individual subjects and the dietary fibers (32). Mean intake was significantly different between the diets for all minerals tested. Karaya gum contained more minerals than any of the other fibers and significantly increased the intake of most minerals during the period it was consumed. Apparent retention of calcium, magnesium, manganese, iron, copper and zinc were calculated from the mineral determinations in the food composites duplicating subject intake, and urinary and fecal excretions of each subject (Table 1). Compared to apparent mineral balance after the basal diet, adding cellulose, carboxymethylcellulose, karaya gum or locust bean gum did not significantly alter the apparent mineral retention of calcium, magnesium, manganese, iron, copper or zinc. A calculated zero balance range was determined for each mineral with the study variance obtained. Based on this range, mean magnesium balance after the basal diet and zinc retention after KG and LBG were

Table 1.

Apparent balance of men after a basal diet and four refined fibers (mean ± SEM*).

Diet	Calcium mmol/d	Magnesium mmol/d	Manganese mmol/d	Iron umol/d	Copper umol/d	Zinc umol/d
Basal	0.8 ± 1.3	4.0 ± 1.3	-7.9 ± 12.7	45.1 ± 17.7	0.6 ± 1.6	4.3 ± 48.6
Cellulose	-1.2 ± 1.1	0.1 ± 1.3	2.1 ± 6.1	-0.9 ± 20.9	0.5 ± 0.7	35.8 ± 50.2
Carboxymethyl-cellulose	-1.6 ± 1.5	0.1 ± 1.7	-17.0 ± 11.5	-21.6 ± 49.2	-0.4 ± 1.4	25.6 ± 32.9
Karaya gum	0.7 ± 2.0	2.6 ± 2.3	0.7 ± 10.5	27.7 ± 37.4	1.0 ± 1.1	47.0 ± 31.7
Locust bean gum	-1.5 ± 1.1	1.5 ± 1.4	-4.8 ± 5.6	22.2 ± 21.0	-0.8 ± 1.4	39.7 ± 24.5
Level of significance	NS	NS	NS	NS	NS	NS

* Mean ± SEM for the 8-day collection period at the end of each feeding period. Pooled SEM from the ANOVA for calcium was 39.0; magnesium, 28.3; manganese, 0.38; iron, 1.32; copper, 0.49; and zinc, .04.
+ Not significant at p < 0.05 by ANOVA.

significantly greater than the zero balance range indicating a positive retention. Mean manganese retention after the CMC diet was below the zero balance range indicating manganese loss. All other mean levels were within the calculated zero balance range.

METHODS STUDY II

The second study investigated the addition of guar gum to the self selected diets of non-insulin dependent diabetic subjects for 6 months. Seven men and nine women completed the 6 month study. The study design was approved by the human studies committees of the Johns Hopkins Hospital and Sinai Hospitals in Baltimore, Maryland. Written informed consent was obtained from the subjects before the study. The subjects consumed either a high carbohydrate-guar gum food bar or the same bar without the guar gum added (34). Baseline studies in a metabolic ward at Johns Hopkins Hospital were started after the volunteers had been stabilized with a fasting blood glucose of 160 mg/dl for at least 1 month. Subjects recorded their self selected diets daily during the month preceding the baseline studies and during the 6 months the food bar was consumed. Subjects were seen weekly by the staff to monitor blood glucose, check food records and food bar consumption. While the subjects were in the metabolic ward, they were fed a weighed constant diet based on their self selected diet records. After 3 days on the constant diet and an overnight fast subjects were given a 4 hour glucose tolerance test using 75 g of glucose with blood drawn before and 1/2, 1, 1 1/2, 2, 3 and 4 hours after the glucose load. Duplicate food samples were collected to match the 3 days of total urine and demarcated fecal collection. Each subject served as his or her own control to compare the level of response prestudy and after consumption of the food bar with or without guar gum.

Total serum cholesterol and triglycerides were analyzed by automated serum analyzers (SMAC, Technicon, Terrytown, NJ) (35). Total HDL cholesterol and HDL_3 were determined by precipitation methods (35). Apoprotein-B was determined by radioimmunodiffusion (35). Serum glucose was determined by an enzymatic method (Beckman Instruments) and plasma C-peptides were determined by a double antibody radioimmunoassay method for synthetic human C-peptides (36). Twelve of the subjects underwent a euglycemic insulin clamp procedure using mode 7 of the biostator artificial beta cell (Life Sciences Instruments, Miles Laboratories, Elkhart, IN). Insulin binding to erythrocytes was measured in the subjects who participated in the euglycemic insulin clamp procedure (36). Mineral intake from the food, fecal and urinary mineral excretion, and apparent mineral balance were collected and calculated (37). Results were analyzed by analysis of variance using SAS general linear model procedure (33).

RESULTS STUDY II

Prestudy the subjects consumed an average of 37% carbohydrate, 21% protein and 41% fat (34). After the addition of the food bars to the diet the guar gum group averaged 53% carbohydrate, 16% protein and 31% fat while the placebo group averaged 44% carbohydrate, 16% protein and 40% fat. Only the increase in carbohydrate was significant. Body weight and total dietary energy intake did not change significantly for either group.

Both the placebo and guar gum groups exhibited a rise (p < 0.025) in serum triglycerides over the 6 month study (35). Most of the rise was due to one individual in each group. Without the one individual in each group the changes were not significant. Total cholesterol, total HDL, HDL_3 and apoprotein-B showed no significant change in either the guar gum or placebo groups from prestudy to the 6 month collection. The area under

11

the glucose tolerance curve decreased 10.5% in the guar gum group and 4.3% in the placebo group from the prestudy to poststudy tolerance (36). The change was not significant in either group. The area under the C-peptide response curves decreased significantly in the guar gum group (45% decrease, p < 0.01) while the response of the placebo group did not significantly change (3% increase). Insulin sensitivity as measured by metabolic clearance rate for glucose increased significantly (p < 0.05) in the guar gum group after the gum was consumed for 6 months, averaging a 73% increase in metabolic clearance rate with an infusion rate of 40 mU insulin/min/m^2. The 12.5% increase observed in the placebo group was not significant. When a glucose infusion rate of 400 mU insulin/min/m^2 was used neither group show a significant change from prestudy clearance rates to the end of the study. The affinity of the erythrocyte receptors for insulin did not change significantly in either group from prestudy to poststudy. Both groups did show a significant increase in receptor number. The guar gum group increased from 84 to 219 picomoles insulin bound/1/10^9 cells and the placebo group from 62 to 101 units. The number of receptors at the end of the study was significantly greater (p < 0.005) in the guar gum group than that in the placebo group.

Calcium, magnesium, iron, copper and zinc intake did not significantly differ between the 2 groups or from the beginning to end of the study (37) (Table 2). Manganese fecal excretion was higher when either bar was supplemented to the diet and was significantly higher in the placebo group. Other variations in mineral excretion in the feces were not significant. The levels of calcium, magnesium and iron in the urine were greater prestudy than poststudy, zinc levels were greater poststudy and copper levels were not different between the two periods. Apparent balance for the 5 minerals were not significantly different between the 2 groups of subjects or prestudy to poststudy within a group.

DISCUSSION

Total cholesterol and LDL cholesterol have generally been reported to decrease in normal subjects (1,13,15,20,38,39), hyperlipidemic subjects (1,3,11,20,40-44) and subjects with diabetes mellitus (1,15-20,26,39,44,45) fed soluble fibers. The results from our study utilizing CMC, LBG and karaya gum are in agreement with the literature. Our longer study feeding guar gum for 6 months did not show a significant reduction in cholesterol or LDL cholesterol. A few other studies longer than 3 months also have reported no reduction in cholesterol (26,40,44). One study (40) reported the maximum reduction of cholesterol levels after 2 months of feeding guar gum with values returning toward prestudy levels after that. The HDL cholesterol levels of our subjects did not significantly change which is in agreement with most other soluble fiber studies (1,13,15,41,42,47). However, Jenkins et al (41) did show a decrease in HDL cholesterol.

Consumption of the soluble gums generally has not affected triglyceride levels (1,13,15,41,42,47). This observation was confirmed with the refined fibers study. However, an increase in triglycerides was observed in the second study when the guar gum and placebo bars were consumed. Gatti et al (44) also reported an increase in triglycerides after guar gum was consumed. It was assumed that some of the diabetic subjects were affected by the increase in carbohydrate in their diet (37% and 38% of the calories prestudy rising to 53% and 45% poststudy in the guar gum and placebo groups respectively) since both groups had at least one individual with markedly increased triglyceride levels. Coulston et al (47) reported significant hypertriglyceridemia when their subjects consumed a 60% carbohydrate diet for 6 weeks compared to levels after a 40% carbohydrate diet.

Consumption of the isolated fibers in the diet did not significantly

Table 2.

Apparent balance after guar gum or placebo bars for six months (mean ± SEM*).

Diet	Calcium mmol/d	Magnesium mmol/d	Iron umol/d	Copper umol/d	Zinc umol/d
Prestudy Placebo	7.0 ± 8.4	0.1 ± 1.8	84 ± 57	3.5 ± 3.5	108.6 ± 68.9
Prestudy Guar Gum	1.3 ± 2.2	1.4 ± 1.2	116 ± 59	2.4 ± 3.6	70.4 ± 41.3
Poststudy Placebo	14.8 ± 5.6	1.9 ± 2.7	139 ± 88	3.0 ± 6.6	70.4 ± 49.0
Poststudy Guar Gum	12.3 ± 5.5	0.4 ± 1.0	174 ± 56	7.3 ± 3.5	101.0 ± 30.3
Analysis of variance					
Sex	p < 0.02	NS	NS	NS	NS
Group	NS	NS	NS	NS	NS
Diet	NS	NS	NS	NS	NS
Sex x diet	NS	NS	NS	NS	NS
Diet x group	NS	NS	NS	NS	NS

* Mean ± SEM for the 8-day collection period at the end of each feeding period.
Pooled SEM from the ANOVA for calcium was 39.0; magnesium, 28.3; manganese, 0.38; iron,
1.32; copper, 0.49; and zinc, 1.04.
+ Not significant at p < 0.05 by ANOVA.

affect fasting or oral glucose tolerance response levels of glucose or insulin in the normal subjects on our study. These results agree with published studies (1,15) in which soluble gums were fed to normal subjects. Diabetic subjects, however, are reported by most studies (1,7,8,15-18,45) to show an improved glycemic response curve when soluble gums are consumed, especially when the gums are consumed with the oral glucose tolerance. Our subjects fed guar gum did not show significant decreases in blood glucose or change in maximal responsiveness to insulin but metabolic clearance rate of glucose at lower levels of insulin and the number of red blood cell insulin receptors increased significantly. Olefsky and Kolterman (51) previously suggested similar changes in carbohydrate metabolism with changes in the number of insulin receptors. Fewer reports are available on the effect of soluble gums on mineral balance than have been published on blood lipids, glucose or insulin. Our results and those available in the literature indicate that soluble gums have less of an effect on mineral balance than do the insoluble fibers. Of the soluble fibers tested, CMC appeared to have the greatest potential for causing a negative apparent mineral balance when fed for longer than the 4 weeks used in the study. When pectin has been added to controlled diets of normal subjects, no significant effect on zinc (48-52), iron (49,50) or copper (48-52), apparent balance has been reported with subjects consuming 15 g of pectin/day for 3 weeks. When ileostomy patients were fed 15 g pectin for 4 days, iron excretion but not zinc or copper excretion was increased over a 7-day control period (52). The effect on iron may have been due to the short duration of the study or to the medical or physical condition of the subjects studied. No change in calcium (24,25,52) or magnesium (24,47,52) apparent balance has been reported with 15 g of pectin fed for 4 days, or 5 weeks or with 36 g pectin fed for 3 weeks. From the results of our studies and those in the literature, it would appear that dietary supplementation of soluble fibers may be an acceptable means of treatment for individuals with hypercholesterolemia or diabetes for extended periods without adversely affecting mineral balance.

Bibliography

1. Pitch SM. ed. Physiological effects and health consequences of dietary fiber. Bethesda, MD:Life Sciences Research Office, Federation of American Societies for Experimental Biology, 1987.
2. Kannel WB, Castelli WP, Gordon T. Cholesterol in the prediction of atherosclerotic disease: new perspectives based on the Framingham study. Ann Intern Med 1979;90:85-91.
3. Vahouny GV. Dietary fiber, lipid metabolism, and atherosclerosis. Fed Proc 1982;41:2801-6.
4. Anderson JW, Gustafson NJ. Dietary fiber and heart disease: Current management concepts and recommendations. Topics in Clin Nutr 1988;3:21-9.
5. Miranda PM, Horwitz DL. High-fiber diets in the treatment of diabetes mellitus. Ann Intern Med 1978;88:482-6.
6. Simpson HCR. High-carbohydrate, high-fiber diets for diabetes. Proc Nutr Soc 1981;40:219-25.
7. Jenkins JDA, Leeds AR, Gussull MA, Cochet B, Alberti KGMM. Decrease in postprandial insulin and glucose concentrations by guar and pectin. Ann Intern Med 1977;86:20-3.
8. Jenkins DJA, Wolever TMS, Bacon S, Nineham R, Lees R, Rowden R, Love M, Hockaday TDR. Diabetic diets: high carbohydrate combined with high fiber. Am J Clin Nutr 1980;33:1729-33.
9. Kirby RW, Anderson JW, Sieling B. Oat-bran intake selectively lowers serum low-density lipoprotein cholesterol concentrations of hypercholesterolemic men. Am J Clin Nutr 1981;34:824-9.
10. Judd PA, Truswell AS. The effect of rolled oats on blood lipids and fecal steroid excretion in man. Am J Clin Nutr 1981;34:2061-7.

11. Zavoral JH, Hannan P, Fields D, Hanson MN, Frantz ID, Kuba K, Elmer P, Jacobs Jr. DR. The hypolipidemic effect of locust bean gum food products in familial hypercholesterolemic adults and children. Am J Clin Nutr 1983;38:285-94.

12. Anderson JW, Chen WL. Plant fiber. Carbohydrate and lipid metabolism. Am J Clin Nutr 1979;32:346-63.

13. Jenkins DJA, Leeds AR, Newton C, Cummings JH. Effect of pectin, guar gum, and wheat fiber on serum cholesterol. Lancet 1975;1:1116-7.

14. Jenkins DJA, Reynolds D, Slavin B, Leeds AR, Jenkins AL, Jepson EH. Dietary fiber and blood lipids: treatment of hypercholesterolemia with guar crispbread. Am J Clin Nutr 1980;33:575-81.

15. Smith U, Holm G. Effect of a modified guar gum preparation on glucose and lipid levels in diabetics and healthy volunteers. Atherosclerosis 1982;45:1-10.

16. Ray TK, Mansell KM, Knight LC, Malmud LS, Owen OE, Boden G. Long-term effects of dietary fiber on glucose tolerance and gastric emptying in noninsulin-dependent diabetic patients. Am J Clin Nutr 1983;37:376-81.

17. Najemnik C, Kritz H, Irsigler K, Laube H, Knick B, Klimm HD, Wahl P, Vollmar J, Brauning C. Guar and its effects on metabolic control in type II diabetic subjects. Diabetes Care 1984;7:215-20.

18. Osilesi O, Trout DL, Glover EE, Harper SM, Koh ET, Behall KM, O'Dorisio TM, Tartt J. Use of xanthan gum in dietary management of diabetes mellitus. Am J Clin Nutr 1085;42:597-603.

19. Anderson JW. Effect of carbohydrate restriction and high carbohydrate diets on men with chemical diabetes. Am J Clin Nutr 1977;30:402-8.

20. Kelsay JL. A review of research on effects of fiber intake on man. Am J Clin Nutr 1978;31:142-59.

21. Ink SL. Fiber-mineral and fiber-vitamin interactions. In:Bodwell CE, Erdman JW, eds. Nutrient interactions. New York, NY: Marcel Deckker, Inc, 1988;253-64.

22. Kelsay JL. Effect of diet fiber level on bowel function and trace mineral balances of human subjects. Cereal Chem 1981;58:2-5.

23. Kelsay JL. Effects of fiber on mineral and vitamin bioavailability. In: Vahouny GV, Kritchevsky D, eds. Dietary fiber in health and disease. New York, NY: Plenum Pub Co, 1982;91-103.

24. Cummings JH, Southgate DAT, Branch WJ, Wiggins HS. The digestion of pectin in the human gut and its effect on calcium absorption and large bowel function. Brit J Nutr 1979;41:477-85.

25. Stasse-Wolthuis M, Albers AFF, van Jeveren JGC, Wil de Jong J, Hautvast JCAJ, Hermus RJJ, Katan MB, Brydon WG, Eastwood MA. Influence of dietary fiber from vegetables and fruits, bran or citrus pectin on serum lipids, fecal lipids, and colonic functions. Am J Clin Nutr 1980;33:1745-56.

26. Jenkins DJA, Wolever TMS, Taylor RH, Reynolds D, nineham R, Hockaday TDK. Diabetic glucose control, lipids and trace elements on long term guar. Brit Med J 1980;280:1353-54.

27. American Association of Cereal Chemists. Approved methods of the AACC. Method 32-30. St Paul, MN: The Association. 1978.

28. Adams D. Nutritive values of American foods in common units. Agriculture Handbook No 456, Washington, DC: Agricultural Research Service, US department of Agriculture, US Government Printing Office, 1975.

29. Recommended Dietary Allowances, Committee on dietary Allowances, Food and Nutrition Board, Commission of Life Sciences, National Research Council. 9th ed. Washington, DC: National Academy Press, 1980.

30. Behall KM, Lee KH, Moser PB. Blood lipids and lipoproteins in adult men fed four refined fibers. Am J Clin Nutr 1984;39:209-14.

31. Behall KM, Scholfield DJ, Lee KH, Moser PB. Blood glucose and hormone levels in adult males fed four refined fibers. Nutr Repts Intern. 1984;30:537-43.

32. Behall KM, Scholfield DJ, Lee K, Powell A, Moser PB. Mineral balance in adult men: effect of four refined fibers. Am J Clin Nutr 1987;46:307-14.

33. SAS Institute Inc. SAS user's guide: statistics version. 5th ed. Cary, NC: SAS Institute Inc, 1985.

34. Van Duyn MAS, Leo TA, McIvor ME, Behall KM, Michnowski JE, Mendeloff AI. Nutritional risk of high-carbohydrate, guar gum dietary supplementation in non-insulin-dependent diabetes mellitus. Diabetes Care 1986;9:497-503.

35. McIvor ME, Margolis S, Behall KM, Michnowski JE, Mendeloff AI. Long-term effects of guar gum on blood lipids. Atherosclerosis 1986;60:7-13.

36. McIvor ME, Cummings CC, Saudek C, Fontan N, Levin P, Mendeloff AI. High carbohydrate high fiber diets increase insulin sensitivity in patients with type II diabetes mellitus. Clin Res 1984;32:403A.

37. Behall KM, Scholfield DJ, McIvor ME, Van Duyn MAS, Leo TA, Michnowski JE, Cummings CC, Mendeloff AI. Effect of guar gum on mineral balances in NIDDM adults. Diabetes Care 1989;12:357-64.

38. Jenkins DJA, Reynolds D, Leeds AR, Waller AL, Cummings JH. Hypocholesterolemic action of dietary fiber unrelated to fecal bulking effect. Am J Clin Nutr 1979;32:2430-5.

39. Khan AR, Khan GY, Mitchel A, Qadeer MA. Effect of guar gum on blood lipids. Am J Clin Nutr 1981;34:2446-9.

40. Simons LA, Gayst S, Balasubramaniam S, Ruys J. Long-term treatment of hypercholesterolemia with a new palatable formulation of guar gum. Atherosclerosis 1982;45:101-8.

41. Jenkins DJA, Reynolds D, Slavin B, Leeds AR, Jenkins AL, Jepson EM. Dietary fiber and blood lipids - Treatment of hypercholesterolemia with guar crispbread. Am J Clin Nutr 1980;33:575-81.

42. Aro A, Uusitupa M, Voutilainen E, Korhonen T. Effects of guar gum in male subjects with hypercholesterolemia. Am J Clin Nutr. 1984;39:911-6.

43. Tuomilehto J, Karttunen P, Vinni S, Kostiainen E, Uusitupa M. A double-blind evaluation of guar gum in patients with dyslipidaemia. Human Nutr Clin Nutr 1983;37C:109-13.

44. Gatti E, Catenazzo G, Camisasca E, Torri A, Denegri E, Sirtori CR. Effects of guar-enriched pasta in the treatment of diabetes and hyperlipidemia. Ann Nutr Metab 1984; 28:1-10.

45. Hagander B, Asp NG, Efendic'S, Nilsson-Ehle P, Schersten B. Dietary fiber decreases fasting blood glucose levels and plasma LDL concentrations in noninsulin-dependent diabetes mellitus patients. Am J Clin Nutr 1988;47:852-8.

46. Anderson JW, Ward K. Long term effects of high-carbohydrate high-fiber diets on glucose and lipid metabolism - A preliminary report on patients with diabetes. Diabetes Care 1978;1:77-82.

47. Coulston AM, Hollenbeck CB, Swislocki ALM, Reaven GM. Persistence of hypertriglyceridemic effects of low-fat high-carbohydrate diets in NIDDM patients. Diabetes Care 1989;12:94-101.

48. Drews LM, Kies C, Fox HM. Effect of dietary fiber on copper zinc and magnesium utilization by adolescent boys. Am J Clin Nutr 1979;32:1893-97.

49. Lei KY, Davis MW, Fang MM, Young MM. Effect of pectin on zinc, copper and iron balances in humans. Nutr Rep Int 1980;22:459-66.

50. Jenkins DJA, Wolever TMS, Taylor RH, Reynolds D, Nineham R, Hockaday TDR. Diabetic glucose control, lipids, and trace elements on long term guar. Brit Med J 1980;280:1353-4.

51. Olefsky JM, Kolterman OG. Mechanisms of insulin resistance in obesity and noninsulin-dependent (Type II) diabetes. Am J Med 1981;70:151-68.

52. Sandberg AS, Alderinne R, Anderson H, Hallgren B, Hatten L. Effect of citrus pectin on the absorption of nutrients in the small intestine. Hum Nutr Clin Nutr 1983;37:171-83.

HYPOCHOLESTEROLEMIC EFFECTS OF OAT PRODUCTS

James W. Anderson

Metabolic Research Group
Veterans Administration Medical Center
University of Kentucky College of Medicine
Lexington, Kentucky

Amy E. Siesel

Metabolic Research Group

The cholesterol-lowering effects of oat products were recognized over a quarter of a century ago[1]. With the possible exception of beans, oat bran lowers serum cholesterol in humans more than any other food[2]. In metabolic ward studies, incorporating either 50 or 100 g of oat bran daily into a typical American diet lowered serum cholesterol of hypercholesterolemic men by 13-19%[2,3]. In ambulatory studies of healthy individuals eating their usual diet, providing 50 g of oat bran daily in the form of muffins decreased serum cholesterol by 12-26%[4,5]. Since every 1% decrease in serum cholesterol reduces estimated risk for coronary heart disease by 2%[6], these oat bran-induced changes in serum cholesterol would theoretically reduce risk for coronary heart disease by approximately 10-35%.

Oat products are rich in a gum, beta-glucan, that appears to be the major cholesterol-lowering component of the oat groat[7]. The mechanisms responsible for these hypocholesterolemic effects are still under intense study. This review will compare the cholesterol-lowering effects of different dietary fibers in rats, summarize previously published reports on oat product effects in humans, briefly present an unpublished study from our group, and review proposed mechanisms for the hypocholesterolemic effects of oat products. Finally the practical implications of oat product use for reducing risk for coronary heart disease will be discussed.

New Developments in Dietary Fiber
Edited by I. Furda and C. J. Brine
Plenum Press, New York, 1990

I. RAT STUDIES: CHOLESTEROL-LOWERING EFFECTS OF DIFFERENT FIBERS

To compare the cholesterol-lowering effects of different fiber sources, we developed a rat model which qualitatively predicts hypocholesterolemic effects of various fibers in humans. As previously described[7-10], when rats are fed the diet outlined in Table 1 which includes 1% cholesterol, 0.1% cholic acid, nutrients and test fibers, the effects of these fibers on serum and liver cholesterol can be assessed.

Table 1. Composition of Diet used in Rat Studies *

	%
Total Carbohydrate	66.5
Sucrose	44.0
Starch	22.5
Total Protein	15.3
Casein or protein in test material	15.0
DL-Methionine	0.3
Total Fat	6.0
Cottonseed Oil or fat in test material	6.0
Total Dietary Fiber	6.0
Cellulose or fiber in test material	6.0
Salt Mix	4.0
Vitamin Mix	1.0
Cholesterol	1.0
Cholic Acid	0.2

* Diet composition is adjusted according to proximate analysis of fiber containing material being tested.

Over the past ten years, our group has tested fiber materials in 18 groups of rats by the above described protocol. In each group of rats, cellulose was used as the control.

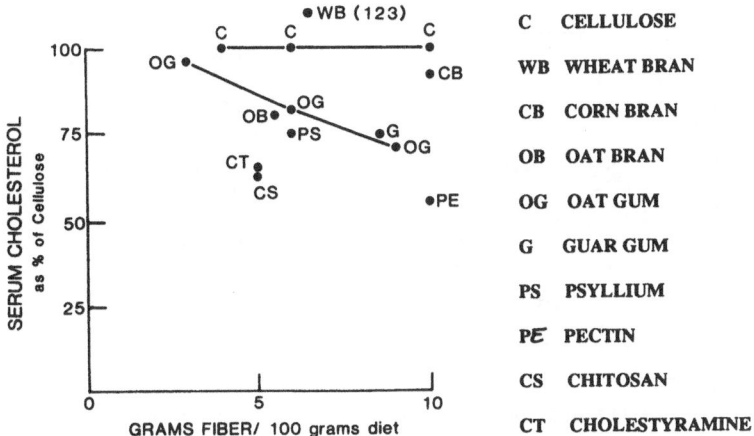

Figure 1. Rat serum cholesterol response to different fibers

Figure 1 summarizes the serum total cholesterol response of rats to the test fibers. In 140 rats fed the cellulose control diet, the serum total cholesterol levels have been remarkably consistent, with a mean value (± SD) of 154 (± 28) mg/dl. The insoluble fibers tested included wheat bran, which increased serum cholesterol slightly[8]; and corn bran, which did not significantly alter serum cholesterol (Anderson, unpublished data).

Several soluble fiber sources also were tested. Oat bran had significant hypocholesterolemic effects[7,9]. Isolated oat gum from oat bran also significantly lowered serum cholesterol[7,10]. Oat gum exhibited a dose response in which higher levels of oat gum in the diet correlated with lower serum cholesterol levels (Anderson, unpublished data). Guar gum[8,9], psyllium (Anderson, unpublished data), and pectin[7,9] had similar significant hypocholesterolemic effects to oat gum. Based on g of fiber/100 g diet, oat gum had greater hypocholesterolemic effects than all other fibers. Chitosan, an aminopolysaccharide, and cholestyramine, a bile acid binding resin, also had significant hypocholesterolemic effects[10].

Figure 2 summarizes the liver cholesterol changes in rats produced by the various fibers. Liver cholesterol averaged a mean (± SD) of 42.6 (± 7) mg/g for the 140 rats fed the cellulose control diet. Wheat bran[8] and corn bran (Anderson, unpublished data) did not significantly alter liver cholesterol. Oat gum significantly reduced liver cholesterol[7,10], and at 3-9% of diet produced a dose-related decrease in

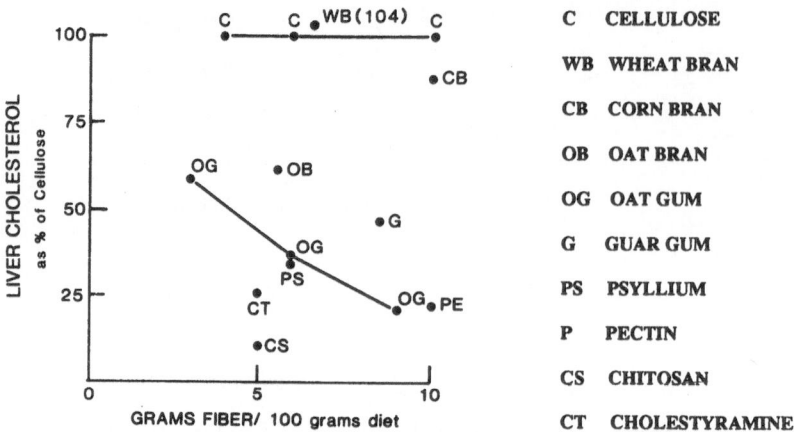

Figure 2.   Rat liver cholesterol response to different fibers

liver cholesterol (Anderson, unpublished data). Oat bran[7,9] and guar
gum[8,9] lowered liver cholesterol less effectively than oat gum, while
pectin[7,9] and psyllium (Anderson, unpublished data) had effects similar
to oat gum. Oat gum lowered liver cholesterol more than all other fibers
tested. Chitosan and cholestyramine had more potent cholesterol-lowering
effects than the soluble fibers studied[10]. The hypocholesterolemic effects
of the materials tested were consistently greater for liver than for serum
in the cholesterol-fed rat animal model.

II. HUMAN STUDIES: CHOLESTEROL-LOWERING EFFECTS OF OAT PRODUCTS

The effects of oat products in humans have been studied extensively in
the past decade. Oatmeal and oat bran are rich in water-soluble fiber.
Approximately one-half of the fiber in the oat groat is soluble B-glucan, or
oat gum[7]. This material has been shown to lower serum cholesterol in animal
studies[7,10] and is thought to be the fraction responsible for the
hypocholesterolemic effects of oats.

In 1963, de Groot and colleagues[1] reported that daily intake of 140 g of
rolled oats in humans reduced serum cholesterol by 12%. Morerecent
metabolic ward and ambulatory studies indicate that without alterations in
fat and cholesterol intake, regular daily use of 50-100 g oat bran lowers
serum cholesterol by 12-19%[2-5]. Additional studies have confirmed and
extended these results[11-16,18].

Our group incorporated 100 g oat bran daily for three weeks into a high-carbohydrate (70% of energy), low-fat (12% of energy) diet for diabetic individuals. Serum total cholesterol and low-density lipoprotein (LDL) cholesterol decreased by 38% and 58% respectively, while high-density (HDL) cholesterol increased 86%[11].

Oat bran intake was tested further in an intensive metabolic ward study in which hypercholesterolemic men ate a typical American diet supplemented with 100 g oat bran daily for 10 days. After 10 days of oat bran supplementation, total serum cholesterol decreased 13% and LDL-cholesterol decreased 14%. HDL-cholesterol did not change significantly[3].

To test the effect of oat bran ingestion for longer periods of time, hypercholesterolemic men received 100 g oat bran daily for 21 days on a metabolic ward. The oat bran diet significantly lowered serum total cholesterol by 19% and LDL-cholesterol by 23%. Again, HDL-cholesterol did not change significantly[2].

In a long-term ambulatory study, hypercholesterolemic men incorporated 50 g oat bran daily into a high-fiber, low-fat diet. After 24 weeks, total serum cholesterol decreased by 26% from initial values and LDL-cholesterol decreased by 24%. Serum cholesterol decreased 22% from initial values, LDL-cholesterol decreased 29%, and HDL-cholesterol increased 9% after 99 weeks[4].

In an ambulatory study, 12 free-living students incorporated either oat bran or wheat bran muffins providing 50 g of bran daily into their usual diet for six weeks. After a seven-week period without muffins, they resumed the use of bran muffins for six more weeks. Each student was on oat bran for six weeks and wheat bran for six weeks. During the study, body weights or serum triglycerides did not change significantly. Wheat bran did not significantly alter serum cholesterol while oat bran significantly lowered serum cholesterol at weeks 3, 4, 5, and 6 producing a reduction of 12% ($p < 0.001$). Thus oat bran muffins lowered serum cholesterol of free living students to a significant degree[5].

Oat products selectively decrease serum LDL-cholesterol while maintaining or increasing HDL-cholesterol levels. A survey of studies indicates oat products reduced LDL-cholesterol an average of 12% and increased HDL-cholesterol an average of 8%[2-4,11-16], thus leading to significant and favorable reductions in the LDL:HDL cholesterol ratio.

Oat bran is a potent, concentrated source of soluble fiber. It can be eaten as a hot cereal or 20-25 g can be incorporated into a muffin. It is now widely available in ready-to-eat cereals, bakery products, cookies, and other new products.

A recent metabolic ward study was designed to determine hypocholesterolemic effects of a processed ready-to-eat cereal containing oat bran[17]. Hypercholesterolemic men were admitted to a metabolic ward for four weeks. The participants ate two oz of the cereal, containing 30 g oat bran with 4 g soluble fiber, as a supplement to a typical American diet for two weeks. Compared to a two-week control period in which the patients ate two oz corn flakes, serum total cholesterol significantly decreased 7% with the oat bran cereal. LDL-cholesterol also significantly decreased 11%, while HDL-cholesterol remained unchanged. This study and a previous rat study with this cereal (Anderson, unpublished data) indicate that processing does not detrimentally affect the hypocholesterolemic properties of oat bran.

III. OAT BRAN COMPARED TO WHEAT BRAN: METABOLIC WARD STUDIES OF
     HYPERCHOLESTEROLEMIC MEN

Apparent differences in hypocholesterolemic effects of soluble and insoluble fibers have not been well documented. An intensive metabolic ward study was undertaken to directly compare these effects in hypercholesterolemic men (Anderson, unpublished data). Oat bran provided the soluble fiber and wheat bran provided the insoluble fiber. Participants entered the metabolic ward for four weeks. Subjects were fed a control diet the first week and then were randomly allocated to either a wheat bran or oat bran diet for three weeks.

Twenty men with hypercholesterolemia were recruited from the patient population at the Veterans Administration Medical Center, Lexington, KY. Total serum cholesterol levels of the subjects ranged from 192-324 mg/dl. Subjects with secondary causes of hyperlipidemia such as hypothyroidism, renal disease, diabetes mellitus, or obesity were excluded from the study. Subjects ranged in age from 38-73 years old.

22

The subjects were fed weight-maintaining diets consisting of commonly available foods. Diets contained approximately 43% carbohydrate, 16% protein, 41% fat, and 450 mg cholesterol. The one-week control diet resembled a typical American diet and contained a standard level of .14 g dietary fiber. The test diets provided 30 g dietary fiber from oat bran or wheat bran. The oat bran diet provided 12 g soluble fiber, and wheat bran diet provided 5 g soluble fiber. The brans were added to the diets in the form of hot cereals and bran muffins.

With oat bran supplementation serum total cholesterol decreased 13% (p<.001), LDL-cholesterol decreased 15% (p<.001), and apolipoprotein B-100 decreased 15% (p<.001) (Table 2). Oat bran slightly reduced triglycerides and apolipoprotein A-1. Wheat bran slightly reduced serum total cholesterol, LDL-cholesterol, apolipoprotein B-100, apolipoprotein A-1, and triglycerides. Both oat bran and wheat bran supplementation lowered HDL-cholesterol 7% (p<.050). The changes in oat bran and wheat bran were significantly different for serum total cholesterol (p<.019) and LDL-cholesterol (p<.022).

Thus, oat bran in direct comparison to wheat bran significantly reduced risk for coronary heart disease by reducing atherogenic blood lipids. Oat bran significantly reduced serum total cholesterol, LDL- cholesterol, and apolipoprotein B-100 while wheat bran did not significantly change these lipid parameters.

Table 2. Oat bran compared to wheat bran: Metabolic ward study serum lipid results. Values are expressed as percentage change from initial values. Significant differences in changes between oat bran and wheat bran are denoted by *.

	Oat	Wheat
Total Cholesterol	-13% *	-4%
Triglycerides	-9%	-9%
HDL-cholesterol	-7%	-7%
Apolipoprotein A-100	-5%	+3%
LDL-cholesterol	-15% *	-2%
Apolipoprotein B-100	-15%	-1%

IV. PROPOSED MECHANISMS FOR THE HYPOCHOLESTEROLEMIC EFFECTS OF OAT PRODUCTS

Various fibers have physico-chemical properties which can affect serum cholesterol by mediating changes in the digestion and absorption of foods. Physico-chemical properties include water-holding capacity, apparent solubility, binding ability, degradability, particle size, and alteration by processing[18]. Soluble fibers alter gastrointestinal function and lower serum cholesterol levels through several possible mechanisms.

A. Modification of Bile Acid Absorption and Metabolism

Certain fibers may lower serum cholesterol by modifying bile acid absorption and metabolism. Early investigators and subsequent studies have documented that bile acids bind strongly to dietary fibers, possibly increasing fecal bile acid excretion[18-20]. The loss of bile acids creates a demand on cholesterol for bile acid synthesis and diverts available cholesterol from lipoprotein synthesis. However, a strong relationship between the ability of fibers to bind bile acids and the ability to lower serum cholesterol is lacking[21,22].

Oat bran and oat gum have important hypocholesterolemic actions. These fibers hydrate during digestion, forming gels and viscous solutions which entrap bile acids. This results in fewer bile acids available for micelle formation, leading to impaired lipid, cholesterol, and bile acid absorption[19].

The relationship between reduced serum cholesterol and increased fecal bile acid excretion, however, is inconsistent. Oat bran and beans both significantly reduce serum cholesterol. However, oat bran increases fecal bile acid excretion, while beans decrease fecal bile acid excretion[2,3]. Wheat bran has no effect on serum cholesterol or fecal bile acid excretion[23]. Bile acid binding resins are potent cholesterol-lowering agents and increase excretion of bile acids[20]. Soluble fiber supplementation decreases cholesterol levels, but the increase in fecal bile acid excretion is minimal compared to the increase with resins. Thus, other mechanisms in addition to increased fecal bile acid excretion probably contribute to the hypocholesterolemic effects of dietary fibers[24].

Changes in the spectrum of biliary and fecal bile acids or changes in bile acid secretion rates may contribute to the effect of dietary fibers on serum cholesterol. It has been postulated that fibers bind bile acids and cholesterol decreasing their intestinal absorption[18]. Bacteria in the colon convert primary bile acids (cholic and chenodeoxycholic) to secondary bile acids (deoxycholic and lithocholic) which are less well absorbed. In humans, significant changes in biliary bile acid concentrations in response to different dietary fibers have been observed[22]. Water soluble fiber (pectin) decreased the ratio of primary to secondary bile acids while insoluble fiber (cellulose) increased the primary to secondary ratio. In studying fecal bile acids, Story[21] has shown that rats fed oat bran and other water soluble fibers increased fecal excretion of chenodeoxycholic acid and its derivatives and decreased fecal excretion of cholic acid and its derivatives. Chenodeoxycholic acid is less well absorbed than cholic acid and may be preferentially synthesized and excreted in larger quantities. Oat bran increases fecal bile acid excretion in humans[2,3]. Oat bran increases primary bile acid excretion, cholic and chenodeoxycholic acid[3] and in one study particularly increased chenodeoxycholic acid excretion[2].

Human subjects were injected with sodium taurocholate-C^{14} to determine the secretion rates of bile acids. Pectin significantly increased secretion of C^{14}-deoxycholate metabolites of sodium taurocholate while cellulose decreased secretion of C^{14}-deoxycholate metabolites[22]. Lignin did not alter any of the C^{14} metabolites. Possible mechanisms for these differences include alterations in bacterial metabolic activity in the colon and in bile acid delivery to the colon.

Few studies have measured the effects of oat bran on both biliary bile acid concentrations and fecal bile acid excretion. Further work is also needed to determine the influence of fiber on bile acid secretion rates before conclusions can be drawn concerning the amount of bile acids secreted, reabsorbed, or excreted.

B. Interference with Lipid Absorption and Metabolism

Dietary fibers may decrease serum cholesterol by altering lipid metabolism and interfering with lipid absorption from the small intestine. Soluble fiber delays gastric emptying in rats and humans, slowing access of nutrients to digestive enzymes and absorptive surfaces of the small intestine[18]. Decreased gastric emptying in response to a diet high in viscous fiber retards lipid digestion[18]. Dietary fibers alter activities of digestive enzymes by changing viscosity and the pH of intestinal contents, thereby decreasing digestion of lipids in the upper small intestine[25].

Fiber binds bile acids and decreases their availability for optimal fat digestion and absorption[19-21]. Bound bile acids prevent micelle formation and absorption, increasing fecal fat excretion. Soluble fiber intake increased fecal fat excretion by 2-4 g/day in some studies[12,26,27]. Rolled oats have been shown to increase fecal fat excretion by 5 g/day, although some of the fecal fat loss may also be due to the low digestibility of certain oat lipid[12]. However, fat losses of this small magnitude would not decrease serum cholesterol by 20%. More likely, viscous fibers sequester the lipid and bile acid components of intestinal contents, delaying their absorption to distal sites of the small intestine[18].

Additional physiologic actions of soluble fiber which alter lipid absorption include decreased motility of the digestive tract, increased thickness of the unstirred water layer of the small intestine, changes in the morphologic and functional properties of the absorptive mucosal cells, and changes in intestinal capillary and lymphatic flow[18,27].

C. Production of Short-Chain Fatty Acids

Colonic fermentation of fiber polysaccharides, particularly soluble fibers, produces short-chain fatty acids (SCFA) as metabolic end-products. Acetate, propionate, and butyrate are the major SCFA produced and absorbed into the portal circulation[30]. Feeding oat bran to rats significantly increased portal vein SCFA concentrations and reduced serum and liver cholesterol levels compared to cellulose[29]. It is postulated that SCFA attenuate hepatic cholesterol synthesis and thereby contribute to the hypocholesterolemic effects of certain dietary fibers[28-30].

Supplementing rats with dietary propionate significantly reduced serum cholesterol[29,30]. At physiologic concentrations, propionate significantly inhibited cholesterol synthesis using isolated rat hepatocytes[31]. Figure 3 illustrates the effects of increasing concentrations of propionate on C^{14} incorporation into cholesterol of isolated rat hepatocytes. Propionate concentrations from 0.25 mM-1.0 mM significantly inhibited cholesterol synthesis.

In humans, dietary fiber increases fecal SCFA excretion[18]. Dietary fiber may also increase serum SCFA and portal vein SCFA, but this is not well documented. Pomare and associates found higher peripheral venous acetate levels in humans which remain elevated for several hours after feeding 20 g of soluble fiber as pectin[32]. The increase in peripheral SCFA may accompany alterations in hepatic cholesterol synthesis and contribute to the hypocholesterolemic effects of oat bran and other water-soluble fibers.

Some investigators question the role of SCFA in mediating the hypocholesterolemic effects of oat bran and soluble fibers[29,33]. Further studies are necessary to determine the potential role of SCFA in regulating serum cholesterol and to elucidate hepatic cholesterol metabolism and the effects of differing peripheral SCFA concentrations on cholesterol metabolism.

D. Hormonal Effects on Lipid Metabolism

Dietary fiber intake affects key hormones such as insulin and glucagon, altering rates of hepatic cholesterol and lipid synthesis[18]. Insulin is a key regulatory hormone which inhibits lipolysis in adipose tissue and stimulates lipid synthesis in liver and other tissues. Glucagon antagonizes these effects on lipid metabolism.

High-carbohydrate, high-fiber diets lower insulin requirements and increase insulin sensitivity in normal and diabetic subjects. These glycemic changes may contribute to reduced hepatic cholesterol and fatty acid synthesis. In addition, serum glucagon and other pancreatic and gastrointestinal hormones are reduced by increased dietary fiber intake[18,34-36]. Oats promote favorable glycemic responses. In normal subjects, plasma glucose and insulin responses to oat baked products were less than 50% of the response to corn or wheat products[36]. The viscous gum content of oats possibly inhibited starch digestion and absorption leading to decreased glycemic response. In addition, the glycemic index of oat flakes is significantly lower than the glycemic index of wheat or corn flakes[37]. Oats seem to exert favorable effects on glycemic factors, thereby possibly contributing to decreased rates of hepatic cholesterol and lipid synthesis.

Figure 4 illustrates the favorable effects of oat bran supplementation on the plasma glucose and insulin responses to a glucose tolerance test in

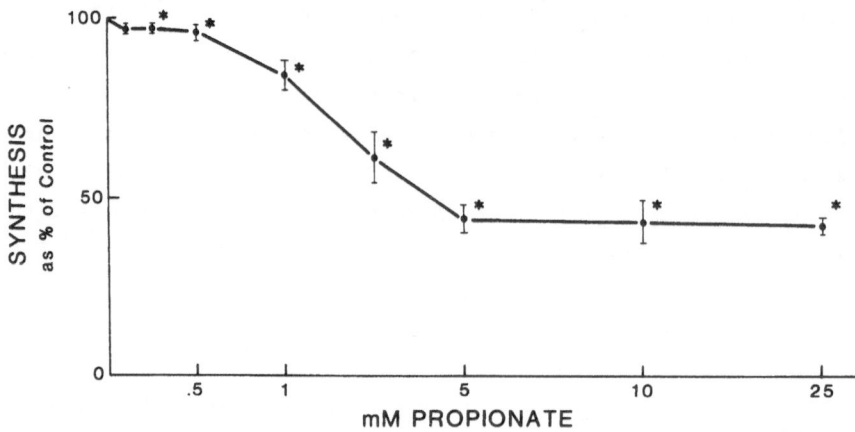

Figure 3.　　Effect of propionate on hepatic cholesterol synthesis

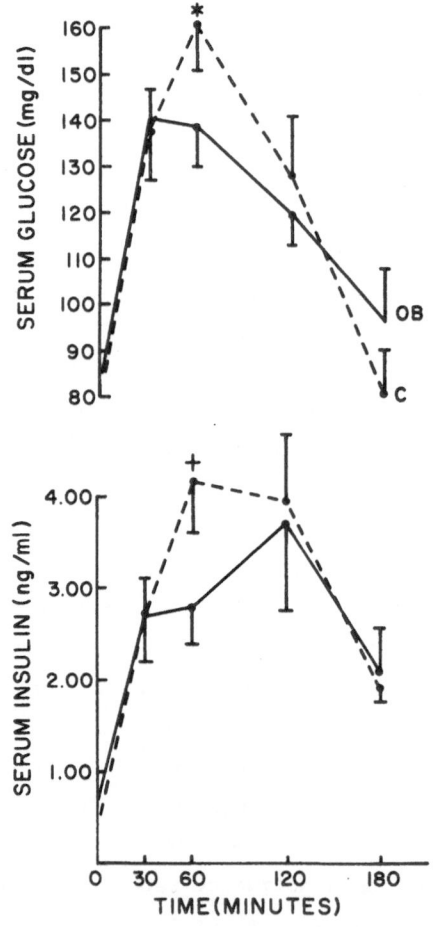

Figure 4.　　Oat bran vs. control. Effect on plasma
　　　　　　　insulin and glucose response

non-diabetic men. As described previously[3], these men ate a control diet for two weeks and an oat-bran- supplemented diet for two weeks. Carbohydrate and fat content of the two diets were identical. Oral glucose tolerance tests (100 g glucose in citrus-flavored drink) were performed after an overnight fast at the end of each diet. Oat bran was not given with the glucose load and subjects had not received oat bran since the previous evening. Both plasma glucose and insulin responses to the glucose load were significantly lower after the oat bran diet than after the control diet. Thus oat bran supplementation for two weeks decreased plasma glucose and insulin responses even when the oat bran was not administered with the glucose load.

Dietary fibers, particularly fiber from oats, have a significant impact on pancreatic and gastrointestinal hormones. Dietary fiber reduces insulin concentration, increases peripheral serum insulin sensitivity, and lowers serum glucose. These hormonal changes possibly contribute to the hypocholesterolemic effect of certain fibers by influencing the regulation of hepatic cholesterol and fatty acid synthesis.

V. PRACTICAL IMPLICATIONS

A. Coronary Heart Disease Risk Factors

Hypercholesterolemia is one of the most important risk factors for coronary heart disease (CHD)[39]. Soluble fibers can lower serum cholesterol levels by 20-30%, significantly reducing CHD risk[6]. Dietary fibers influence other lipid risk factors for coronary heart disease, including serum HDL and LDL cholesterol, triglycerides, andapolipoproteins A-1 and B-100[6]. Risk for CHD is closely linked to increased LDL-cholesterol and apolipoprotein B-100 levels[39]. CHD risk is inversely linked to HDL -cholesterol[40] and apolipoprotein A-1 levels[41]. In general, dietary fiber effects on triglycerides and HDL-cholesterol are variable for different dietary fibers. Most soluble fiber sources generally reduce LDL-cholesterol, while oat bran selectively lowers LDL-cholesterol and maintains or increases HDL-cholesterol[2-4,11-16].

The effects of dietary fiber on apolipoproteins have not been established. Durrington and colleagues[42] noted decreased apolipoprotein B-

100 after three weeks of soluble fiber supplementation. Oat bran significantly reduced apolipoprotein B-100 by 15% (Anderson, unpublished data). The effects of dietary fibers on various apolipoprotein fractions require further study.

In addition to lipid risk factors, dietary fiber also influences non-lipid risk factors for CHD. These include clotting factors, hormonal and glycemic factors, obesity, and hypertension[6]. The impact of fiber on other risk factors for CHD is not clearly characterized[6].

Increased levels of clotting factors such as fibrinogen and von Willebrand factor may be associated with CHD[6]. Dietary fiber supplementation decreases fibrinogen levels and has an uncertain effect on von Willebrand factor[6].

Hormonal and metabolic changes mediated by hyperinsulinemia and hyperglycemia can increase the risk for CHD[43]. Insulin resistance leads to chronic hyperglycemia and increased circulating insulin which may promote hepatic cholesterol and fatty acid synthesis[43]. Dietary fiber reduces serum insulin levels and fasting and postprandial blood glucose[18,34-36]. Insulin and glucose responses are favorably influenced by an oat-containing meal in comparison to meals containing wheat or corn[37,38].

Obesity is an independent risk factor for CHD and contributes to other risk factors such as chronic hyperglycemia, hyperinsulinemia, and hypertension[6]. In weight reduction trials, a hypocaloric diet, high in carbohydrate and dietary fiber promoted weight loss[44]. Fiber generally promotes long-term weight loss and maintenance[45]. Few well-controlled studies examining the effects of fiber on obesity have been published. Further work is necessary to systematically document the relationship.

Hypertension is linearly correlated to CHD risk[46]. Increased serum cholesterol levels often accompany increased blood pressure[46]. Alternate approaches to drug therapy in the treatment of hypertension are desirable due to lipid-raising side effects of many anti-hypertensive agents[47]. Dietary fiber supplementation reduces serum lipids as well as blood pressure levels. Documented studies clearly demonstrate increased fiber intake beneficially affects blood pressure[48]. Dietary intervention, including generous amount of fiber intake provides non-pharmacologic treatment for hypertension and other CHD risk factors.

B. Nutrition Management of Hypercholesterolemia

Lifestyle changes, including diet and exercise modifications, are recommended prior to and in addition to drug therapy to permanently reduce serum cholesterol levels[49]. Nutrition intervention is the primary treatment recommended for hypercholesterolemia. Approximately 60% of individuals in the United States have elevated serum cholesterol levels, and three-quarters of these elevations are diet-related[6]. Consequently, most individuals with hypercholesterolemia respond well to dietary intervention.

Traditional diets used in hyperlipidemia management stipulate decreased total and saturated fat intake, decreased dietary cholesterol intake, and adjusted caloric intake to achieve or maintain desirable body weight. Combining a high-fiber, low-fat diet with exercise and a healthy life style effectively reduces serum cholesterol and offers other health benefits, including improved glycemic control, easier weight management, and lower blood pressure[50].

Our group developed a high-carbohydrate, high-fiber (HCF) diet which improves serum lipid status and general metabolic well-being. The HCF diet provides 55% of energy as carbohydrate, 25% fat, 20% protein, 150 mg cholesterol, and 50 g dietary fiber daily[48]. These levels are suitable and practical for home use. Caloric intake is adjusted to achieve or maintain desirable body weight[51].

To meet the HCF diet recommendations, individuals must decrease their intake of high-fat meats and dairy products, eggs, and fats. Foods high in refined sugar and fat are also limited because they provide little fiber, vitamins, or minerals. To increase fiber intake, consumption of foods such as whole grains, legumes, fruits, and vegetables are recommended. Fruits, oats, barley and beans contain substantial amounts of cholesterol-lowering soluble fiber, thereby enhancing the metabolic effects of the HCF diet. Individuals with elevated serum cholesterol levels should include at least an additional 5 g of soluble fiber daily from oat products, beans, or psyllium[51].

Nutritional intervention with the HCF plan significantly improves health status. Lipid-lowering effects of the HCF diet are maximized when the diet is implemented in conjunction with other life-style changes, such as increased exercise, smoking cessation, avoidance or reduction of alcohol use, and appropriate stress management. Nutritional counseling is crucial to diet modification to assess current eating habits, reinforce positive aspects and implement the HCF diet.

C. Hypocholesterolemic Effects of Foods and Supplements

Soluble fiber includes pectin, gums, certain hemicelluloses, and storage polysaccharides[52]. Foods high in soluble fiber include oats, barley, fruits, and legumes[52]. Oat bran and beans are particularly rich in soluble fiber[52]. To increase soluble fiber intake, the following should be included in the HCF diet: one bowl of oatmeal, oat bran, or dry unsweetened oat cereal and two oat bran muffins; one-half to one cup of cooked dried beans (pinto, navy, kidney); three to four servings a day of fruit (prunes or raisins); or a combination of these foods daily. Dietary supplements, including psyllium or soy fiber preparations, can be used in the management of hyperlipidemia by persons unable to meet the soluble fiber requirements through their diet[51].

Supplementing the diet with concentrated soluble fiber sources significantly lowers serum cholesterol by about 15% and selectively lowers LDL-cholesterol. Serum total cholesterol can be reduced 20-25%, LDL-cholesterol reduced 25-30%, and HDL-cholesterol increased 10-20% by long-term supplementation with concentrated soluble fiber sources.

D. Role of Fiber in Reducing Risk for Coronary Heart Disease

CHD is the major cause of death in the United States and in most western countries. Hyperlipidemia, particularly hypercholesterolemia, has moved to the forefront as a major risk factor for CHD[39]. Research shows that a 1% reduction in serum cholesterol results in a 2% reduction in incidence of CHD[6]. Considering that more than one-half the middle-aged men and women in the United States have serum cholesterol values exceeding 200 mg/dl, reducing serum cholesterol levels in Americans should significantly reduce mortality and morbidity from CHD in the United States[6].

Dietary intervention is the treatment of choice to reduce serum cholesterol levels. Drugs currently available to lower cholesterol are associated with deleterious side effects[39,53]. A high-fiber intake combined with a high-carbohydrate and low-fat intake effectively reduces hypercholesterolemia as well as reduces other diet-induced CHD risk factors including hypertension, diabetes, and obesity[54]. Evidence suggests that dietary fiber may also be an independent risk factor for CHD[6]. Numerous studies have shown a negative correlation between fiber intake and CHD incidence and mortality, independent of changes in serum cholesterol levels[6].

Oat bran is more cost-effective than cholestyramine or colestipol in treating individuals with hypercholesterolemia[55]. Nutrition intervention using oat bran to treat hypercholesterolemic individuals contributes significantly to CHD risk reduction. Oat bran is accessible, practical, and cost-effective for general population use.

VII. CONCLUSIONS

Oat products have more potent cholesterol-lowering action than virtually any other widely-used food product. Oat bran has the potential to lower serum cholesterol by 10-20% while oatmeal has similar but less potent effects. Importantly, oat products selectively lower atherogenic LDL-cholesterol without significantly affecting the protective HDL-cholesterol. The mechanisms responsible for these effects are still under study. Oats and other soluble fiber sources may lower serum cholesterol by altering bile acid secretion or metabolism, by modifying lipid absorption or lipoprotein metabolism, through fermentation to short chain acids which may influence hepatic cholesterol synthesis, or through changes in insulin secretion or hormonal effects. Because of the substantial hypocholesterolemic effects of oat bran, it is widely used in the dietary management of hypercholesterolemia. The regular use of oats as part of a low-fat, low-cholesterol diet should reduce the risk for coronary heart disease.

REFERENCES

1. A. P. de Groot, R. Luyken, and N. A. Pikaar, Cholesterol -lowering effect of rolled oats, Lancet. 2:303 (1963).
2. J. W. Anderson, L. Story, B. Sieling W-J. L. Chen, M.S. Petro, and J. Story, Hypocholesterolemic effects of oat bran or bean intake for hypercholesterolemic men, Am J Clin Nutr. 40: 1146 (1984).
3. R. W. Kirby, J. W. Anderson, B. Sieling, E. D. Rees, W-J. L. Chen, R. E. Miller, and R. M. Kay, Oat bran intake selectively lowers serum low density lipoprotein cholesterol concentrations of hyperchlesterolemic men, Am J Clin Nutr. 34:824 (1981).
4. J. W. Anderson, L. Story, B. Sieling, and W-J. L. Chen, Hypocholesterolemic effects of high-fiber diets rich in water-soluble fibers, J Can Diet Assoc. 45:140 (1984).

5.   K. Storch, J. W. Anderson, and V. R. Young, Oat-bran muffins lower serum cholesterol of healthy young people, Clin Res. 34:447 (1984).

6.   J. W. Anderson, D. A. Deakins, T. L. Floore, B. M. Smith, and S. E. Whitis, dietary fiber and coronary heart disease, in: "CRC Reviews," F. M. Clydesdale, ed., CRC Press, Boca Raton, FL. in press.

7.   W-J. L. Chen, J. W. Anderson, and M. R. Gould, Effects of oat bran, oat gum, and pectin on lipid metabolism of cholesterol fed rats, Nutr Rep Int. 24:1093 (1981).

8.   W-J. L. Chen and J. W. Anderson, Effects of guar gum and wheat bran on lipid metabolism of rats, J Nutr. 109:1028 (1979).

9.   W-J. L. Chen and J. W. Anderson, Effects of plant fiber in decreasing plasma total cholesterol and increasing high-density lipoprotein cholesterol, Proc Soc Exp Biol Med. 162:310(1979).

10. C. D. Jennings, K. Boleyn, S. R. Bridges, P. Wood, and J. W. Anderson, A comparison of the lipid-lowering and intestinal morphological effects of cholestyramine, chitosan and oat gum in rats, Proc Soc Exp Biol Med. 189:13 (1988).

11. M. R. Gould, J. W. Anderson, J.W., and S. O'Mahoney, Biofunctional properties of oats, in: "Cereals for Food and Beverages," G. E. Inglett and L. Munck, eds., Academic Press Inc., New York (1980).

12. P. A. Judd and A. S. Truswell, The effects of rolled oats on blood lipids and fecal steroid excretion in man, Am J Clin Nutr. 34:2061 (1981).

13. L. V. Van Horn, K. Liu, D. Parker, L. Emidy, Y. Liao, W. H. Pan, D. Giumetti, J. Hewitt, and J. Stamler, Serum lipid response to oat product intake with a fat-modified diet, J Am Diet Assoc. 86:759 (1986).

14. W. H. Turnbill and A. R. Leeds, Reduction of total and LDL-cholesterol in plasma by rolled oats, J Clin Nutr Gastroenterol. 2:177 (1987).

15. L. V. Van Horn, L. A. Emidy, K. Liu, Y. Liao, C. Ballew, J. King, and J. Stamler, Serum lipid response to a fat-modified, oatmeal-enhanced diet, Prev Med. 17:377 (1988).

16. K. V. Gold and D. M. Davidson, Oat bran as a cholesterol-reducing dietary adjunct in a young, healthy population, West J Med. 148:299 (1988).

17. C. C. Hamilton, J. Tietyen, D. B. Spencer, and J. W. Anderson, Serum lipid responses of hypercholesterolemic men to a ready-to-eat oat bran cereal, Abstract, American Dietetic Association. 72nd Meeting, October 23-27. 1989.

18. J. W. Anderson, D. A. Deakins, S. R. Bridges, Hypocholesterole- mic effects of soluble fibers: mechanisms, in: "Dietary Fiber: Chemistry, Physiology, and Health Effects," D. Kritchevsky and J. W. Anderson, eds., Plenum Press, New York, in press.

19. J. A. Story and D. Kritchevsky, Comparison of the binding of various bile acids and bile salts in vitro by several types of dietary fiber, J Nutr. 106:292 (1976).

20. G. V. Vahouny, R. Tombes, M. M. Cassidy, D. Kritchevsky, and L. L. Gallo, Dietary fibers V: binding of bile salts, phospholipids and cholesterol from mixed micelles by bile acid sequestrants and dietary fibers, Lipids, 15:1012 (1980).

21. J. A. Story, Dietary fiber and lipid metabolism, Proc Soc Exp Biol Med. 180:447 (1985).

22. L. C. Hillman, S. G. Peters, C. A. Fisher, and E. W. Pomare, The effects of the fiber components pectin, cellulose, and lignin on bile salt metabolism and biliary lipid composition in man, Gut. 27:29 (1986).

23. J. H. Cummings, M. J. Hill, D. J. A. Jenkins, J. R. Pearson, and H.S. Wiggins, Changes in fecal composition and colonic function due to cereal fiber, Am J Clin Nutr. 29:1468 (1976).

24. R. M. Kay and A. S. Truswell, Dietary fiber: Effects on plasma and biliary lipids in man, in: "Medical Aspects of Dietary Fiber," G.A. Spiller, ed., Plenum Press, New York, (1980).

25. B. O. Schneeman and D. Gallaher, Effects of dietary fiber on digestive enzyme activity and bile acids in the small intestine, Proc Soc Exp Biol Med. 180:409 (1985).

26. R. M. Kay and A. S. Truswell, Effects of citrus pectin on blood lipids and fecal steroid excretion in man, Am J Clin Nutr. 30:171 (1977).

27. M. M. Cassidy, F. G. Lightfoot, and G. V. Vahouny, Dietary fiber, bile acids, and intestinal morphology, in: "Dietary Fiber in Health and Disease," G.V. Vahouny and D. Kritchevsky, eds., Plenum Press, New York, (1982).

28. J. H. Cummings, E. W. Pomare, W. J. Branch, C. P. E. Naylor, and G. T. Macfarlane, Short-chain fatty acids in human large intestine, portal, hepatic, and venous blood, Gut, 28:1221 (1987).

29. R. J. Illman and D. L. Topping, Effects of dietary oat bran on faecal steroid excretion, plasma volatile fatty acids and lipid synthesis in rats, Nutr Res. 5:839 (1985).

30. W-J. L. Chen, J. W. Anderson, and D. Jennings, Propionate may mediate the hypocholesterolemic effects of certain soluble plant fibers in cholesterol-fed rats, Proc Soc Exp Biol Med. 175:215 (1984).

31. W-J. L. Chen and J. W. Anderson, Hypocholesterolemic effects of soluble fibers, in: "Dietary Fiber (Basic and Clinical Aspects)," G. V. Vahouny and D. Kritchevsky, eds. Plenum Press, New York (1986).

32. E. W. Pomare, W. J. Branch, and J. H. Cummings, Carbohydrate fermentation in the human colon and its relation to acetate concentrations in venous blood, J Clin Invest. 75:1448 (1985).

33. R. J. Illman, D. L. Topping, G. H. McIntosh, R. P. Trimble, G. B. Storer, M. N. Taylor, and B. Q. Cheng, Hypocholesterolemic effects of dietary propionate: studies in whole animals and perfused rat liver, Ann Nutr Metab. 32:97 (1988).

34. J. W. Anderson and K. Ward, High-carbohydrate, high-fiber diets for insulin treated men with diabetes mellitus, Am J Clin Nutr. 32:2312 (1979).

35. J. W. Anderson, J. Zeigler, and D. Deakins, High carbohydrate, high fiber diets increase insulin sensitivity and carbohydrate disposal in type I diabetic individuals, Diabetes. 37(suppl I):11A (1988).

36. J. W. Anderson, Dietary fiber and diabetes, in: "Dietary Fiber in Health and Disease," G. V. Vahouny and D. Kritchevsky, eds., Plenum Press, New York (1982).

37. K. W. Heaton, S. N. Marcus, P. M. Emmett, and C. H. Bolton, Particle size of wheat, maize, and oat test meals: Effects on plasma glucose and insulin response, Am J Clin Nutr. 47:675 (1988).

38. D. J. A. Jenkins, T. M. S. Wolever, and R. H. Taylor, H. Barker, H. Fielden, J. Baldwin, A. C. Bowling, H. C. Newman, A. L. Jenkins, D. V. Goff, Glycemic index of foods: a physiologicalbasis for carbohydrate exchange, Am J Clin Nutr. 34:362 (1981).

39. A. M. Gotto, Interactions of the the major risk factors for coronary heart disease, Am J Med. 80(suppl 2A):48 (1986).

40. W. B. Kannel, W.P. Castelli, T. Gordon, Cholesterol in the prediction of atherosclerotic disease. New perspectives based on the Framingham study, Ann Inter Med. 90:85 (1979).

41. J. J. Maciejko, D. R. Holmes, and B. A. Kottke, Apolipoprotein

A-I as a marker of angiographically assessed coronary-artery disease, N Engl J Med. 309:385 (1983).

42. P. N. Durrington, C. H. Bolton, A. P. Manning, and M. Hartog, Effect of pectin on serum lipids and lipoproteins, whole-gut transit-time, and stool weight, Lancet. 2:394 (1976).

43. J. W. Anderson, Hyperlipidemia and diabetes: Nutrition considerations, in: "Nutrition and Diabetes," L. Jovanovic and C. Peterson, eds., Alan R. Liss, New York, (1985).

44. J. W. Anderson, High-fiber, hypocaloric vs. very-low-calorie diet effects on blood pressure of obese men, Am J Clin Nutr. 43:695 (1986).

45. J. W. Anderson, "Nutrition Management of Metabolic Conditions," The HCF Diabetes Foundation, Lexington, KY. (1986).he evolution of current hypertension therapy, Am J Med. 85(suppl 3B):14 (1988).

46. W. P. Castelli and K. Anderson, A population at risk. Prevalence of high cholesterol levels in hypertensive patients in the Framingham study. Am J Med. 80(suppl 2A):23 (1986).

47. R. H. McDonald, The evolution of current hypertension therapy, Am J Med. 85(suppl 3B):14 (1988).

48. J. W. Anderson, Plant fiber and blood pressure, Ann Intern Med. 98:842 (1983).

49. Report of the National Cholesterol Education Program Expert Panel on detection, evaluation, and treatment of high blood cholesterol in adults, Arch Intern Med. 148:36 (1988).

50. J. W. Anderson, Fiber and health: an overview, Am J Gastroenterol. 81:892 (1986).

51. J. W. Anderson, "The HCF Guidebook," The HCF Diabetes Foundation, Lexington, KY. (1987).

52. J. W. Anderson and S. R. Bridges, Dietary fiber content of selected foods, Am J Clin Nutr. 47:440 (1988).

53. C. J. Lavir, G. T. Gau, R. W. Squires, B. A. Kottke, Management of lipids in primary and secondary prevention of cardiovascular disease, Mayo Clinic Proc. 63:605 (1988).

54. J. W. Anderson and N. J. Gustafson, Dietary fiber and heart disease: current management concepts and recommendations, Top Clin Nutr. 3:21 (1988).

55. B. P. Kinosian and J. M. Eisenberg, Cutting into cholesterol. Cost-effective alternatives for treating hypercholesterolemia, J Am Med Assoc. 259:2249 (1988).

GASTROINTESTINAL RESPONSES TO DIETARY FIBER

Barbara O. Schneeman

Department of Nutrition
University of California, Davis
Davis, CA 95616

Various sources of dietary fiber can slow the process of digestion and absorption of macronutrients. It is likely that the physical properties of fiber sources such as particle size, viscosity, water-holding capacity and gel formation, and bile acid binding capacity are important in determining the effect of a fiber source on gastrointestinal function and nutrient absorption. The effects of fiber on gastrointestinal function have recently been reviewed (Schneeman, 1987, 1989; Vahouny and Cassidy, 1985) and several mechanisms have been suggested by which the physical properties of fiber slow digestion and absorption. In reviewing the experimental evidence it is clear that dietary fibers can affect the functioning of all of the gastrointestinal organs. For example, viscous polysaccharides slow gastric emptying and hence the rate of nutrient delivery to the small intestine. Within the small intestine various types of fiber can interfere with digestive enzyme activity, slow diffusion and mixing, and bind components from micelles.

The ability of dietary fiber to bind micellar components such as bile acids and phospholipids and to slow the absorption of lipid from the small intestine could contribute to the ability of certain dietary fibers to lower plasma cholesterol. Most of the fat consumed in the diet consists of triglyceride, and the predominant fatty acids consumed are in the C_{16} and C_{18} families. Pancreatic lipase, which hydrolyzes fatty acids from triglycerides, is a water-soluble enzyme and can only hydrolyze fats at the surface between the oil and water phase. Hence the availability of an emulsifying system in the small intestine is essential for lipid digestion and absorption to occur. Fat malabsorption leading to steatorrhea has been associated with impairment of lipid emulsification. Within the small intestinal contents the emulsifying agents available include fatty acids, monoglycerides, cholesterol, lecithin and lysolecithin, protein and bile salts. Vahouny and Cassidy (1985) reported that guar gum bound 36% of sodium taurocholate, 22% of lecithin, 23% of cholesterol, 23% of monolein, and 33% of fatty acids from an in vitro micellar solution. In contrast, wheat bran and cellulose absorbed much lower percentages of micellar components. They did not report the ability of dietary fibers to bind lysolecithin which is probably a more important emulsifying agent in the intestine than lecithin. In addition to the potential ability to bind micellar components, certain fibers can alter the physical properties of the intestinal contents which might lead to entrapment of micellar components.

In 1986 Gallaher and Schneeman reported an in vivo method to determine bile acid and phospholipid binding within the small intestinal contents.

New Developments in Dietary Fiber
Edited by I. Furda and C. J. Brine
Plenum Press, New York, 1990

Rats are fed a test meal containing the source of fiber to be tested, and after 2 hours their small intestinal contents are removed and pancreatic lipase activity inhibited. The intestinal contents are separated into an oily layer, aqueous phase, and an insoluble phase by ultracentrifugation. The aqueous phase is obtained by slicing the centrifuge tube. The concentration of micellar components in the aqueous phase is compared to the concentration in the total intestinal contents. Hence the ratio of aqueous phase concentration to total intestinal contents indicates the proportion of micellar component that is present in the aqueous phase. In the first study two control treatments were included -- a fiber free test diet to establish the baseline ratio without any fiber treatment, and a positive control with cholestyramine, which is a bile-acid binding resin. As will be discussed later, because cellulose did not bind intestinal bile acids or phospholipids, it also is referred to as a control treatment. A bile acid or phospholipid binding ratio significantly less than the control treatment indicates binding or entrapment of components in the insoluble intestinal phase, whereas a ratio significantly higher than the control treatment indicates that the components have been partitioned into the aqueous phase of the intestinal contents.

Using this experimental approach two studies have been conducted with fiber sources that have a unique range of chemical and physical properties (Gallaher and Schneeman, 1986; Ebihara and Schneeman, 1989). These fiber preparations include wheat bran, which does not have much in vitro bile acid binding capacity but has been reported to increase fecal bile acid excretion; lignin, which binds bile acids in vitro; chitosan, which is an aminopolysaccharide that has been reported to lower plasma cholesterol; guar gum and konjac mannan, which are viscous, soluble polysaccharides that have been reported to lower plasma cholesterol; and oat bran, which contains betaglucan, a soluble polysaccharide.

The combined results of these studies are shown in Figures 1 and 2. The bile acid binding ratio is shown in Figure 1. As expected, the bile acid-binding resin, cholestyramine, significantly lowered the ratio compared to the fiber free and cellulose groups. Among the fiber treatments, lignin had a significantly lower ratio than the fiber free or cellulose groups, and the chitosan and wheat bran groups had significantly lower ratios than the cellulose group. Consequently all four of these treatments led to an interaction with the bile acids that reduced the amount in the aqueous phase and enhanced the proportion in the insoluble phase. The cholestyramine and these three fiber sources are insoluble at the pH of the small intestine, indicating that they bind or entrap bile acids. In contrast, the bile acid binding ratio was significantly higher in the guar gum and konjac mannan groups than the fiber free and cellulose groups. Both of these fiber sources are soluble in the aqueous phase of the small intestinal contents, and the higher ratio suggests that the bile acids were associated with the fiber-containing phase. The ratio did not change significantly for the oat bran group. The implications of this lack of difference for the oat bran group is difficult to interpret because oat bran contains both soluble and insoluble components.

The phospholipid binding ratios are shown in Figure 2. Interestingly, the ratios did not follow identical patterns as the bile acid binding ratio for all of the treatments, suggesting that the interactions are specific for each fiber and the individual micellar components. For cholestyramine and chitosan the ratio was lower than the fiber free and cellulose groups, as observed for the bile acid ratio. However, for the groups fed wheat bran, lignin, oat bran, guar gum, or konjac mannan the ratio did not differ significantly from the fiber free or cellulose groups. The cholestyramine and chitosan were the only treatments that bound both bile acids and phospholipids. It is likely that chitosan entraps micelles as it precipitates at the pH of the small intestine (Ebihara and Schneeman, 1989; Furda, 1983).

38

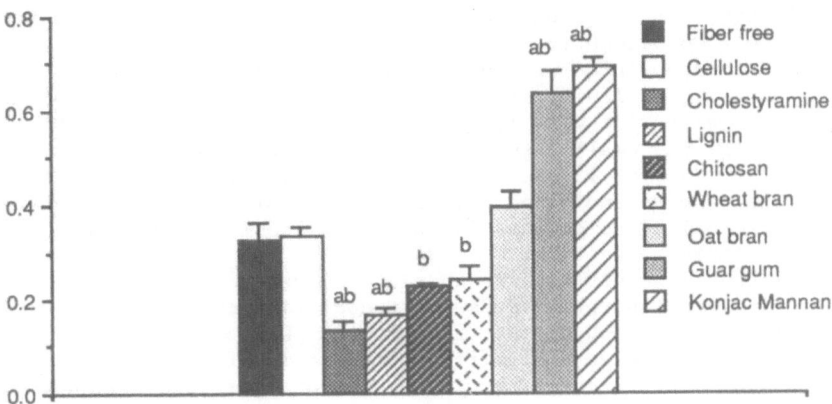

Figure 1. Bile Acid Binding Ratio. Each value is the ratio of bile acids in the aqueous phase of the small intestinal contents. The letter "a" indicates a significant difference from the fiber free group and "b" a difference from the cellulose group.

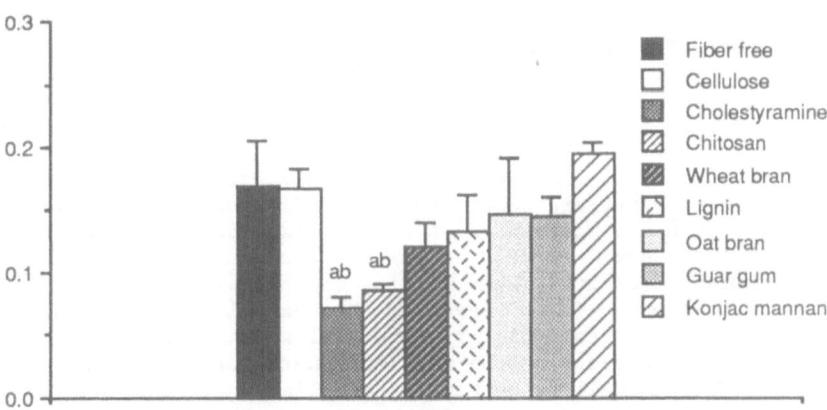

Figure 2. Phospholipid Binding Ratio. Each value is the ratio of phospholipids in the aqueous phase of the small intestinal contents. The letter "a" indicates a significant difference from the fiber free group and "b" a difference from the cellulose group.

Although the phospholipid binding ratio for guar gum and konjac mannan indicate a lack of interaction between the fiber and phospholipids, these dietary treatments affect the concentration of phospholipid in the aqueous phase. Because both of these fibers have a high water-holding capacity, they significantly increase the aqueous volume of the intestinal contents by 30-50%. Hence the concentration of phospholipid in the aqueous phase actually is lower than the control treatments because of the lack of interaction yet greater volume.

An important aspect of these studies is to consider the effect of these varying treatments that range in specificity for binding bile acids and phospholipids on the disappearance of lipid from the small intestine. In the first study only cholestyramine significantly lowered the amount of lipid which was solubilized in the intestinal contents; however, Gallaher and Schneeman (1986) demonstrated that the total amounts of phospholipid and bile acid in the aqueous phase are significant predictors of the amount of lipid solubilized in the aqueous phase and available for absorption from the small intestine. Consequently the interaction of dietary components with bile acids and phospholipids will contribute significantly to their effect on the rate and site of lipid absorption. In the second study ^{3}H-cholesterol and ^{14}C-triglyceride were included in the test diet to estimate lipid disappearance from the small intestine. Both konjac mannan and guar gum, but not chitosan, delayed disappearance of the isotopes at 2 hours after the test meal. The results indicate that in the case of a drug, cholestyramine, the binding of micellar components is sufficiently strong to have a significant impact on lipid solubilization. However, in the case of dietary fibers the interactions are weaker and the absolute amount of lipid solubilized during digestion and absorption does not necessarily change. However, under these conditions the effect of the fiber on the physical properties of the intestinal contents becomes a significant factor in affecting the rate of lipid absorption. Specifically, konjac mannan and guar gum will both increase the viscosity of the intestinal contents, whereas wheat bran, lignin and chitosan are insoluble and unlikely to be viscous at the pH of the intestinal contents. Although all treatments have some bile acid binding capacity, only the viscous polysaccharides delay the disappearance of lipid from the small intestine.

It has been demonstrated in vitro that viscosity caused by guar gum will slow the diffusion of micelles (Phillips, 1986) and this factor can have an important effect on the rate of lipid absorption. Studies in patients with ileostomies also support the observation that viscous fiber sources are more likely to reduce lipid absorption. Sandberg et al. (1983) reported that addition of pectin to a low fiber meal given to ileostomy patients increased the amount of fat in the ileostomy fluid by 96%. In contrast, the amount of fat in the ileostomy fluid was not changed when wheat bran was added to the meal (Sandberg et al., 1981). In patients with pancreatic insufficiency that are receiving pancreatic enzyme replacements, pectin appears to cause a malabsorption of fat (Isaksson et al., 1984). Vahouny et al. (1988) have reported, using lymph-cannulated rats, that viscous polysaccharides are more likely to slow the appearance of oleic acid and cholesterol in lymph. Taken together, this experimental evidence clearly indicates that certain dietary fibers slow the digestion and absorption of fats by their effects on the physical characteristics of the small intestinal contents, as well as by binding micellar components and interfering with digestive enzyme activity. Interestingly, the effect of fiber appears to be on the rate and perhaps the site of lipid digestion and absorption, not on the overall fat digestibility. Vahouny and Cassidy (1986) have reported that apparent fat digestibility as measured by fecal fat excretion is changed no more than 4% by a high fiber diet.

The experimental evidence that indicates that certain fibers can alter the pattern of lipid disappearance from the small intestine has raised questions about the potential effect of fiber supplemented meals on postprandial lipid appearance in the plasma. However, only a limited number of studies have been conducted in human subjects to determine whether the effect of soluble fiber sources on lipid digestion and absorption will attenuate the appearance of triglycerides in the plasma following consumption of a meal. In general the use of plasma triglyceride appearance has not resulted in a clear-cut picture of the effect of fiber on lipid absorption. Anderson et al. (1980) reported that plasma triglycerides did not increase postprandially in subjects given a high carbohydrate, high fiber meal. However, the low fat content of this meal undoubtedly contributed to the lack of postprandial increase. In contrast Jenkins (1978) reported that both pectin and guar gum raise the serum triglycerides above control during the first 3 hours after consuming a Lundh test meal. We have also reported a larger increase in postprandial triglycerides in female subjects in response to a test meal when the meal was supplemented with guar gum and oat bran (Redard et al., 1988). It is interesting to note that treatment with cholestyramine, which binds micellar components, has also been reported to enhance postprandial triglyceridemia in Type IIa patients (Weintraub et al., 1987). It is possible that, although the absorption of lipid may be delayed by viscous fiber sources or cholestyramine treatment, the clearance of triglycerides from plasma may be slowed because of a modification in insulin release. The limited studies conducted to date on the effect of fiber on postprandial lipemia have not attempted to separate the intestinal and hepatic contributions to the triglyceride-rich particles that appear during the postprandial period. Supplementation with viscous polysaccharides may affect the appearance of triglyceride-rich particles by prolonging the intestinal contribution to this fraction in the plasma after a meal.

Because the human diet is typically omnivorous, it is important to study the functions of the gastrointestinal tract in the presence of the nondigestible polysaccharides associated with dietary fiber. The current experimental evidence suggests that the presence of fiber has an important function in regulating the rate and site of nutrient absorption from the small intestine. Further research is needed to understand the physiological importance of this modulating, regulatory role.

Acknowledgments: This work has been supported in part by NIH grant DK 20446. The author is grateful to B. Diane Richter, Daniel Gallaher, Paul Davis, Kiyoshi Ebihara, and Carol Redard for their excellent research contributions.

REFERENCES

Anderson, J. W., Chen, W.-J. L. and Sieling, B., 1980, Hypolipidemic effects of high-carbohydrate, high-fiber diets, Metabolism, 29:551.

Ebihara, K., and Schneeman, B. O., 1989, Interaction of bile acids, phospholipids, cholesterol and triglyceride with dietary fibers in the small intestine of rats, J. Nutr., in press.

Furda, I., 1983, Aminopolysaccharides -- Their potential as dietary fiber, in "Unconventional Sources of Dietary Fiber," I. Furda, ed., ASC Symposium Series 214, Washington, DC, pp. 105-122.

Gallaher, D., and Schneeman, B. O., 1986, Intestinal interaction of bile acids, phospholipids, dietary fibers, and cholestyramine, Am. J. Physiol., 250:G420.

Isaksson, G., Lundquist, I., Akesson, B., and Ihse, I., 1984, Effects of pectin and wheat bran on intraluminal pancreatic enzyme activities and on fat absorption as examined with the triolein breath test in patients with pancreatic insufficiency, Scand. J. Gastroenterol., 19:467.

Jenkins, D. J. A., 1978, Action of dietary fiber in lowering fasting serum cholesterol and reducing postprandial glycemia: Gastrointestinal mechanisms, in "International Conference on Atherosclerosis," L. A. Carlson, ed., Raven Press, New York, pp. 173-182.

Phillips, D. R., 1986, The effect of guar gum in solution on diffusion of cholesterol mixed micelles, J. Sci. Food Agric., 37:548.

Redard, C. L., Schneeman, B. O., and Davis, P. A., 1988, Differences in postprandial lipemia due to gender and dietary fiber, FASEB J., 2:A1418.

Sandberg, A. S., Andersson, H., Hallgren, B., Hasselblad, K., Isaksson, B., and Hulten, L., 1981, Experimental model for in vivo determination of dietary fibre and its effect on the absorption of nutrients in the small intestine, Br. J. Nutr., 45:283.

Sandberg, A. S., Ahderinne, R., Andersson, H., Hallgren, B., and Hulten, L., 1983, The effect of citrus pectin on the absorption of nutrients in the small intestine, Human Nutr.:Clin. Nutr., 37C:171.

Schneeman, B. O., 1987, Dietary fiber and gastrointestinal function, Nutr. Rev., 45:129.

Schneeman, B. O., 1989, Macronutrient absorption, in "Proceedings of the Third International Symposium on Dietary Fiber," In press.

Vahouny, G. V., and Cassidy, M. M., 1985, Dietary fibers and absorption of nutrients, Proc. Soc. Exp. Biol. Med., 180:432.

Vahouny, G. V., and Cassidy, M. M., 1986, Effect of dietary fiber on intestinal absorption of lipids, in "Handbook of Dietary Fiber in Human Nutrition," G. A. Spiller, ed., CRC Press, Boca Raton, FL, pp. 121-128.

Vahouny, G. V., Satchitanandam, S., Chen, I., Tepper, S. A., Kritchevsky, D., Lightfoot, F. G., and Cassidy, M. M., 1988, Dietary fiber and intestinal adaptation: effect on lipid absorption and lymphatic transport in the rat, Am. J. Clin. Nutr., 47:201.

Weintraub, M. S., Eisenberg, S., and Breslow, J. L., 1987, Different patterns of postprandial lipoprotein metabolism in normal, type IIa, type III, and type IV hyperlipoproteinemic individuals, J. Clin. Invest., 79:1110.

DIETARY FIBER AND BILE ACID METABOLISM

Jon A. Story, Julia J. Watterson, Hugh B. Matheson
and Emily J. Furumoto

Department of Foods and Nutrition
Purdue University
West Lafayette, IN 47907

INTRODUCTION

Physical properties possessed by dietary fiber suggest several ways in which interactions of dietary fiber and bile acids might result in alterations of bile acid metabolism which would influence sterol balance, a major determinant of serum cholesterol which, an important risk factor for coronary heart disease. The first property of interest is the ability of some sources of dietary fiber to absorb water, diluting the contents of the intestine, or to form gels which hold water and may interfere with mixing or transport within the intestine. Second, some sources of dietary fiber interact with organic compounds (e.g. bile acids) or with ions, binding or adsorbing them and potentially interfering with their absorption from the intestine. Fermentability of some source of dietary fiber may also contribute to their effects through changes in the environment within the intestine or through absorption of metabolites which may have secondary effects in the liver or at other sites.

OBSERVATIONS

We first became aware of the relationship between some sources of dietary fiber and bile acid excretion as a result of the work of Cookson et al. (1967). They reported an increase in steroid excretion in response to dietary alfalfa at a level which also prevented hypercholesterolemia in rabbits fed a cholesterol-containing diet. We subsequently measured the adsorption of bile acids by a variety of sources of dietary fiber in vitro and found that alfalfa was capable of binding a significant amount of bile acids and bile salts in comparison to cellulose or other materials used as dietary fiber which did not have an effect on serum cholesterol level (Kritchevsky and Story, 1974; Story and Kritchevsky, 1976). More recently we have modified our original method to enable measurement of adsorption of bile acids by water soluble components of dietary fiber and found that a wide variety of dietary fibers adsorb bile acids, but that this adsorption was not always correlated with the ability of that source to alter cholesterol levels (Story and Lord, 1985).

New Developments in Dietary Fiber
Edited by I. Furda and C. J. Brine
Plenum Press, New York, 1990

This observation, that cholesterol lowering was often related to an ability to adsorb bile acids and that in some instances increases in bile acid excretion had also been observed, led us to suggest that these sources of dietary fiber increased bile acid excretion by adsorbing bile acids in the small and large intestines which resulted in negative sterol balance (Story and Kritchevsky, 1976). We have subsequently tested this hypothesis by simultaneously measuring effects of several sources of dietary fiber on liver cholesterol accumulation and steroid excretion in cholesterol-fed rats (Story and Thomas, 1982; Story, 1985). In comparison to cellulose, oat bran and alfalfa were found to reduce accumulation of liver cholesterol in these animals while corn bran did not. Interestingly, steroid excretion was not significantly altered by

Table 1. Fecal Steroid Excretion in Response to
Various Sources of Dietary Fiber in Rats

Dietary Fiber	Daily Steroid Excretion	Bile Acid Concentration
Cellulose	Control	Control
Alfalfa	+ 9	+ 80
Corn Bran	+ 2	+ 60
Oat Bran	−17	+220
Wheat Bran (white)	− 5	+ 83
Corn Bran	+17	+ 3
Barley Bran	+24	+ 44
Oat Bran	−26	+225
Corn Bran I	+11	+ 35
Corn Bran II	+16	+ 90
Wheat Bran (white)	−15	+101
Wheat Bran (red)	−13	+ 53

any of these sources of dietary fiber with the greatest change a 17% reduction in response to oat bran. This suggests that some other mechanism was responsible for the observed changes in cholesterol accumulation in the rat. We have examined steroid excretion in response to several other sources of dietary fiber (Table 1). Of the sources of dietary fiber examined in this model, only oat bran and alfalfa reduce cholesterol accumulation. Steroid excretion does not seem to be consistently related to the ability to prevent cholesterol accumulation.

In human studies a quite different picture emerges, but still without a clear relationship. Oat bran has been shown to reduce serum cholesterol levels in hypercholesterolemic men and to simultaneously increase bile acid excretion (Kirby et al., 1981; Anderson et al., 1984). However the magnitude of this increase, as well as that reported for other sources of dietary fiber and polysaccharides (e.g. pectin), has never been sufficient to be considered the sole reason for the observed reduction in cholesterol levels. In addition, as can be seen in Table 2, several other sources of dietary fiber have been shown to alter daily bile acid excretion in human subjects with a variety of effects on serum cholesterol levels. For example, cellulose, white and red wheat bran and psyllium have been shown to increase bile acid excretion but, of these, only psyllium and red wheat bran have ever been shown to reduce serum cholesterol.

Table 2. Bile Acid Excretion by Humans in Response to Dietary Fiber

Dietary Fiber	Daily Bile Acid Excretion	Bile Acid Concentration	Reference
	(% change)		
Cellulose	+ 25	NA	Stanley et al., 1973
Wheat Bran (white)	0	− 39	Eastwood et al., 1973
	+ 90	− 35	Cummings et al., 1976
Wheat Bran (red)	+ 65	− 12	Spiller et al., 1986
Oat Bran	+ 51	+ 25	Kirby et al., 1981
	+ 65	+ 14	Anderson et al., 1985
Rolled Oats	+ 62	+ 37	Judd & Truswell, 1981
Psyllium	+ 70	NA	Stanley et al., 1973

NA = Not available

Feeling that some other difference in bile acid excretion might be a key to this lack of agreement, we examined the spectrum of bile acids excreted by rats in response to some of the sources of dietary fiber mentioned above. We observed (Story and Thomas, 1982; Story, 1985) a change in the spectrum of bile acids excreted. Dietary fiber sources that reduced cholesterol accumulation (e.g., oat bran and alfalfa) increased the proportion of chenodeoxycholic acid derivatives excreted.

Previous work would suggest that a change of this nature would alter sterol balance. Increases in the pools of derivatives of chenodeoxycholic acid would tend to reduce cholesterol absorption (Ponz de Leon et al., 1979), reduce cholesterol synthesis (Coyne et al., 1976; Cooper, 1976) and reduce the proportion of bile acids reabsorbed in the small intestine (Beher et al., 1967) possibly increasing bile acid excretion. All these changes would result in a negative sterol balance.

In an effort to examine this effect more precisely we have measured bile acid pool sizes in rats in response to some of these dietary fiber sources. We have employed a method in which a bile duct cannula is inserted and the bile drained allowing measurement of the pool size for all bile acids, secretion rate and basal synthesis after the pool has been washed out. Feeding of pectin or psyllium causes an increase in the bile acid pool size and daily secretion rate with a smaller effect on basal synthetic rate (Table 3). Within the pool, the percentage of chenodeoxycholic acid derivatives increase from 31% when cellulose was fed to 43% and 41% when pectin or psyllium were fed. This was consistent with results from a similar experiment in which oat bran and corn bran were fed. In that case, derivatives of chenodeoxycholic acid increased in response to oat bran (32%) but were reduced by corn bran (15%) in comparison to cellulose (28%). It would appear that this change in the

Table 3. Changes in Bile Acid Pools in Response
to Pectin and Psyllium

Dietary Fiber	Bile Acid Pool Size	Basal Synthetic Rate	Daily Secretion Rate
Cellulose	Control	Control	Control
Pectin	+ 169	+ 101	+ 178
Psyllium	+ 170	+ 42	+ 170

spectrum of bile acids might be a key to understanding the mechanism for the effects of dietary fiber on cholesterol metabolism.

CONCLUSIONS

How might the physical properties of dietary fiber cause some of the observed effects on bile acid excretion and pool size? Gel formation is thought to be an important part of the effects of water soluble sources of dietary fiber or polysaccharides. Gels may interfere with transport and absorption of lipids in the small intestine as well as with bile acid absorption in the ileum. Our data with pectin and psyllium indicate that gel formation might be responsible for increased pool size by reducing reabsorption of bile salts in the small intestine. This reduced return to the liver would stimulate synthesis of an expanded pool. Absorption of water by dietary fiber would seem to have little effect on bile acid metabolism except in reducing concentrations of bile acids in the colon and feces. This could have implications for risk for colon cancer (Hill, 1985).

Adsorption seems to be an important phenomena but its direct implications in altering bile acid excretion is less well understood. Care needs to be taken in assuming that in vitro adsorption is related to in vivo interactions between dietary fiber and bile acids. Our experiments would suggest that pH and other variables can be adjusted to alter adsorption in a manner unrelated to what would be observed in vivo. As we gain a better understanding of the relationship between bile acid excretion and changes in sterol balance with dietary fiber sources the place of adsorption will become more apparent.

Fermentation of dietary fiber in the cecum and colon is an area of great interest in improving our understanding of the effects of dietary fiber on sterol balance. It has been suggested that increased production of short chain fatty acids by fermentation of the polysaccharides of dietary fiber in the colon and their subsequent absorption is responsible for the observed effects. Some evidence has been presented that suggests that propionate, the primary short chain fatty acid reaching the liver, inhibits cholesterol synthesis by rat hepatocytes (Chen et al., 1984). More recent data suggest that this effect is very dependent on level of the fatty acid used and on the substrate used to measure cholesterol synthesis (R. A. Freedland, personal communication). A great deal of additional information is needed to confirm or deny this mechanism.

46

Other potential effectors of changes in sterol balance should be considered. We have observed a reduction in the activity of HMG CoA reductase, the rate limiting step in cholesterol synthesis, in response to oat bran in rats within a short period of time (one to four hours) after consuming a diet containing this source of fiber for the first time (Kelley et al., 1987). This would suggest that there may be other, rapidly absorbed, components involved.

A great deal of additional information is needed in order to delineate the mechanism responsible for the action of the various sources of dietary fiber in lowering serum cholesterol. However we have sufficient knowledge and experience with several of these sources which are effective in lowering cholesterol levels in human to allow prudent use of them as a part of a hypocholesterolemic diet. Meanwhile our recommendation to the general public, concerning their intake of dietary fiber, needs to include the suggestion that we include a variety of dietary fiber sources in our diets but make and effort to increase our intake above current levels.

REFERENCES

Anderson, J. W., Story, L., Sieling, B., Chen, W-J. L., Petro, M. S. and Story J. A., 1984, Hypocholesterolemic effects of oat-bran or bean intake for hypercholesterolemic men. Am. J. Clin. Nutr. 40:1146.
Cookson, F. B., Altschul, R and Fedoroff, S., 1967, The effects of alfalfa on serum cholesterol and in modifying or preventing cholesterol-induced atherosclerosis in rabbits. J. Atheroslcer. Res. 7:69.
Cummings, J. H., Hill, M. J., Jenkins, D. J. A., Pearson, J. R. and Wiggins, H. S., 1976, Changes in fecal composition and colonic function due to cereal fiber. Am. J. Clin. Nutr. 29:1468.
Eastwood, M. A. and Mitchell, W. D., 1976, Physical Properties of fiber: a biological evaluation, in: "Fiber in Human Nutrition," G. A. Spiller and R. J. Amen, eds., Plenum Press, New York.
Eastwood, M. A., Krikpatrick, J. R., Mitchell, W. D., Bone, A. and Hamilton, T., 1973, Effects of dietary supplements of wheat bran and cellulose on faeces and bowel function. Br. Med. J. 4:392.
Hill, M. J., 1985, Bile acids and human colorectal cancer, in: "Dietary Fiber in Health and Disease," G. V. Vahouny and D. Kritchevsky, eds., Plenum Press, New York.
Judd, P. A. and Truswell, A. S., 1981, The effect of rolled oats on blood lipids and fecal steroid excretion in man. Am. J. Clin. Nutr. 34:2061.
Kelley, M. J. and Story, J. A., 1987, Short-term changes in hepatic HMG-CoA reductase in rats fed diets containing cholesterol or oat bran. Lipids 22:1057.
Kirby, R. W., Anderson, J. W., Sieling, B., Rees, E. D., Chen, W-J. L., Miller, R. E. and Kay, R. M., 1981, Oat-bran intake selectively lowers serum low-density lipoprotein cholesterol concentrations of hypercholesterolemic men. Am. J. Clin. Nutr. 34:824.
Kritchevsky, D. and Story, J. A., 1974, Binding of bile salts in vitro by nonnutritive fiber. J. Nutr. 104:458.
McConnell, A. A., Eastwood, M. A. and Mitchell, W. D., 1974, Physical characteristics of vegetable food stuffs that could influence bowel function. J. Sci. Food Agric. 25:1457.
Spiller, G. A., Story, J. A., Wong, L. G., Nunes, J. D., Alton, M., Petro, M. S., Furumoto, E. J., Whittam, J. H. and Scala, J., 1986, Effect of increasing levels of hard wheat fiber on fecal weight, minerals and steroids and gastrointestinal transit time in healthy young women. J. Nutr. 116:778.

Stanley, M. M., Paul, D., Gacke, D. and Murphy, 1973, Effects of
    cholestyramine, Metamucil and cellulose on fecal bile salt excretion
    in man. Gastroenterol. 65:889.
Story, J. A., 1985, Modification of steroid excretion in response to
    dietary fiber, in: "Dietary Fiber," G. V. Vahouny and D. Kritchevsky,
    eds., Plenum Press, New York.
Story, J. A. and Kritchevsky, D., 1976, Comparison of the binding of
    various bile acids and bile salts in vitro by several types of fiber.
    J. Nutr. 106:1292.
Story, J. A. and Kritchevsky, D., 1976, Dietary fiber and lipid
    metabolism, in: "Fiber in Human Nutrition," G. A. Spiller and R. J.
    Amen, eds., Plenum Press, New York.
Story, J. A. and Lord, S. L., 1987, Bile salts: in vitro studies with
    fibre components. Scand. J. Gastroenterol. 22(Suppl. 129):174.
Story, J. A. and Thomas, J. N., 1982, Modification of bile acid spectrum
    by dietary fiber, in: "Dietary Fiber in Health and Disease," G. V.
    Vahouny and D. Kritchevsky, eds., Plenum Press, New York.

PHYSIOLOGICAL EFFECTS AND PHYSICO-CHEMICAL

PROPERTIES OF SOY COTYLEDON FIBER

Grace S. Lo

Protein Technologies International
Checkerboard Square
St. Louis, MO 63164

INTRODUCTION

Soy fiber derived from dehulled and defatted soybean cotyledon is a
unique dietary fiber source. This fiber tends to react physically as an
insoluble fiber, and yet it has many physiological soluble fiber
properties.

It is a common misconception that soluble and insoluble dietary fiber
values obtained from analytical assays directly correspond to the
physiological soluble and insoluble fiber properties. Due to the
dependency of analytical methods used for measuring chemical soluble and
insoluble dietary fibers, the values obtained from chemical analyses vary
extensively. Table 1 summarizes the soluble and insoluble fiber analyzed
by different methods using soy cotyledon fiber as an example. The soluble
fiber value varies, ranging from 2% to 60% of total dietary fiber (TDF)
depending on the assay method (1-6).

Soy fiber derived from soybean cotyledon is comprised of both
cellulosic and non-cellulosic dietary fiber components. It contains 75%
total dietary fiber as analyzed by AOAC TDF method (1). The major
non-cellulosic portions of soybean cotyledon fiber are neutral
arabinogalactan and pectin-related acidic polysaccharides. Acidic
polysaccharides are highly branched, made of a backbone of D-galacturonic
acid and D-galactose interspersed with rhamnose, arabinose, xylose and
fucose (7-8). The complete structures of these fibers have not been
elucidated.

Based on Englyst's fractionation procedure, the sugar components of
soluble and insoluble non-starch polysaccharides (NSP) are summarized in
Table 2. The total NSP of soy cotyledon fiber is 78%, of which 27% is
soluble NSP and 73% is insoluble NSP. Although Englyst's NSP method
measures both neutral and acid soluble NSP, it still does not measure the
viscosity and fermentability of fiber in the human gut (2).

PHYSIOLOGICAL EFFECTS OF SOY COTYLEDON FIBER

The physiological effects of soy cotyledon fiber have been studied
extensively in several laboratories since 1977. Research from both animal
and human clinical studies has demonstrated that soy fiber derived from

Table 1

Dietary Fiber Values of Fibrim[R] Soy Fiber
Determined by Various Methods

Assay Method[a]	TDF g	SDF g	IDF g	% of TDF SDF	% of TDF IDF
AOAC[1]	75.9	4.7	71.1	6	94
Englyst & Cummings[2]	78.7	21.9	56.8	27	73
Li[3]					
121°C	73.7	18.6	55.1	25	75
130°C	77.4	47.6	29.8	61	39
Mongeau[4]					
100°C	72.5	5.6	67.1	8	92
121°C	66.0	40.0	26.0	61	39
Southgate[5]	71.4	1.5	69.9	2	98
Theander & Westerlund[6]	78.2	4.8	73.4	6	94

[a]Parentheses indicate reference cited in reference section.

TDF: total dietary fiber
SDF: soluble dietary fiber
IDF: insoluble dietary fiber

cotyledon meal has multiple physiological functions (9-20). It has a direct effect on the gastrointestinal tract by regulating nutrient utilization and bowel function, and an indirect effect on blood lipids and glucose metabolism by normalizing blood sugar and lipids. The potential health benefits of soy cotyledon fiber can be summarized as follows.

Cholesterol Lowering Effects

Clinical trials have shown that supplementing soy cotyledon fiber (Fibrim[R] Soy Fiber) to a fat modified and low cholesterol diet can further reduce blood cholesterol in people having elevated plasma cholesterol levels (9-10). Shorey et al (9) first demonstrated that supplementing 25 grams Fibrim[R] Soy Fiber in the form of croutons or cookies to a relatively low cholesterol (app. 380 mg/day) and low P/S ratio (P/S = 0.5) diet significantly lowered plasma cholesterol levels by 5 to 11% in 31 volunteers with mildly elevated plasma cholesterol levels while participating in a cardiovascular fitness program in Austin, Texas. Table 3 summarizes the hypocholesterolemic effect of Fibrim[R] Soy Fiber in a blind crossover designed study. For those receiving the fiber during the first 4-week interval, mean cholesterol value declined significantly ($p < 0.001$) from 252 to 224 mg/dl, a 28 mg reduction or 11% change. In the next 4-week period on the starch placebo, mean cholesterol values regressed 4% to 233 mg/dl. For those receiving starch during the first 4-week interval, the value changed insignificantly from 247 to 241 mg/dl, but decreased significantly ($p < 0.05$) on Fibrim[R] Soy Fiber treatment to 230 mg/dl, an 11 mg or 5% reduction.

Table 2

Dietary Fiber and Sugar Components of Soluble and Insoluble
Non-Starch Polysaccharide (NSP)

Dietary Fiber	Non-starch polysaccharide, % db		
Component	Soluble	Insoluble	Total
Cellulose	--	11.6	11.6
Non-cellulosic polysaccharide:			
Rhamnose	0.7	1.1	1.8
Fucose	0.5	1.3	1.8
Arabinose	3.7	9.2	12.9
Xylose	1.2	3.5	4.7
Mannose	0.3	0.9	1.2
Galactose	7.5	19.5	27.0
Glucose	0.4	0.4	0.8
Uronic Acid	6.8	9.0	15.8
Total	21.1	56.5	77.6

Analyzed by Englyst's NSP Method (2)

Supplementing 25 grams Fibrim[R] Soy Fiber per day also benefited
Type IIA hypercholesterolemic and Type IV hypertriglyceridemic patients
who had already reduced their plasma cholesterol levels by adherence to a
National Institutes of Health (NIH) low-saturated fat, low-cholesterol
diet (10). Fibrim[R] Soy Fiber significantly lowered mean total
cholesterol by 13 mg/dl ($p < 0.04$) and mean LDL cholesterol by 12 mg/dl
($p < 0.05$) in the 11 patients with Type IIA hypercholesterolemia (Table 4).
The effects were more variable in the 9 patients with Type IV
hypertriglyceridemia. Although the mean cholesterol reductions were
comparable in magnitude to those in the Type IIA hypercholesterolemic
patients, these were not statistically significant. Plasma HDL
cholesterol did not change in either Type IIA hypercholesterolemic or Type
IV hypertriglyceridemic patients.

When the data from the Type IIA and IV hyperlipidemic patients were
combined, the results indicated that dietary supplementation of Fibrim[R]
Soy Fiber significantly reduced plasma total cholesterol by $5.0 \pm 1.7\%$
($p < 0.01$) and LDL cholesterol by $4.7 \pm 1.9\%$ ($p < 0.03$). Further analysis
of the data revealed that Fibrim[R] Soy Fiber significantly lowered mean
plasma total cholesterol, LDL cholesterol and LDL triglyceride levels in
hyperlipidemic patients with impaired glucose tolerance, but did not
significantly change plasma total or LDL cholesterol levels in patients
with normal glucose tolerance, although changes in mean levels were in the
same direction (Table 5). There were no significant changes in plasma

Table 3

Effect of Fibrim[R] Soy Fiber on Plasma Cholesterol in Humans
with Mildly Elevated Plasma Cholesterol Levels[a]

Period	Treatment	Total Cholesterol, mg/dl Mean ± SE	Percent Change From Previous Period[b]
Initial Group A	None	252 ± 8	
Group B	None	247 ± 5	
Period I Group A	Soy Fiber	224 ± 7	−11*
Group B	Starch	241 ± 6	−2
Period II Group A	Starch	233 ± 8	+4**
Group B	Soy Fiber	230 ± 5	−5**

[a]Shorey et al, 1985 (9)

[b]Significance of difference between initial or treatment values by paired "t"
test: * $p < 0.001$, ** $p < 0.05$

Table 4

Effect of Fibrim[R] Soy Fiber on Lipoprotein Lipids
in Patients with Primary Hypercholesterolemia (Type IIA)
and Primary Hypertriglyceridemia (Type IV)[a]

	Type IIA Patients			Type IV Patients		
	Placebo I	Soy Fiber[b]	Placebo II	Placebo I	Soy Fiber	Placebo II
No. of Subjects	11	11	11	9	9	9
Total Chol (mg/dl)	330	317*	326	239	225	250
VLDL-Chol (mg/dl)	15	13	18	87	80	99
LDL-Chol (mg/dl)	264	252*	261	123	116	127
HDL-Chol (mg/dl)	46	45	46	32	31	33

[a]Lo et al, 1986 (10)
[b]Significantly different at $p < 0.05$

Table 5

Effect of Fibrim[R] Soy Fiber on Lipoprotein Lipids
of Subjects with Normal and Impaired Glucose Tolerance[a,b]

	Normal glucose tolerance		Impaired glucose tolerance	
	Placebo	Soy Fiber	Placebo	Soy Fiber
No. of Subjects	12 (mg/dl)	12	8 (mg/dl)	8
Total Chol	274 + 25	266 + 26	314 + 25	300 + 22*
VLDL-Chol	47 + 12	46 + 12	46 + 17	44 + 19
LDL-Chol	191 + 30	182 + 29	216 + 3	203 + 29***
HDL-Chol	36 + 3	36 + 3	44 + 3	43 + 3
Total TG	279 + 45	203 + 30	272 + 83	285 + 107
VLDL-TG	149 + 36	139 + 33	170 + 74	194 + 93
LDL-TG	34 + 3	31 + 3	50 + 6	38 + 5**
HDL-TG	14 + 1	13 + 1	16 + 2	15 + 1

[a]Lo et al, 1986 (10)

[b]Data represent mean. Statistical comparison between placebo and soy fiber supplementation. *$p < 0.03$, **$p < 0.01$, ***$p < 0.005$

cholesterol levels of college students having mean fasting plasma cholesterol values of 160 mg/dl (11).

Early animal experimental results (12-13) support the efficacy of Fibrim[R] Soy Fiber in lowering plasma cholesterol levels observed in human clinical trials (9-10). Two experiments were designed to determine the effect of Fibrim[R] Soy Fiber and other dietary fiber sources on rats fed control diets or cholesterol-containing diets (12). In the first experiment, seven male Sprague-Dawley weanling rats per group were fed nutritionally adequate diets containing 0, 5, 10, or 20% soy fiber with or without 1% cholesterol for 42 days. Results revealed that supplementation of soy fiber to cholesterol-containing diets lowered serum cholesterol and triglyceride levels. The rats fed cholesterol-containing diets with supplementation of 10 or 20% soy fiber had significantly lower serum cholesterol levels. Serum cholesterol and triglycerides were the lowest in the rats fed cholesterol-free diets with 20% soy fiber supplementation, but were not statistically significant (Figure 1).

Table 6 shows the effect of Fibrim[R] Soy Fiber on liver fat and cholesterol content of rats. Cholesterol feeding significantly increased liver fat and cholesterol levels. Supplementing soy fiber significantly lowered liver fat and cholesterol of rats fed cholesterol containing diets. This study suggested that soy fiber is capable of lowering serum

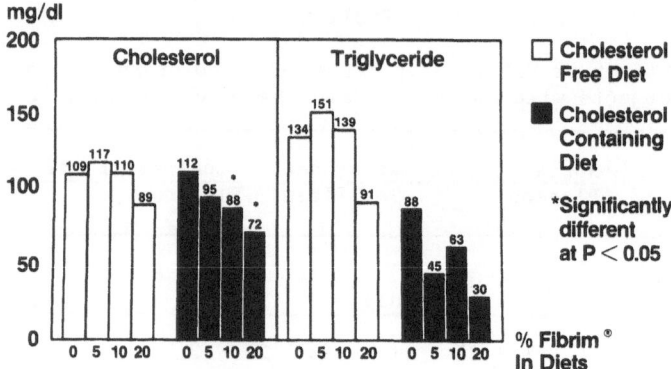

Fig. 1   Effect of different levels of Fibrim[R] Soy Fiber on serum cholesterol and triglyceride values of rats fed cholesterol-free and cnolesterol-containing diets (Experiment No. 1), [Lo et al, 1980 (12)].

Table 6

Effect of Fibrim[R] Soy Fiber on Liver Fat
and Liver Cholesterol Content of Rats
Fed Cnolesterol-Free and Cholesterol-Containing Diets[a]
(Experiment 1)

Dietary Level[b]		Liver Data[c]	
Soy Fiber (%)	Cholesterol %	Fat (%)	Cholesterol (mg/g)
0	0	3.8a	3.40a
5	0	3.3a	3.57a
10	0	2.8a	3.32a
20	0	2.9a	3.49a
0	1	17.8c	45.74d
5	1	15.1bc	39.33c
10	1	14.2b	35.54bc
20	1	12.3b	31.05b

[a]Lo et al, 1980 (12)

[b]Test diets contain no cholic acid.

[c]Means not followed by tne same letter are significantly different at $p < 0.05$.

and liver lipids. It can prevent dietary cholesterol from being absorbed by an unknown mechanism, but may not be able to remove endogenous cholesterol from the metabolic pool.

The second experiment was designed to compare the hypocholesterolemic effect of two levels of Fibrim[R] Soy Fiber, AACC wheat bran and pectin. Ten male weanling rats per group were fed the same basal diet as Experiment 1, but contained 1% cholesterol and 0.25% cholic acid with 0, 5, or 10% Fibrim[R] Soy Fiber, 10% AACC wheat bran or 5% pectin for 42 days. The effects of these dietary fiber sources on serum cholesterol and triglycerides are summarized in Figure 2.

Rats fed diets containing 10% soy fiber had significantly lower serum cholesterol and triglyceride levels. Rats fed diets containing 5% soy fiber had lower serum cholesterol and triglycerides, but were not statistically different from those rats fed the low fiber basal diets. Rats fed diets containing 5% pectin had significantly lower serum cholesterol, but not lower serum triglycerides. Rats fed diets containing 10% wheat bran had neither lower serum cholesterol nor serum triglyceride levels.

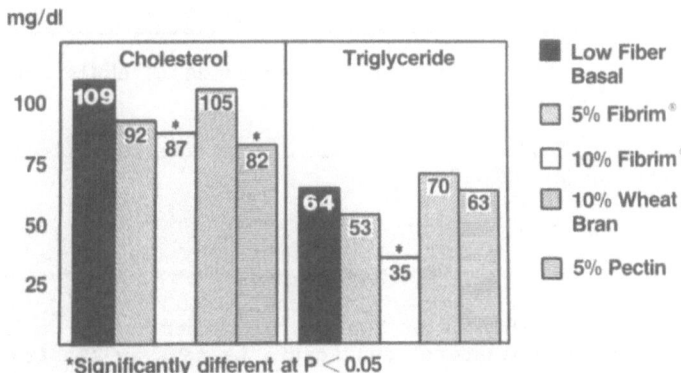

Fig. 2   Effect of different dietary fiber sources on serum cholesterol and triglyceride values in rats fed hyperlipidemic diets (Experiment No. 2), [Lo et al, 1980 (12)].

The effects of Fibrim[R] Soy Fiber, wheat bran, and pectin on liver fat and cholesterol content of rats are summarized in Table 7. The data indicate that rats fed cholesterol-containing diets with pectin had significantly lower liver fat and cholesterol content than other treatments. Liver fat and cholesterol were not reduced by soy fiber or wheat bran supplementation. Results from this study suggest that pectin is capable of lowering liver fat and cholesterol possibly by depressing the absorption of dietary cholesterol and/or inhibition of bile acid reabsorption. Fibrim[R] Soy Fiber is capable of reducing serum cholesterol levels possibly by preventing the absorption of dietary cholesterol.

A rabbit study was conducted to investigate the mechanism of the cholesterol lowering effects of Fibrim[R] Soy Fiber (13). Male New Zealand white rabbits at the age of 6 months were randomly assigned into 2 groups. Rabbits were fed casein based high-fat diets with either cellulose or soy fiber as the dietary fiber source. Data summarized in Table 8 show that soy fiber prevented plasma cholesterol elevation and

Table 7

Effect of Dietary Fiber on Liver Fat
and Liver Cholesterol Content of Rats
Fed Hyperlipidemic Diets[a]
(Experiment 2)

| Treatment | Liver Data[b] | |
	Fat (%)	Cholesterol (mg/g)
Low Fiber Basal	24.8 b	46.1 bc
5% Fibrim[R] Soy Fiber	25.0 b	50.8 bc
10% Fibrim[R] Soy Fiber	23.8 b	51.7 c
10% AACC Wheat Bran	22.9 b	42.7 b
5% Pectin	18.5 a	23.5 a

[a]Lo et al, 1980 (12); test diet contains 1% cholesterol and 0.25% cholic acid.

[b]Means not followed by the same letter are significantly different at $p < 0.05$.

Table 8

Effect of Fibrim[R] Soy Fiber
on Plasma Cholesterol and Aortic Lesions in Rabbits[a]

| Parameters | Treatment[b] | |
	Cellulose	Soy Fiber
Initial Plasma TTL-Cholesterol (mg/dl)	23	31
Final Plasma TTL-Cholesterol (mg/dl)	173	46*
Final Plasma LDL-Cholesterol (mg/dl)	96	15*
Final Plasma HDL-Cholesterol (mg/dl)	42	26*
HDL/LDL-Cholesterol Ratio	0.44	1.73*
Stained Lesions (no. of rabbits)	8	1*

[a]Lo et al, 1987 (13)

[b]Data represented as mean of 9 rabbits per group. Significantly different at $p < 0.05$.

reduced the incidence of aortic lesions in rabbits fed high-fat, cholesterol-free diets for 36 weeks.

The mechanism of some type of dietary fiber effect on lipid metabolism is not fully understood. However, these animal studies support the hypocholesterolemic effects observed in humans having elevated plasma cholesterol (9-10).

Effect on Diabetic Control

The possible benefits of supplementing Fibrim[R] Soy Fiber in the dietary management of diabetic patients were first suggested by Tsai et al (11). Figure 3 demonstrates that adding 15 grams of Fibrim[R] Soy Fiber to a 100 gram glucose solution reduced postprandial reactive hypoglycemia at 180 minutes after consumption in 14 healthy young male subjects.

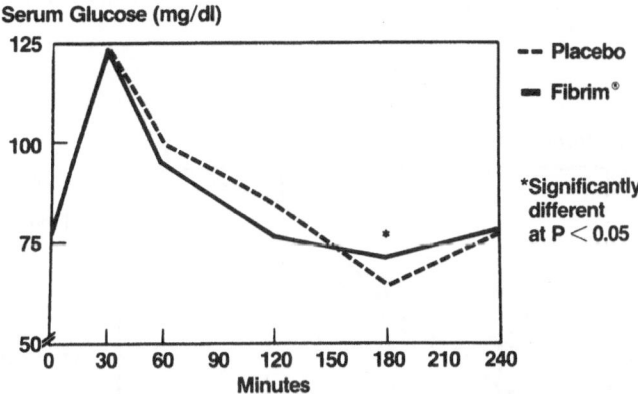

Fig. 3   Effect of Fibrim[R] Soy Fiber on serum glucose levels during glucose tolerance test in healthy young male subjects [Tsai et al, 1983 (11)].

Later, Tsai et al (14) evaluated the effect of Fibrim[R] Soy Fiber supplementation on postprandial blood glucose levels of 7 obese, non-insulin dependent, adult onset diabetic volunteers in a controlled, crossover designed experiment. Each subject was asked to consume two meals, one a low-fiber control meal, and the other a soy fiber-supplemented meal. Compared to the low-fiber meal, the soy fiber-supplemented meal significantly lowered postprandial serum glucose levels at 180 and 240 minutes (Figure 4), and significantly reduced the rise of postprandial plasma triglyceride levels (Figure 5).

Improvements in glucose tolerance (Figure 6) and insulin responses (Figure 7) also were observed in Type IIA and Type IV hyperlipidemic patients with impaired glucose tolerance tests (10). In this study, 25 grams of Fibrim[R] Soy Fiber were added to a NIH-low saturated fat, low-cholesterol diet for nine weeks. Fasting glucose levels were reduced by 8.5% and glucose area by 13%. Insulin areas also were decreased by 28% when standard glucose tolerance tests were performed before and after Fibrim[R] Soy Fiber supplementation.

Effect on Bowel Function

Regulating bowel function was the most consistent benefit of increased fiber consumption. However, not all dietary fiber sources have

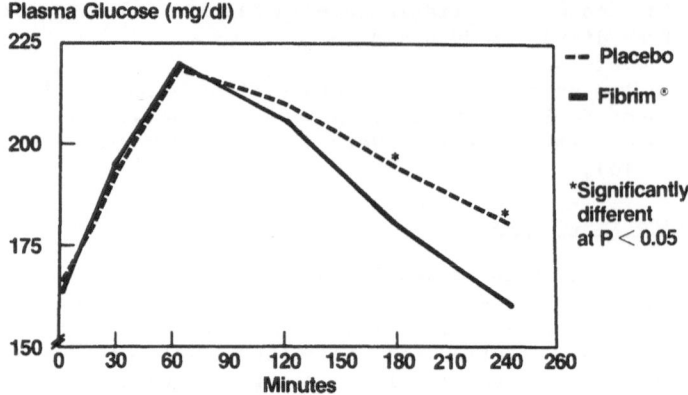

Fig. 4   The effect of Fibrim^R Soy Fiber on postprandial plasma glucose responses to ingestion of mixed meal by obese diabetic patients [Tsai et al, 1987 (14)].

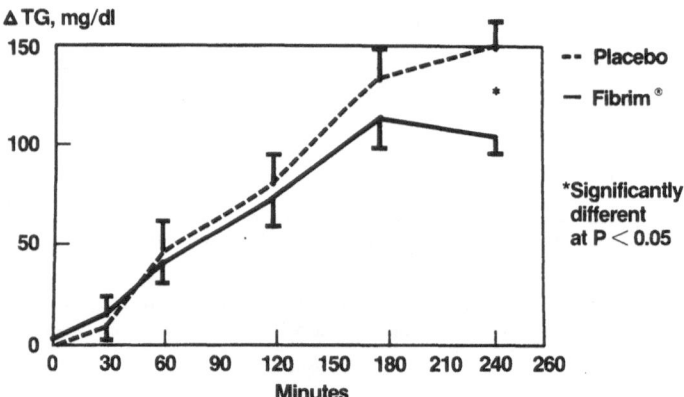

Fig. 5   Effect of Fibrim^R Soy Fiber on postprandial plasma triglyceride responses to ingestion of mixed meal by obese diabetic patients [Tsai et al, 1987 (14)].

equal ability in regulating bowel function. Research has demonstrated that Fibrim^R Soy Fiber can increase fecal bulk, primarily by increasing fecal moisture content (11,15).

When equal weights of Fibrim^R Soy Fiber, AACC wheat bran, and alpha-cellulose were substituted for corn starch in pig diets, fecal wet weight and fecal moisture were increased significantly only in pigs fed diets containing Fibrim^R Soy Fiber and wheat bran (15), (Figure 8). Fecal transit time was decreased significantly in pigs fed diets containing all three dietary fiber sources (Figure 9). However, in pigs fed diets using cellulose as the dietary fiber source, fecal moisture was lower than in pigs fed the low-fiber diet (Figure 10).

Human studies demonstrated that supplementing 13 to 60 g of Fibrim^R Soy Fiber per day to humans consuming a low-fiber solid diet (11, 20) or liquid diets (16-19) significantly increased fecal weight and normalized fecal moisture and transit time.

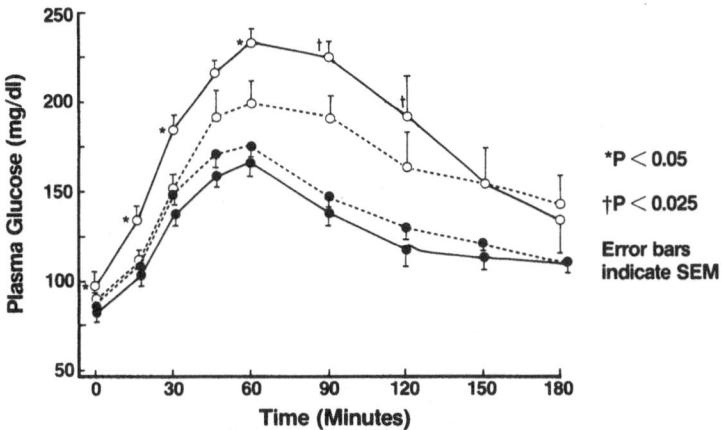

Fig. 6   The effect of Fibrim^R Soy Fiber on plasma glucose levels of
          patients with impaired (IGT) and normal glucose tolerance (NGT)
          during standard oral glucose tolerance test. 0____0 IGT on
          placebo; 0----0 IGT on Fibrim^R; ●____● NGT on placebo; ●----●
          on Fibrim^R [Lo et al, 1986 (10)].

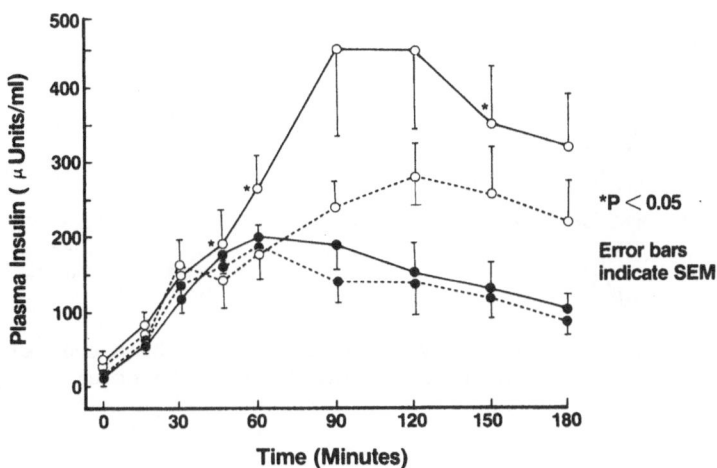

Fig. 7   The effect of Fibrim^R Soy Fiber on plasma insulin levels of
          patients with impaired (IGT) and normal glucose tolerance (NGT)
          during standard oral glucose tolerance test. 0____0 IGT on
          placebo; 0----0 IGT on Fibrim^R; ●____● NGT on placebo; ●----●
          on Fibrim^R [Lo et al, 1986 (10)].

Tsai et al (11) demonstrated that supplementation of 25 g of
Fibrim^R Soy Fiber per day to healthy young male volunteers resulted in
an increase in fecal weight, mainly by increasing fecal water content
(Figure 11).

Slavin et al (16) compared the bowel function of healthy men
consuming liquid diets with and without Fibrim^R Soy Fiber. The average
wet stool weights were increased from an average of 67 g per day on a
low-fiber liquid diet to 115 g and 150 g/day when liquid diets were
supplemented with 30 g and 60 g/day, respectively (Figure 12). They also

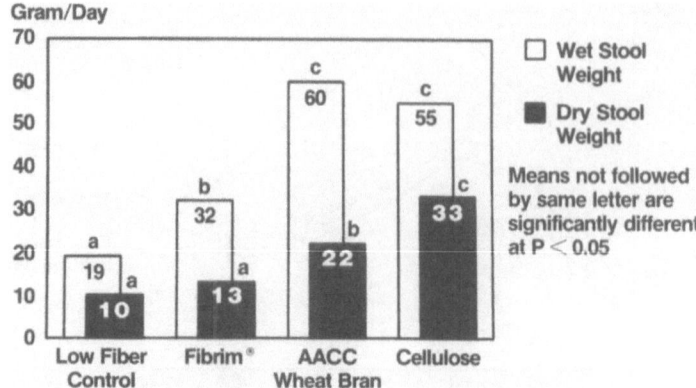

Fig. 8   Effect of Fibrim[R] Soy Fiber, AACC wheat bran and cellulose on fecal outputs in pigs [Lo et al, 1979 (15)].

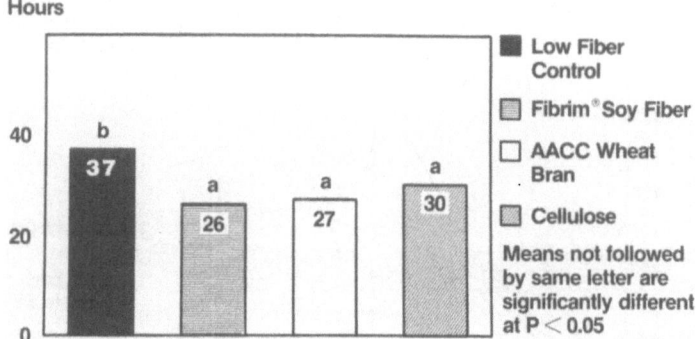

Fig. 9   Effect of Fibrim[R] Soy Fiber, AACC wheat bran and cellulose on gastrointestinal transit time in pigs [Lo et al, 1979 (15)].

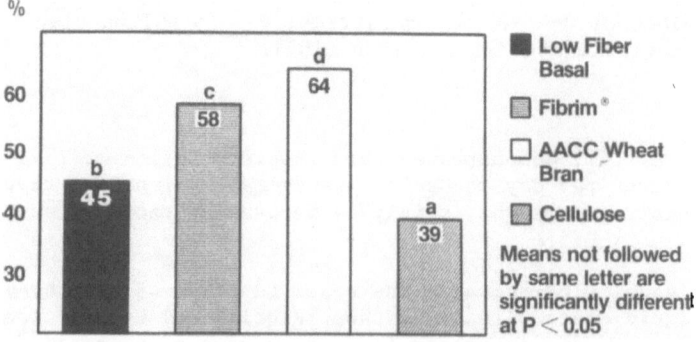

Fig. 10   Effect of Fibrim[R] Soy Fiber, AACC wheat bran and cellulose on fecal moisture contents in pigs [Lo et al, 1979 (15)].

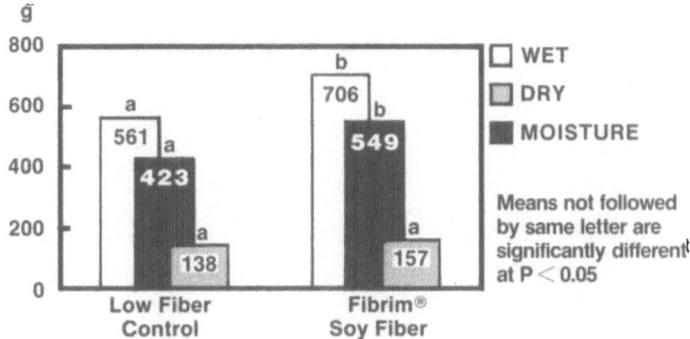

Fig. 11   Effect of Fibrim^R Soy Fiber on fecal outputs in healthy young
male subjects [Tsai et al, 1983 (11)].

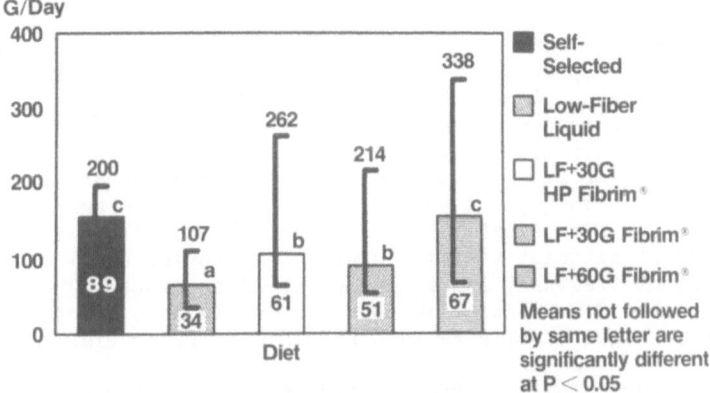

Fig. 12   Effect of daily fecal wet weights (g/day) of healthy men
consuming liquid diets with and without Fibrim^R Soy Fiber
[Slavin et al, 1985 (16)].

reported that the physiological effect of soy fiber on laxation was not
changed by heat processing. Also, although stool weights were increased
significantly on a 60 g soy fiber per day supplementation, transit rates
were similar on all the fiber containing diets (Figure 13).

Bowen et al (17) had similar observations to Slavin et al (16), and
demonstrated that supplementing 20 g, 30 g, or 40 g soy fiber per day in a
liquid formula diet significantly increased fecal wet weight and frequency
of stools in 20 young men. The authors also reported that nitrogen
balance was unaffected by fiber augmentation in both of their studies
(17). Fischer et al (18) showed that even a low level of dietary fiber
supplementation of soy fiber to a low-fiber liquid formula (range from
11.5 to 20.1 g/d) increased stool weight and improved stool consistency in
11 retarded epileptic patients. Similar findings were reported by
Heymsfield et al (19) that a low level of soy fiber supplementation
(12.4 g soy fiber/2000 Kcal per day) also increased stool wet weight from
124 ± 63 g/day to 180 ± 106 g/day in 13 undernourished patients.

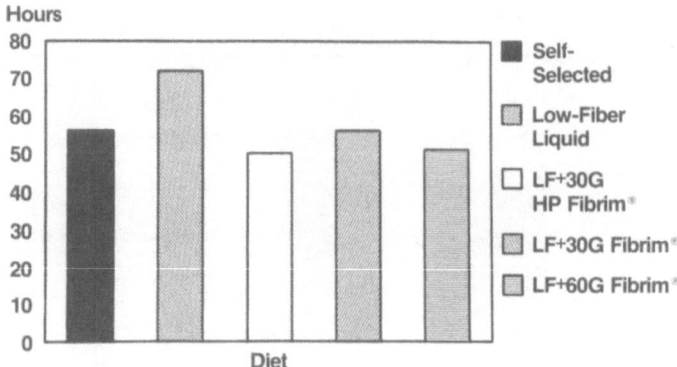

Fig. 13   Effect of transit time on healthy men consuming liquid diets
with and without Fibrim[R] Soy Fiber [Slavin et al, 1985 (16)].

The efficacy of Fibrim[R] Soy Fiber and Metamucil[R] in regulating
bowel habits was compared in a blind crossover designed study (20). A
total of 21 subjects, 14 normal subjects and 7 subjects with irregular
bowel habits, were randomly assigned into 2 groups. Subjects were asked
to consume their customary diet plus 9 g dietary fiber from either 12 g
Fibrim[R] Soy Fiber or 21 g Metamucil[R] for 14 consecutive days. After a
rest period of one month, the subjects were switched to test product for
another 14 days. Both Fibrim[R] Soy Fiber and Metamucil showed
statistically significant positive signs of improving stool morphology,
regulating stool frequency and relieving the discomfort of rectal
evacuation (Table 9). Metamucil[R] caused slightly more urgency and
significantly larger stool size than Fibrim[R] Soy Fiber (Table 10).

Based on physiological effects discussed above, soy cotyledon fiber
(Fibrim[R] Soy Fiber) performs as both soluble and insoluble dietary fiber
in vivo. This conclusion is based on the observation made from both
animal feeding studies (12, 13, 15), and clinical trials conducted with
human subjects (9, 10, 14, 16-20). However, several analytical methods
suggest that soy cotyledon fiber is primarily composed of "insoluble"
dietary fiber.

ANALYTICAL ISSUES

The analytical solubility of dietary fiber is affected by several
factors such as chemical structure, molecular weight, pH, temperature,
reaction time and particle size.

The characteristics of the physiological "soluble dietary fibers"
commonly described by nutritionists include: (1) form a gel in the
stomach, delaying stomach emptying; (2) ferment in the colon to produce
gas and short chain fatty acids (SCFA); (3) increase stool weight, mainly
by increasing moisture content of stool, and (4) affect lipid and
carbohydrate metabolism.

The characteristics of physiological "insoluble dietary fibers" are:
(1) less or non-fermentable; (2) increase stool weight, mainly by
increasing dry solids; (3) decrease transit time, and (4) affect nutrient
utilization.

Table 9

Effect of Metamucil[R] and Fibrim[R] Soy Fiber
on Gastrointestinal Responses in Humans
(Compared to Baseline Data)

Observations	Probability[a] Metamucil[R]	Fibrim[R]
Number of Defecations	0.0009	0.0009
Stool Size	0.0001	0.0059
Stool Morphology	0.0714	0.0461
Regularity of Bowel Habits	0.0481	0.0245
Comfortable & Completeness	0.0577	0.0154
Urgency	0.0193	0.1875
Abdominal Gas	0.0032	0.01

[a]Sign Test (Binomial Test was the Statistical Method Used)

It gradually becomes evident that soluble/insoluble dietary fiber values determined by the analytical assay are only an operational definition, and do not directly reflect physiological responses of the human gut. This is because solubility represents only one physico-chemical property of dietary fiber. Viscosity and fermentability also play major roles in the gastrointestinal tract to determine total physiological responses of dietary fiber in humans.

The concept of classifying dietary fiber sources as "soluble" and "insoluble" was originated at an early stage of human feeding studies to differentiate common dietary fiber sources like wheat bran, alpha-cellulose, pectin, and gums. These fiber sources are largely soluble in water and increase viscosity in the stomach, like pectin and gums, lower plasma cholesterol and improve glucose tolerance. Insoluble dietary fiber sources such as wheat bran and alpha-cellulose are found to have little or none of these effects, but they do affect gastrointestinal responses by decreasing transit time and increasing fecal bulk primarily by increasing fecal dry matter. Thus, it was thought that "soluble" and "insoluble" classifications would simplify the selection of a fiber source to provide particular physiological benefits. It hasn't.

As often happens in a new field of investigation, further work has revealed that other fiber sources such as Fibrim[R] Soy Fiber, psyllium seeds, oat bran, and other legume fibers have physiological properties attributed to both "soluble" and "insoluble" fibers.

The difficulty of correlating analytical data to physiological response is primarily due to the water insoluble non-cellulosic polysaccharides, such as some types of hemicellulose and pectic

Table 10

Comparison of Gastrointestinal Responses Between Metamucil[R] and Fibrim[R] Soy Fiber

Observations	Metamucil[R]	Fibrim[R]	Significance Level[a]
Number of Defecations	13/21	13/21	NS
Stool Size	17/21	10/21	0.02
Stool Morphology	12/21	10/21	NS
Regularity of Bowel Habits	13/21	13/21	NS
Comfortable & Completeness	14/21	14/21	NS
Urgency	10/21	4/21	0.05
Abdominal Gas	11/21	9/21	NS

[a]Chi Square Test was the Statistical Method Used. NS = No significant difference.

substances. These fiber components may solubilize or change viscosity in different physiological conditions to produce unique physico-chemical properties, resulting in different physiological responses in the gastrointestinal tract.

Fermentation occurring in the large intestine increases the complexity. The colon provides favorable conditions for the slow breakdown of most complex carbohydrates for bacterial growth and maintains the cellular function of the large intestine. Fermentation end-products alter the chemical environment of the colon by the formation of volatile fatty acids and various gases, thus further affecting the physiological responses of individual dietary fiber sources.

There has been an overemphasis on the generalization of characteristics of soluble/insoluble dietary fiber components in relation to health claims. As Dr. Jon Story once said, "We need to think seriously about whether this generalization is appropriate, whether it is adequate, and whether or not it is correct in all situations." (21) Therefore, one should not draw any conclusions on physiological responses of dietary fiber based on any given analytical data. Relative to health benefits, only through clinical testing of volunteers can the physiological effects of any given fiber be determined.

REFERENCES

1. Total dietary fiber in foods by enzymatic gravimetric procedure, JAOAC 68 (2):339, 43.A14 - 43.A20 (1985).

2. H. N. Englyst and J. H. Cummings, Improved method for measurement of dietary fiber as non-starch polysaccharides in plant foods, AOAC 71:808 (1988).

3. B. W. Li, Simplified method for the determination of total dietary fiber and its soluble and insoluble fraction in foods, ACS Symposium on Dietary Fiber (1989).

4. R. Mongeau, Rapid gravimetric method for analysis of water soluble and insoluble dietary fiber. Health Protection Branch Laboratory, Bureau of Nutritional Sciences, Ottawa, Canada. Laboratory procedure LPFC-142, unpublished personal communication (1988).

5. D. A. T. Southgate, Measurement of unavailable carbohydrates: structural and non-structural polysaccharides, in: "Determination of Food Carbohydrate," D. A. T. Southgate, ed., Applied Science Publishers, Ltd., London, (1976).

6. O. Theander and E. A. Westerlund, Studies in dietary fiber: 3. Improved procedures for analysis of dietary fiber, J. Agric. Food Chem, 34:330 (1986).

7. G. O. Aspinall, R. Begbie, and J. E. McKay, Polysaccharide components of soybeans, Cereal Sci. Today 12:225 (1967).

8. G. O. Aspinall, Chemistry of soybean carbohydrates, in: Soybean Utilization Alternatives, a symposium sponsored by the Center for Alternative Crops and Products, Univ. of Minnesota, (1988).

9. R. L. Shorey, P. J. Day, R. A. Willis, G. S. Lo, and F. H. Steinke, Effects of soybean polysaccharide on plasma lipids, J. Am. Diet. Assoc. 85:1461 (1985).

10. G. S. Lo, A. P. Goldberg, A. Lim, J. J. Grundhauser, C. Anderson, and G. Schonfeld, Soy fiber improves lipid and carbohydrate metabolism in hyperlipidemic subjects, Atherosclerosis 62:239 (1986).

11. A. C. Tsai, E. L. Mott, G. M. Owen, M. R. Bennick, G. S. Lo, and F. H. Steinke, Effects of soy polysaccharide on gastrointestinal functions, nutrient balance, steroid excretions, glucose tolerance, serum lipids, and other parameters in humans, Am. J. Clin. Nutr. 38:504 (1983).

12. G. S. Lo, S. L. Settle, F. H. Steinke, and D. T. Hopkins, Effects of soy polysaccharide fiber on lipid metabolism in rats, Fed. Proc. 39:784 (1980).

13. G. S. Lo, R. H. Evans, K. S. Phillips, R. R. Dahlgren, and F. H. Steinke, Effect of soy fiber and soy protein on cholesterol metabolism and atherosclerosis in rabbits, Atherosclerosis 64:47 (1987).

14. A. C. Tsai, A. I. Vinik, A. Lasichak, and G. S. Lo, Effects of soy polysaccharide on postprandial plasma glucose, insulin, glucagon, pancreatic polypeptide, somatostatin, and triglyceride in obese diabetic patients, Am. J. Clin. Nutr. 45:596 (1987).

15. G. S. Lo, S. L. Settle, F. H. Steinke, and D. T. Hopkins, Effect of transit time and fecal output of soy polysaccharides in pigs, Fed. Proc. 38:548 (1979).

16. J. L. Slavin, B. A. Nelson, E. A. McNamara, and K. Cashmere, Bowel function of healthy men consuming liquid diet with and without dietary fiber, JPEN 9:317 (1985).

17. P. E. Bowen, M. McCallister, F. W. Thye, L. J. Taper, P. A. Sherry-Scanman, and A. L. Hecker, Bowel function and macronutrient absorption using fiber-augmented liquid formula diets, JPEN 6:588 (1982).

18. M. Fischer, B. Liebl, S. Van Calcar, and J. Marlett, Bowel performance and mineral balances in a nonambulant, profoundly retarded epileptic population receiving fiber containing enteral feedings, JPEN 11 (1):175 (1987).

19. S. B. Heymsfield, C. Roongspisuthipong, M. Evert, K. Casper, P. Heller, and S. S. Akrabawi, Fiber supplementation of enteral formulas: effects on the bioavailability of major nutrients and gastrointestinal tolerance, JPEN 12:265 (1988).

20. G. S. Lo, Subjective clinical observations on the comparison of gastrointestinal responses between Metamucil and Fibrim[R] Soy Fiber, Unpublished observation, 1984.

21. J. Story, Report on the results of the discussion and recommendations on the efficacy of Fibrim[R] Soy Fiber supplementation in lowering plasma cholesterol, in: "Conference on Soy Fiber, Abstracts and Discussions," Protein Technologies International, St. Louis, Mo., 1987.

INTERACTION OF DIETARY FIBER WITH LIPIDS -

MECHANISTIC THEORIES AND THEIR LIMITATIONS

Ivan Furda

General Mills, Inc.
Minneapolis, Minnesota 55427

INTRODUCTION

It is now well established that dietary fiber is not an inert entity. On the contrary, it can effectively interact with other food components, whether they are macro or micro nutrients. Among the macronutrients, dietary lipids have been studied most extensively in conjunction with dietary fiber. The inclusion of dietary fiber in diet not only displaces lipids, but it also frequently alters or diminishes their physiological and nutritional contributions. This is usually demonstrated in reduced calorie density of the diet, in reduction of blood and hepatic lipids, and in alteration and output of fecal lipids. These changes are of a major significance since they are believed to be beneficial in preventing or reducing serious diseases which include obesity, coronary heart disease as well as certain types of cancer. In order to explain these changes, it is important to understand the interactions and reactions between different dietary fibers and lipids on the molecular basis. Although a number of excellent theories on how dietary fiber reacts or interacts with dietary lipids and how the hypolipidemic effects are achieved have been proposed, our understanding of the mechanism of these interactions is still unsatisfactory. The main reason for this lies in great chemical diversity of existing dietary fibers, and to a lesser degree of lipids, and in extreme difficulty to monitor the reactions which take place in the intestine and in the other parts of the human body. It is likely that better physicochemical characterization of dietary fibers and lipids and correlation of these attributes with corresponding physiological effects will inevitably improve our understanding of these vital interactions.

The following paragraphs discuss the possible consequences of excessive fat intake and primarily, they focus on the main theories and hypotheses of fiber-lipids interactions and resulting hypolipidemic effects that have been proposed and investigated. A special attention is paid to some of their weaknesses and limitations.

POSSIBLE HEALTH CONSEQUENCES OF EXCESSIVE FAT INTAKE

The incidence of serious diseases existing in affluent societies that are related to increased fat intake is dangerously high. Table 1.

New Developments in Dietary Fiber
Edited by I. Furda and C. J. Brine
Plenum Press, New York, 1990

illustrates this problem as it exists in the United States[1,2,3].
Although a significant improvement in the area of coronary heart disease
has been recently achieved, the obesity and the certain types of cancer
that are associated with the high fat intake are still growing. It is
therefore of a paramount importance to reduce the total daily fat intake
and to understand how dietary fiber may diminish the negative
physiological and nutritional consequences of the high fat diet.

Table 1

Risks of Excessive Fat Intake

Coronary Heart Disease

Hypercholesterolemic Population in U.S.A. (Adults)[1]
    25% have blood cholesterol > 240 mg/dl -- high risk
    50% have blood cholesterol > 200 mg/dl -- moderate risk

Obesity Affects approximately 34 million adults, ages 20 to 74

       years in the United States.[2]

Cancer
    Colon - 5% of Americans will have it in their lifetime (15%
        of total cancers).[3]

    Breast - 1 in 10 women will have it in her lifetime.[3]

MAJOR TYPES OF DIETARY FIBER-LIPID INTERACTIONS

Most of the interactions between the dietary fibers and lipids
result in reduced lipid absorption. The possible mechanisms of dietary
fiber influences on lipid absorption were summarized by Vahouny[4] and the
proposed theories are shown in Table 2. While the inhibition of lipid
absorption through different mechanisms is probably a dominant mode of
fiber action, other mechanisms have been proposed and investigated as
well. Some of these mechanisms focus on specific body organs and
enzymes which they produce[5]. The activities of these enzymes may be
altered by the interference of some fibers or by the specific compounds
closely associated with them[6]. The other proposed mechanisms zero in on
products of fiber fermentation and their effects on hepatic cholesterol
metabolism[7]. Yet another mechanistic hypothesis implicates enhanced
lipoprotein metabolism in peripheral tissues as the final effect of
fiber ingestion[8]. In summary, these mechanistic theories are:

- Inhibition of pancreatic lipase by dietary fiber;

- Inhibition of HMG CoA reductase by isoprenoid compounds
  associated with dietary fiber;

- Effect of volatile short chain fatty acids on hepatic
  cholesterol synthesis;

- Increased rate of LDL catabolism in lipoprotein metabolism.

Table 2

Possible Mechanisms of Dietary Fiber Influences on Lipid Absorption

Direct effects
        Gastric emptying
        Altered transit times
        Interference with bulk phase diffusion and availability to
        intestinal surface
        Binding of fiber to intestinal surface coat
        Sequestration of bile acids and other micellar components

Indirect effects
        Effects on bile acid pool size and composition
          a. Increased fecal excretion of acidic and neutral
              steroids
          b. Increased alpha-hydroxylation of cholesterol
        Altered responses of gut glucagon and pancreatic insulin
        Adaptive changes in intestinal structure and function

---

Adapted from Vahouny (1982)[4]

BILE ACIDS BINDING THEORY

Among all the theories which have been proposed to explain hypocholesterolemic effect of specific fibers, this theory received the greatest attention and most extensive evaluation. Its principle is based on the observation that circulating bile acids can be bound, adsorbed or sequestered by specific dietary fibers and subsequently excreted rather than reabsorbed[9]. In other words, the effective fibers act as cholestyramine which interrupts the enterohepatic circulation of bile acids. To compensate bile acids losses, more cholesterol in the liver is converted to bile acids which eventually leads to decrease in total serum cholesterol and low density lipoproteins. The incremental loss of bile acids associated with fiber ingestion is typically 30-70% which is usually accompanied by small incremental loss of neutral sterols amounting to 2-18%[10].

Although entirely logical in principle, this theory has a few weaknesses and limitations:

- Chemically, there is no base for "ionic binding" of bile acids by neutral or acid fibers/polysaccharides. Only basic polymers such as cholestyramine or chitosan can bind bile acids by ionic bonds. Physical entrapment or adsorption, rather than chemical binding of bile acids by neutral or acid fiber is a more likely type of interaction.

- The changes in bile acids excretion alone are not large enough to account for changes in serum cholesterol[11]. Simulated threefold increase in fecal bile acids output achieved by administration of low doses of cholestyramine (0.05g/kg/d) failed to reduce cholesterol in normolipidemic volunteers[12]. The effective doses of cholestyramine increase the excretion of bile acids usually by more than sixfold[13].

- These changes are also inconsistent with serum cholesterol lowering. For example, beans or soy polysaccharide reduce or do not change bile acids excretion and yet they reduce serum cholesterol[14,15]. On the other hand, insoluble fibers which are generally not effective in lowering of serum cholesterol, show sometimes elevated excretion of bile acids[15].

-   Little is known about the effects of changes in the pool size of the specific bile acids on overall sterol balance.

THEORY OF FIBER INTERFERENCE WITH STABILITY OF MIXED MICELLES

It is well known that lipids can be absorbed only from the emulsion which is formed of mixed micelles. If the mixed micelles disintegrate, the micellar components cannot be effectively absorbed and utilized. Specific dietary fibers exert different chemical and physical forces which tend to destabilize mixed micelles leading sometimes to their disintegration or entrapment. Although seemingly similar, there are substantial differences between the bile acids binding theory and this theory. Unlike in the bile acids binding theory, the entrapment, adsorption or binding of the portion of micelle is not specific predominantly to bile acids. The whole micelles or several of their components can be entraped or bound by specific fibers. This usually leads to reduced lipid absorption. Also, in the bile acids binding theory the interruption of the enterohepatic circulation of bile acids includes increased hepatic cholesterol oxidation as opposed to the destabilization of mixed micelles which is limited to interactions in the gut. Some fibers, primarily those which are positively charged, destabilize mixed micelles by forming ionic bonds with the negatively charged micellar components, or they may entrap the whole micelles[16,17]. The neutral and negatively charged fibers may also destabilize mixed micelles, however, the mechanism in this case is less clear.

Destabilization through Fiber - Mixed Micelles Binding

This mechanism was proposed by Nagyvary[18] for micelles binding to aluminum pectinate. He had shown that in a series of pectic acid, sodium pectinate and aluminum pectinate, the aluminum salt was most hypolipidemic in rats. Nagyvary proposed the formation of a ternary complex between aluminum pectinate and mixed micelles in which the free positive charge on aluminum cation forms ionic bond with the negatively charged components of mixed micelles. Similar interactions between aluminum and ferric pectinaceous fibers, and negatively charged components of mixed micelles were also described[19].

Table 3

Hypolipidemic Effects of Different Chemical Forms of Pectin

Different Chemical Form

Polygalacturonic Acid <<Low Methoxyl Pectin <High Methoxyl Pectin <Al^{3+} Pectinates
DM=0　　　　　　　DM<50%　　　　　　DM>50%

Different Molecular Size

Galacturonic Acid << Low MW Pectin <High MW Pectin
  (no effect)

DM = Degree of Methoxylation
MW = Molecular weight

Hoagland[20] proposed that even divalent cations such as calcium can participate in formation of calcium pectinate - mixed micelle complexes and explain the hypocholesterolemic effect of pectinaceous fibers present in carrots. The relative hypolipidemic efficacy of different chemical forms of pectin fibers, based on these and other numerous studies, is illustrated in Table 3.

Entrapment of Whole Micelles

This mode of action has been proposed for chitosan by several investigators[16,17,21]. Among the natural fibers, chitosan is the only polysaccharide which is soluble at gastric pH and which precipitates in neutral environment. During this stage, the molecules of chitosan start to associate and entrap other, preferentially negatively charged molecules in the small intestine. Specifically, chitosan can entrap negatively charged mixed micelles in toto after they have been attracted and associated with chitosan molecules by ionic linkages[16,17].

The structural differences between chitin and chitosan and their relevance to the hypolipidemic activity in one case and inactivity in chitin's case are illustrated in Figure 1. Since chitin lacks the high density of the positive charge in its molecule as well as the ability to solubilize in the stomach and subsequently precipitate in the small intestine, it cannot be hypolipidemic[16,22]. When chitosan is hydrolized to small oligomers which do not precipitate in the small intestine, in spite of the high density of the positive charge, they also become ineffective[23].

Destabilization of Mixed Micelles by Neutral and Negatively Charged Fibers

Unlike fibers which have an excess of positive charge in their molecules, it is difficult to envision chemical interactions between neutral or negatively charged fiber polymers and predominantly negatively charged mixed micelles. It is more likely that these interactions are of the physical nature in which the dispersed fibers act as barriers to uninhibited diffusion and absorption of organic molecules and mixed micelles[24,26].

While the general validity of the interference of numerous fibers with the stability of micelles containing hydrolized dietary lipids is accepted, the precise understanding of the involved mechanisms is not complete. The gray areas and questions to this mechanistic theory could be the following:

- By which mechanism neutral and negatively charged fibers destabilize mixed micelles?

- Since the fecal acid sterols are usually low, it is not clear whether positively charged fibers bind or entrap bile acids in the upper small intestine only temporarily and later release them for reabsorption.

- Unlike cholestyramine, nothing is known about the behavior of the positively charged polysaccharides in humans since only animal studies were conducted so far.

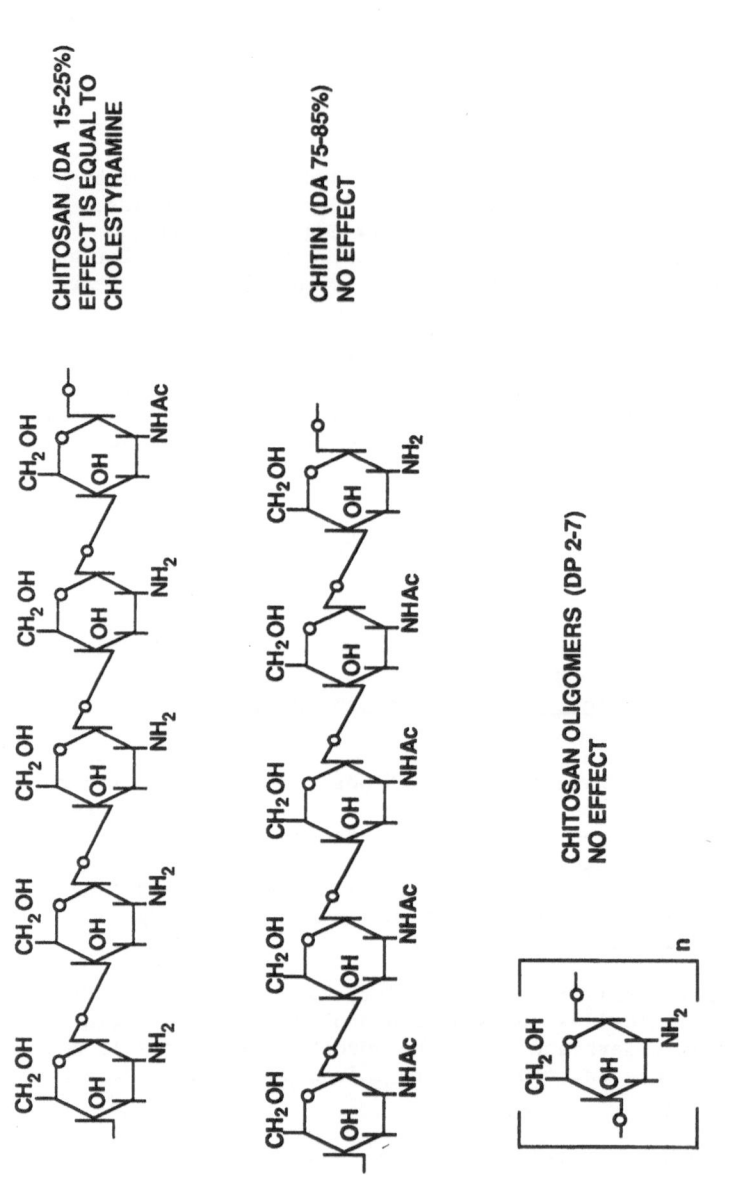

CHITOSAN (DA 15-25%)
EFFECT IS EQUAL TO
CHOLESTYRAMINE

CHITIN (DA 75-85%)
NO EFFECT

CHITOSAN OLIGOMERS (DP 2-7)
NO EFFECT

n = 2-7

DA = Degree of Acetylation

DP = Degree of Polymerization

Figure 1. Hypocholesterolemic effect of different chemical forms of aminopolysaccharides

- Destabilization of mixed micelles is reflected in increased excretion of other lipids besides acid and neutral sterols. The relative effectiveness of different fibers and type of excreted lipids are not known; they need elucidation.

- Incremental fecal fat excretion (2-5 g/day), which is caused primarily by soluble fibers, is not large enough to explain typical 10-25% reductions in serum cholesterol.

THEORY OF ADAPTATION OF INTESTINAL STRUCTURE AND FUNCTION TO FIBER FEEDING

Conceived and studied by Vahouny[25], this theory postulates that prolonged fiber feeding can modify not only the morphological characteristics of the intestine but also nutrient absorptive functions of the intestinal epithelium. These structural changes may be correlated, at least in part, with transport characteristics of the small intestine even in the absence of the fiber source.

Figure 2 shows the effect of prefeeding diets containing specific fibers on subsequent cholesterol absorption in fasted rats. All fibers, whether soluble or insoluble, had a profound effect on the absorption of cholesterol into lymph during the first four hours of collection. The overall 24-hour recoveries improved primarily with the insoluble fibers, while with the soluble fibers, the cholesterol absorption was delayed and impaired.

There are some inconsistencies and limitations with this theory as well:

- The indirect effects of short term (3-day) and long term (4-6 weeks) prefeeding of different fibers on subsequent absorptions of cholesterol, triglycerides and oleic acid in rats showed their delayed and impaired absorption not only by viscous soluble fibers pre-feedings but also by cellulose and other insoluble fibers which usually do not reduce serum lipids[27].

- In rat studies lasting 30 days, no evidence was obtained to suggest that the consumption of a diet containing guar gum (20g/kg), which is similar to levels used in human studies, leads to any adaptive reduction in their rates of cholesterol or glucose absorption[26].

- The requirements for reversal of intestinal adaptive changes induced by fiber feedings as well as translation of these findings to humans are presently not known.

THEORY OF INTERFERENCE WITH BULK PHASE DIFFUSION AND AVAILABILITY TO INTESTINAL SURFACE

The basic premise of this theory is the observation that viscous soluble fibers act as antimotility agents which reduce the diffusion of nutrients, and as intestinal surface coating agents which increase the thickness of unstirred water layer[24,26,28]. These attributes slow down and reduce the absorption of nutrients including lipids and cholesterol (Figure 3). This theory is supported by a number of animal and human studies in which hypolipidemic effects have been observed in conjunction with soluble viscous fibers but not with insoluble fibers[29].

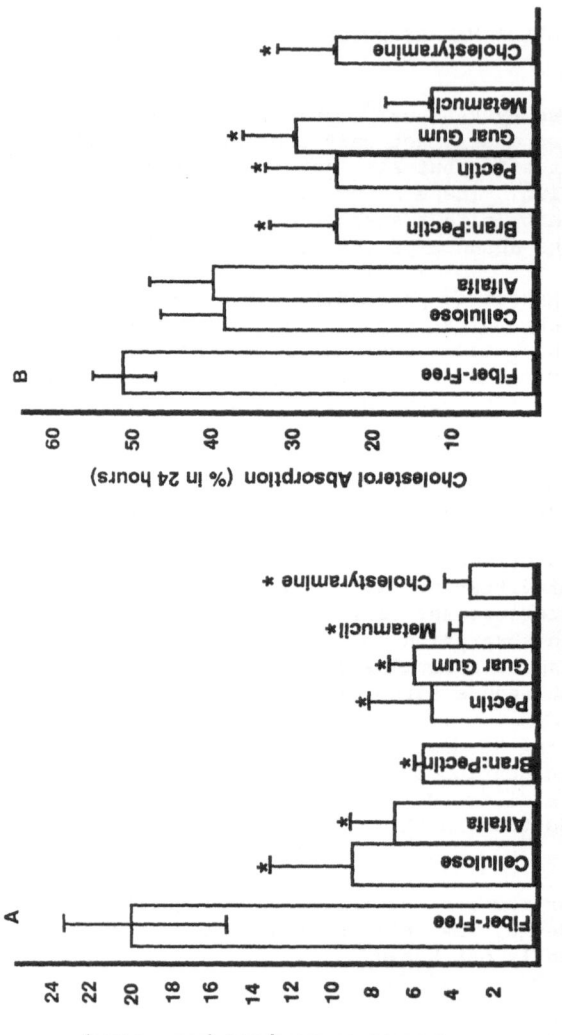

Figure 2. Absorption of cholesterol (25 mg) in the thoracic duct lymph of fasted rats fed for 4 weeks *ad libitum* on defined diets containing: no fiber; 10% cellulose, alfalfa, or a bran-pectin (3:1) mixture; 5% pectin, guar gum, or Metamucil; or 2% cholestyramine.

(A) Recovery during the first 4 hours after an intraduodenal lipid test dose. (B) Recovery by 24 hours after the lipid dose. Asterisks indicate significant differences ($p < 0.05$) from fiber-free group.

Adapted from Vahouny (1986)[25]

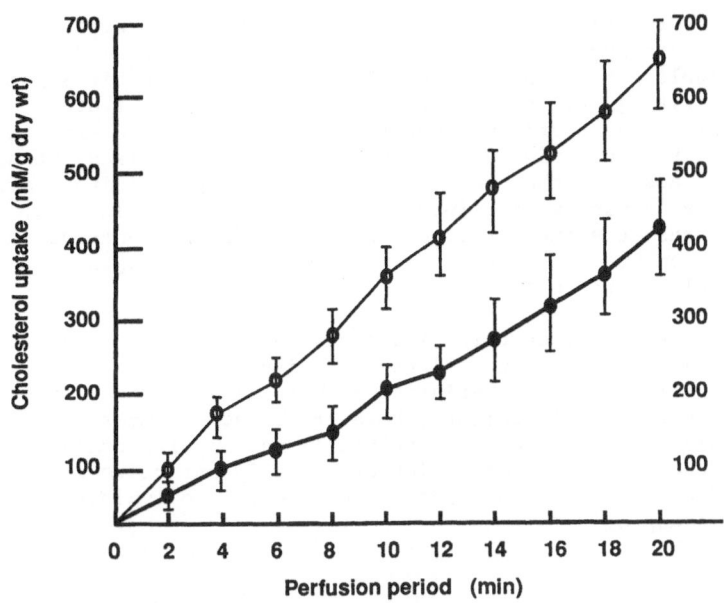

Figure 3. Uptake of cholesterol (nM/g dry weight) v. period of perfusion (min) by loops of rat intestine, pre-perfused with Ringer solution only (o) or with Ringer solution containing guar gum (•).
(Rate of uptake by test loops reduced by approx. 36%).

Adapted from Gee et al. (1983)[26]

The weaknesses and inconsistencies of this theory could be illustrated by the following examples:

- While some soluble viscous and gel forming fibers are effective serum cholesterol reducers (guar gum, beta glucan, psyllium, etc.), the other soluble viscous fibers (alginates, agar, hydroxypropylcellulose) have not been associated with appreciable effects[30,31].

- Low viscosity gum arabic was shown to be also effective[32].

- According to intestine adaptation theory, the presence of the antimotility agent is not required since reduction of lipid and cholesterol absorption can be achieved by fiber prefeeding alone[25].

- Experiments with chitosans having wide spectrum of viscosities (17-1600 cps) showed impressive and virtually identical cholesterol-lowering effects in rat studies[23].

THEORY OF INHIBITION OF PANCREATIC LIPASE ACTIVITY

The basis of this theory is an assumption that some fibers may inhibit the activity of pancreatic lipase. This would lead to reduced degree of hydrolysis and absorption of dietary triglycerides and subsequently to hypolipidemic effects. Schneeman, et al[33] had shown that insoluble fibers such as cellulose and xylan inhibit significantly lipase activity _in vitro_ while pectin does not. On the other hand, neither cellulose, pectin nor wheat bran inhibited pancreatic lipase activity in rat intestine or in the pancreas[5,34]. Lairon, et al[35,36] have shown, however, that wheat bran has a strong inhibitory effect on the activity of pancreatic lipase _in vitro_ as well as in rats. They postulated that the inhibition may be due to the presence of soluble proteic components and to a lesser degree to the adsorption of lipase on the insoluble particles[37].

It appears that at this time it is not clear how significant is the inhibition of pancreatic lipase activity by dietary fiber, and which fibers are effective. The existing results indicate that some insoluble fibers are effective inhibitors, however, based on numerous animal and human studies, these fibers are generally not hypolipidemic. On the contrary, the soluble fibers which are usually effective in reducing hyperlipidemia had not been associated with lipase inhibition to any appreciable degree.

INHIBITION OF HEPATIC CHOLESTEROL SYNTHESIS-PROPIONATE THEORY

Fermentable soluble fibers such as beta glucan from oat bran increase concentrations of short chain fatty acids (SCFA) in colon and subsequently in portal vein and hepatic blood[38]. It has been suggested that propionate specifically, inhibits hepatic cholesterol synthesis and contributes to the cholesterol lowering of oat bran or other fermentable soluble fibers[7].

Although the propionate theory in principle appears quite logical, there are some questions and weak points associated with it:

- Concentrations of propionate in the hepatic portal vein, although raised by oat bran are less than 2% of those at which inhibition of hepatic cholesterogenesis had been observed _in vitro_[39].

- Recent studies with rats fed soluble fibers with different digestibilities increased the level of SCFA in portal vein but the highest concentrations did not parallel the lowest serum cholesterol levels[40].

- Soluble fibers with low fermentability such as psyllium are not necessarily less hypocholesterolemic then beta glucan from oat bran[41, 52, 53].

- Inhibition of cholesterol synthesis in the liver and intestines is not responsible for plasma cholesterol lowering induced by dietary supplementation of sodium propionate in rats and pigs[43].

- Oligosaccharides of hydrolized beta glucan[42], or chitosan with no residual viscosity[23], which are not absorbable in small intestine but fermentable in large intestine did not show cholesterol lowering effects in experimental animals.

- Experiments involving the bypass of small intestine showed no effect.

THEORY OF INCREASED RATE OF LDL CATABOLISM IN LIPOPROTEIN METABOLISM

Chen, et al[8] had observed that oat bran feeding lowers rapidly LDL without altering their precursors VLDL. They proposed that the explanation could be in increased rate of LDL catabolism. The increased LDL catabolism may be the result of soluble fiber induced elevated peripheral serum acetate concentrations which may inhibit cholesterol synthesis in peripheral tissues resulting in increase in peripheral LDL receptors and subsequent increase in LDL clearance.

The inhibition of the cholesterol synthesis by acetate and lactate in isolated rat hepatocytes was studied by Beynen, et al[44] and it is shown in Fig. 4.

This theory needs further verification since the increased rate of LDL clearance has not been demonstrated yet after soluble fiber or oat bran feeding. It is rather a hypothesis at this time.

THEORY OF ALTERED RESPONSES OF GASTROINTESTINAL HORMONES

Another indirect effect of dietary fiber on lipid absorption could be through altered responses of gastrointestinal hormones[4]. Components of dietary fiber alter the rate of nutrient absorption and may thereby influence the release of these hormones. The release of the gastrointestinal hormones which stimulate insulin release may be modified by specific fibers in the gut contents[45]. In general, dietary fibers reduce the levels of circulating gastrointestinal hormones, namely gastric inhibitory polypeptide (GIP), glucagon and enteroglucagon[46]. Although there is no sharp distinction between the effectiveness of individual fibers, the soluble viscous fibers tend to suppress the release of these hormones more consistently. Whether the suppression of release of these hormones and reduction of serum insulin influence the regulation of hepatic cholesterol and fatty acid synthesis and subsequent hypolipidemic effects is not entirely clear. Some animal studies[47,48] showed however that the administration of glucagon either inhibited the intestinal absorption of cholesterol in hypercholesterolemic and normocholesterolemic animals, or it increased the bile flow[49]. In other animal studies[50] plasma enteroglucagon was elevated significantly more with soluble nonstarch polysaccharides than with cellulose. As in previous cases, this mechanistic theory raises a number of questions and issues:

-    Is there any relationship between the hormonal changes and the hypolipidemic effects induced by specific fibers?

-    Some studies reported an increase or no change in levels of insulin and other circulating hormones followed by specific fiber feedings[46].

-    There is no clear understanding about the physical and chemical parameters of specific fibers that are responsible for altered hormonal responses.

-    Is there any linkage between the reduction of serum insulin induced by viscous fibers, or by nibbling, and the possible inhibition of HMG CoA reductase?[51]

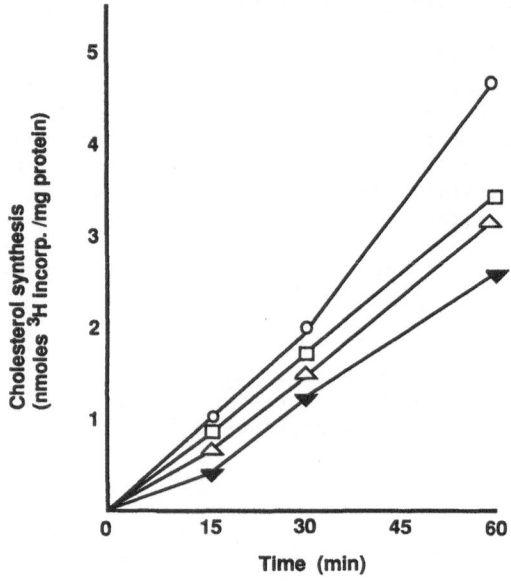

Figure 4. Effects of lactate and acetate on the rate of cholesterol synthesis by isolated rat hepatocytes. Control (o); lactate, 10mM (△); acetate, 10mM (☐); lactate plus acetate (▼). All incubations contained 10mM glucose and [^{3}H]H$_2$O (1mCi/ml).

Adapted from Beynen, et al. (1982) [44]

CONCLUSIONS

Although a number of excellent theories attempting to explain the mechanism of interactions of dietary fiber with lipids and the resulting hypocholesterolemic and hypolipidemic effects have been proposed and investigated, they are neither consistent nor complete. It is still not clear whether there is a dominant mechanism involved, or whether these physiological effects are achieved by the combination, or by the simultaneous interplay of several mechanisms. It is also not known whether each type of fiber has its own mechanism.

In order to better understand this area, an approach which proposes a more comprehensive characterization of investigated fibers may prove to be useful. As it was shown in some examples, it is possible to correlate the magnitude of the hypocholesterolemic effects with the chemical structure of the specific fibers or polysaccharides. Thus, better physical, chemical and analytical characterization of the investigated fibers used in in vivo studies should be helpful in better understanding of involved mechanisms. Such characterization should include attributes and parameters which have been used previously, as well as those which have not been used yet, or they were used only sporadically.

The necessary information should include:

-    origin and source of fiber,
-    its preparation and processing,
-    fiber analysis (TDF, IF, SF),
-    gross chemical characterization of the major fiber polysaccharide (type of sugars, degree of esterification or etherification),
-    water binding and cation exchange capacity,
-    viscosity,
-    estimation of the major accompanying components (starch, protein, lipids, sugars, ash).

When appropriate, additional data and analyses, such as:

-    isolated individual polysaccharide fractions,
-    type of linkages in the main polysaccharide component,
-    polysaccharide side chains attached to main polysaccharide chain;
-    molecular weight of the major polysaccharide component, and
-    detailed analysis of accompanying non-fiber components

will provide extremely useful information for elucidation of *in vivo* interactions. The research on dietary fiber is now entering a more advanced phase in which the explanation of these interactions will require thorough correlation of the physico-chemical attributes of individual fibers with observed nutritional and physiological efficacy.

REFERENCES

1. The Expert Panel, Report of the National Cholesterol Education Program (NCEP) expert panel on detection, evaluation, and treatment of high blood cholesterol in adults, Arch. Intern. Med. 148:36 (1988).
2. 1987 Anthropometric reference data and prevalence of overweight, USA 1976-1980. National Health Survey, series 11, No. 238. DHHS publication No. (PHS) 87-1688. Hyattsville, MD: National Center for Health Statistics.
3. Cancer Facts and Figures - 1989. American Cancer Society, Inc., Atlanta, GA, Statistics, p. 10, 1989.
4. G. V. Vahouny, Dietary fibers and intestinal absorption of lipids, in: "Dietary Fiber in Health and Disease," G. V. Vahouny and D. Kritchevsky, eds. Plenum Press, New York and London (1982).
5. B. A. Schneeman, Pancreatic and digestive function, in: "Dietary Fiber in Health and Disease," G. V. Vahouny and D. Kritchevsky, eds. Plenum Press, New York and London (1982).
6. A. A. Qureshi, W. C. Burger, D. M. Peterson, C. E. Elson, The structure of an inhibitor of cholesterol biosynthesis isolated from barley, J. Biol. Chem. 261: 10544 (1986).
7. W-J. L. Chen. J. W. Anderson, D. Jennings, Propionate may mediate the hypocholesterolemic effects of certain soluble plant fibers in cholesterol-fed rats, Proc. Soc. Exp. Biol. Med. 175:215 (1984).
8. W-J. L. Chen, J. W. Anderson, Hypocholesterolemic effects of soluble fibers, in: "Dietary Fiber, Basic and Clinical Aspects," G. V. Vahouny and D. Kritchevsky, eds. Plenum Press, New York (1986).

9.   J. A. Story and D. Kritchevsky, Dietary fiber and lipid metabolism, in: "Fiber in Human Nutrition," G.A. Spiller and R. J. Amen, eds. Plenum Press, New York (1976).

10. J. W. Anderson and J. Tietyen-Clark, Dietary fiber: Hyperlipidemia, hypertension and coronary heart disease, Am. J. Gastroenterology 81:907 (1986).

11. J. A. Story, Modification of steroid excretion in response to dietary fiber, in: "Dietary Fiber, Basic and Clinical Aspects," G. V. Vahouny and D. Kritchevsky, eds. Plenum Press, New York (1986).

12. R. M. Kay, O. D. Rotstein, K. P. Vasal et al., Induction of a threefold increase in fecal bile acid output improves bile composition but does not alter plasma cholesterol concentration in man. Gastroenterology 78:1192 (1980).

13. C. D. Moutafis, L. A. Simons, N. B. Myant, P. W. Adams and V. Wynn, The effect of cholestyramine on the fecal excretion of bile acids and neutral steroids in familial hypercholesterolemia, Atherosclerosis. 26:329 (1977).

14. J. W. Anderson, L. Story, B. Sieling, W-J. L. Chen, M. S. Petro, and J. Story, Hypocholesterolemic effects of oat bran or bean intake for hypercholesterolemic men, Am. J. Clin. Nutr. 40:1146 (1984).

15. Bile acid excretion, in: "Physiological Effects and Health Consequences of Dietary Fiber," S. Pilch, ed. Life Sciences Research Office FASEB, Bethesda, MD (1987).

16. I. Furda, Aminopolysaccharides - their potential as dietary fiber, in: "Unconventional Sources of Dietary Fiber," I. Furda, ed. Amer. Chem. Soc. Symposium Series 214, Washington, D. C. (1983).

17. J. L. Nauss, J. L. Thompson and J. Nagyvary, The binding of micellar lipids to chitosan, Lipids 18:714 (1983).

18. J. J. Nagyvary, J. D. Falk, M. L. Hill, M. L. Schmidt, A. K. Wilkins and E. L. Bradbury, The hypolipidemic activity of chitosan and other polysaccharides in rats, Nutr. Rep. Int. 20: No. 5, 677 (1979).

19. I. Furda, Interaction of pectinaceous dietary fiber with some metals and lipids, in: "Dietary Fibers Chemistry and Nutrition," G. E. Inglett, S. I. Falkehag, eds. Academic Press, New York (1979).

20. P. D. Hoagland and P. E. Pfeffer, Cobinding of bile acids to carrot fiber, J. Agric. Food Chem. 35:316 (1987).

21. K. Ebihara and B. O. Schneeman, Interaction of bile acids, phospholipids, cholesterol and triglycerides with dietary fibers in the small intestine of rats, J. Nutr. 119:1100 (1989).

22. M. Sugano, T. Fujikawa, Y. Hiratsuji, K. Nakashima, N. Fukuda, Y. Hasegawa, A novel use of chitosan as a hypocholesterolemic agent in rats, Amer. J. Clin. Nutr 33:787 (1980).

23. M. Sugano, S. Watanabe, A. Kishi, M. Izume and A. Ohtakara, Hypocholesterolemic action of chitosans with different viscosity in rats, Lipids 23:187 (1988).

24. G. A. Gerencser, J. Cerda, C. Burgin, M. M. Baig and R. Guild, Unstirred water layers in rabbit intestine: Effects of pectin, Proc. Soc. Exp. Biol. Med. 176:183 (1984).

25. G. V. Vahouny and M. M. Cassidy, Dietary fiber and intestinal adaptation, in: "Dietary Fiber Basic and Clinical Aspects," G. V. Vahouny and D. Kritchevsky, eds. Plenum Press, New York (1986).

26. J. M. Gee, N. A. Blackburn and I. T. Johnson, The influence of guar gum on intestinal transport in the rat, Brit. J. Nutr. 50:215 (1983).

27. G. V. Vahouny, T. Roy, L. L. Gallo, J. A. Story, D. Kritchevsky and M. M. Cassidy, Dietary Fibers III. Effects of chronic intake on cholesterol absorption and metabolism in the rat, Am. J. Clin. Nutr. 33:2182 (1980).

28. B. Elsenhans, U. Sufke, R. Blume and W. F. Caspary, The influence of carbohydrate gelling agents on rat intestinal transport of monosaccharides and neutral amino acids in vitro, Clinical Science 59:373 (1980).

29. Effects of dietary fiber on normal physiological functions - Serum lipids, in: "Physiological Effects and Health Consequences of Dietary Fiber," S. M. Pilch, ed. Life Sciences Research Office FASEB, Bethesda, MD (1987).

30. D. M. W. Anderson, Exudate and other gums as forms of soluble dietary fibre, in "Nutritional and Toxicological Aspects of Food Processing," R. Walker and E. Quattruci, eds. Taylor and Francis, N. Y. (1988).

31. S. Kiriyama, Y. Okazaki and A. Yoshida, Hypocholesterolemic effect of polysaccharides and polysaccharide-rich foodstuffs in cholesterol-fed rats, J. Nutr., 97:382 (1969).

32. A. H. M. Ross, M. A. Eastwood, W. G. Brydon, J. R. Anderson, D. M. W. Anderson, A study of the effects of dietary gum arabic in humans, Am. J. Clin. Nutr. 37:368 (1983).

33. G. Dunaif and B. O. Schneeman, The effect of dietary fiber on human pancreatic enzyme activity in vitro, Am. J. Clin. Nutr. 34:1034 (1981).

34. B. O. Schneeman and D. Gallaher, Changes in small intestinal digestive enzyme activity and bile acids with dietary cellulose in rats, J. Nutr. 110:584 (1980).

35. D. Lairon, P. Borel, E. Termine, R. Grataroli, C. Chabert, J. C. Hauton, Evidence for a proteinic inhibitor of pancreatic lipase in cereals, wheat bran and wheat germ, Nutr. Rep. Int. 32:1107 (1985).

36. P. Borel, D. Lairon, M. Senf, M. Chautan, H. Lafont, Wheat bran and wheat germ: effect on digestion and intestinal absorption of dietary lipids in the rat, Am. J. Clin. Nutr. 49:1192 (1989).

37. D. Lairon, H. Lafont, J. L. Vigne, G. Nalbone, J. Leonardi, J. C. Hauton, Effects of dietary fibers and cholestyramine on the activity of pancreatic lipase in vitro, Am. J. Clin. Nutr. 42:629 (1985).

38. J. H. Cummings, E. W. Pomare, W. J. Branch, C.P.E. Naylor and G. T. Macfarlane, Short-chain fatty acids in human large intestine, portal, hepatic, and venous blood, Gut 28:1221 (1987).

39. J. R. Illman and D. L. Topping, Effects of dietary oat bran on fecal steroid excretion, plasma volatile fatty acids and lipid synthesis in rats, Nutr. Res. 5:839 (1985).

40. M. K. Rudd and R. D. Reeves, Effects of soluble dietary fiber on cholesterol metabolism in the rat. Presented at the Annual FASEB meeting, Abstract No. 4890, New Orleans (1989). Prepared for press.

41. J. W. Anderson, N. Zettwoch, T. Feldman, J. Tietyen Clark, P. Oeltgen, C. W. Bishop, Cholesterol-lowering effects of psyllium hydrophilic mucilloid for hypercholesterolemic men, Arch. Intern. Med. 148:292 (1988).

42. J. G. Fadel, R. K. Newman, C. W. Newman and A. E. Barnes, Hypocholesterolemic effects of beta glucans in different barley diets fed to broiler chicks, Nutr. Rep. Int. 35: No. 5, 1049 (1987).

43. R. J. Illman, D. L. Topping, G. H. McIntosh, R. P. Trimble, G. B. Storer, M. N. Taylor, B.Q. Cheng, Hypocholesterolemic effects of dietary propionate: studies in whole animals and perfused rat liver, Ann. Nutr. Metab. 32:97 (1988).

44. A. C. Beynen, K. F. Buechler, A. J. van der Molen, M. J .H. Geelen, The effects of lactate and acetate on fatty acid and cholesterol biosynthesis by isolated rat hepatocytes, Int. J. Biochem. 14:165 (1982).

45. A. R. Leeds, Modification of intestinal absorption by dietary
     fiber and fiber components, in "Dietary Fiber in Health and
     Disease," G. V. Vahouny and D. Kritchevsky," eds. Plenum Press,
     New York and London (1982).
46. Postprandial serum glucose and hormone levels, in "Physiological
     Effects and Health Consequences of Dietary Fiber," S. Pilch, ed.
     Life Sciences Research Office FASEB, Bethesda, MD (1987).
47. M. Friedman, S. O. Byers and R. H. Rosenman, Effect of glucagon on
     blood cholesterol levels in rats, Lancet ii, 464 (1971).
48. S. O. Byers, M. Friedman and S. R. Elek, Further studies
     concerning glucagon-induced hypocholesterolemia, Proc Soc. Exp.
     Biol. Med. 149:151 (1975).
49. O. O. Thomsen, J. A. Larsen, The effects of glucagon, dibutyrilic
     cyclic AMP and insulin on bile production in the intact rat and
     the perfused rat liver, Acta Physiol. Scand. 111:23 (1981).
50. I. T. Johnson, J. M. Gee and J. C. Brown, Plasma enteroglucagon
     and small bowel cytokinetics in rats fed soluble non-starch
     polysaccharides, Am. J. Clin. Nutr. 47:1004 (1988).
51. D. J. A. Jenkins, T. M. S. Wolever, V. Vuksan, et al., Nibbling
     versus gorging: Metabolic advantages of increased meal frequency,
     N. Engl. J. Med. 321:929 (1989).
52. L. P. Bell, K. J. Hectorne, H. Reynolds, T. Balm, D. B.
     Hunninghake, Cholesterol-lowering effects of psyllium hydrophilic
     mucilloid as an adjunct to a prudent diet for patients with mild
     to moderate hypercholesterolemia, J. Am. Med. Assoc. 261:3419
     (1989).
53. L. P. Bell, K. J. Hectorne, H. Reynolds, D. B. Hunninghake,
     Cholesterol-lowering effects of soluble fiber cereals as part of a
     prudent diet for patients with mild to moderate
     hypercholesterolemia, Am. J. Clin. Nutr. Prepared for press.

CHELATING PROPERTIES OF DIETARY FIBER AND PHYTATE. THE ROLE FOR MINERAL

AVAILABILITY

Wenche Frølich

MATFORSK, Norwegian Food Research Institute
Osloveien 1, 1430 Ås, Norway

INTRODUCTION

Chemical determination of the content of minerals present in the food do not indicate the amount of the minerals which are available for absorption and utilization. Bioavailability of minerals and trace elements is shown to be affected by both intrinsic and extrinsic factors. Intrinsic factors are those internal to the human or animal and include among others nutritional and health status, sex and age. Components in the diet are the extrinsic factors and could also be described as interactions between the components and the minerals, with either positive or negative effects.

In vivo studies are the most reliable way to study bioavailability of minerals, and both radiolabeled absorption tests, effects on mineral levels in different body tissues and metabolic balance techniques have been used.

Much less is known about mineral requirement and bioavailability in humans than in animals. Only limited studies on minerals connected with defienciency states can be carried out in humans. It is also difficult to make long-term experiments and studies that might cause risks to humans would be ethically wrong, e.g. studies on children and pregnant women. Evidence suggest, however, that many observations carried out on animals, do have relevance for man. However, direct extrapolations from animals to humans do not give correct answers in all cases, and before any definite conclusions can be drawn, human studies are needed.

A lot of controversis exists in the litterature when it comes to mineral bioavailability. To some degree this could be explained by the differences in the experimental design. Most human studies have short term duration. It has, however, been shown that the body seems to adapt to a dietary regimen with time. In experiments performed with whole grain cereals no change in mineral absorption could be seen after three to four weeks, in spite of a percent inhibition in the beginning of the experiments.

In in vivo experiments unphysiological dosages of the chelating agent have often been used. The interpretation of the results from these experiments are difficult, and extrapolation to normal diets often impossible.

New Developments in Dietary Fiber
Edited by I. Furda and C. J. Brine
Plenum Press, New York, 1990

It is also important to bear in mind that when studying mineral bio-availability, the composite meal should be taken into consideration, and not only one single component or fraction of the food. Studies on a single food item can on the other hand be used to identify the dietary components that might influence the degree of absorption.

In vitro studies, on the other hand, do not give definite answers, when it comes to bioavailability of minerals. They could, however, eluci-date some of the factors that may be important for determining in vivo bindings between minerals and factors in the complex diet.

WHOLE GRAIN CEREALS IN THE DIET

Whole grain cereals are some of the best sources, not only for dietary fiber, but also for minerals and trace elements. An increased con-sumption of whole grain flour is therefore a nutritional aim in many countries.

Most minerals and trace elements in the cereals are closely related to the outer layers, especially the aleurone cells, where also dietary fiber and phytic acid are recovered.

The complexity of interactions that may take place between minerals and different components in the dietary fiber complex, makes it very diff-icult to predict the bioavailability of the minerals in whole grain cereals by the amount of minerals present. Due to the chelating properties of the different dietary fiber components and the phytic acid, there has been a lot of concern about the effect of an unrefined high-fiber diet on mineral availability. Great controversies, in terms of mineral interac-tions, have been in connection with the chelating properties of dietary fiber versus phytic acid. In earlier studies it was claimed that phytic acid was the component chiefly responsible for the chelating of divalent minerals. More recently it has been suggested that phytic acid is not the component that is solely responsible for the decreased mineral availabi-lity, as dietary fiber itself also might be of importance.

It is, however, important to stress that there is a considerably higher amount of minerals present in whole grain products. Therefore an inhibition of the percent uptake of minerals does not necessarely mean a decreased absolute absorption and decrease in the mineral pool size of the body.

DIETARY FIBER

Dietary fiber is often divided into a soluble and an insoluble frac-tion due to relative solubilities and other chemical properties. The solu-bility of the minerals may be decreased as a result of chelation to fiber components. Dietary fibers could increase the viscosity and reduce the rate of migration of the minerals. This in addition to reduced transit time could result in changes in bioavailability of the minerals.

Several in vitro studies have been published on different fiber com-ponents affecting mineral association. This has been the case both with food as wheat, corn, rice and soy, but also with isolated components from the fiber complex as phytate, cellulose and lignin. These studies have shown that several factors must be taken into consideration when interpre-ting the results: (a) the presence of different chelators as phytate, tannins, citrate and amino acids, (b) pH in the solution, (c) concentra-tion of the mineral, but also other minerals that may compete for a given

binding site, (d) heat treatment that may effect the binding, (e) fermentability of the fiber components in the colon and the potential for uptake
of the minerals in the gut.

ASSOCIATION OF MINERALS TO DIFFERENT CEREALS

The binding behaviour of different minerals to cereal fibers seems to
be extremely variable, probably due to the various chelating properties of
the different fiber components. No agreement exists between the different
minerals studied.

The various cereals have also quite different dietary fiber composition with different chemical structure and binding capacity, making the
chelating properties of the same mineral special for each cereal. As fiber
and phytic acid most often occur together in the food, it is difficult to
distinguish between the effect of the different fiber components and
phytic acid.

In many of the in vitro studies, isolated dietary fiber components
and phytate are added to study the effect of these components on mineral
bioavailability. It is important to realize that the addition of isolated
components to the diet may not reflect the effect of the same amount of
the native fiber or phytate present in the original food item, even if it
may give indications. Therefore model systems with only one single binding
component cannot be extrapolated directly to food where more than one
binding component exists.

DIETARY FIBER

Analysis of dietary fiber in foods is a subject under discussion. In
this study an enzymatic, gravimetric method (Asp et al., 1983) was used to
determine soluble and insoluble components of dietary fiber.

The dietary fiber content of whole grain wheat flour was 12.7 % and
for white wheat flour with an extraction rate of 80 %, 3.6 %. The soluble
fiber content was 1.8 %, independent of the degree of extraction. The
content of insoluble fiber components increased rapidly when the the
extraction rate exceeded 80 %. The relative composition of monosaccharides
differed little between the two extraction rates investigated, the main
components being glucose, arabinose and xylose (Nyman et al., 1984). The
soluble fraction contained mainly of arabinoxylans.

The dietary fiber content of whole grain rye flour was 16.8 % and for
rye flour with an extraction rate of 70 %, 7.5 %. The soluble fiber fraction accounted for 3.8 %, and like wheat independent of extraction rate.
Rye consisted mainly of glucose, arabinose and xylose.

Barley contained of 10.5 % dietary fiber at an extraction rate of
50 %, which is the product available on the Norwegian marked. The soluble
fraction was 4.2 %, which has been shown to be the same also for other
extraction rates. Barley had the highest content of β-glucans in the
soluble fraction (Nyman et al., 1984). The main components of the total
dietary fiber complex were glucose, arabinose and xylose.

The oat kernel (accounting for 70 % of the oat grain) had a dietary
fiber content of 11.5 %, of which 2.7 % was soluble fiber components. Like
barley, the soluble fiber components consisted mainly of β-glucans. In
contrast to the other cereals, the amount of soluble fiber components was
distributed throughout the whole kernel (Frølich and Nyman, 1988). The

insoluble fiber components consisted mainly of arabinoxylans and cellulose. The oat husk, accounting for the remaining 30 % of the oat grain,
had a dietary fiber content of around 85 %. The husk consisted mainly of
insoluble fiber components, containing cellulose, xylans and lignin
(Frølich and Nyman, 1988).

PHYTIC ACID

Phytic acid (myo-inositol hexaphosphate) occours naturally in many
foods derived from plants. It is the storage form of phosphorus in most
seeds and is found in significant quantities in cereal grains. Of the
total amount of phosphorus in plants, 60 - 90 % is found as phytate (Asada
et al., 1969, Reddy et al., 1982, Frølich et al., 1985).

Phytic acid forms strong complexes (phytates) with many essensial bi-
and trivalent metal ions in foods as well as in the intestine. The presence of phytic acid may, therefore, have an effect on the bioavailability
of minerals.

The content of phytate in wheat, rye, oat and barley was measured by
using the method of Holt (1955); which is based on complex formation of
phytic acid and ferric ions at pH 1 - 2. The content of phytate in the
four cereals were 1.0 %, 0.6 %, 1.1 % and 0.5 %, based on dry weigth. For
wheat, rye and oat this is in accordance with what has been found earlier
(Maga 1982, Fretzdorff et al., 1986). For barley, however, a content of
0.97 - 1.08 % (Lolas et al., 1976) has been reported earlier. The explanation for this difference is most probably that the extraction rate of the
barley of 50 % used in our study is lower than in other studies.

In all the four cereals nearly all the phytate, recovered together
with the dietary fiber, were found in the soluble fiber fraction.

DEGRADATION OF PHYTATE

Knowlegde of the degradation of phytate in the food is limited. The
degree of breakdown depends on many variables and is not well understood.

Most methods for determining phytate are phrone with the same limita-
tion, namely defining the species of phytate that are measured. In foods
which have been exposed to hydrolysis, this will not give an exact amount
of the content of phytate as it has been shown that inositol polyphos-
phates other than hexaphosphate may coprecipitate with phytate under conditions used in the methods. Also traces of inorganic phosphate may be
precipated, resulting in a too high estimate of the phytate content.

Hydrolysis of phytate was studied by using high resolution ^{31}P - NMR-
spectroscopy (Frølich et al., 1986). With this method the fate of phytate
could be followed both in an isolated system and in a breaddough during
fermentation. When using this system, a spectrum with resonances of the
intensity ratio 1:2:2:1 was obtained, which is in accordance with the sym-
metry of the phytate molecule. Figure 1 shows the time dependence in the
^{31}P-NMR spectra during hydrolysis of commercial, pure phytate and phytase
isolated from wheat. The hydrolysis of the phytate molecule showed a very
distinct pattern, and seemed to proceed in a step-wise-manner.

Each of the lower inositol phosphates was the dominant species present at different stages during hydrolysis. The rate of hydrolysis of the different inositol phosphates decreased with increasing degree of phosphorylation, and was twice as fast for the phytate as for the penta-phosphate. After 5 - 6 hours only the tri-, di- and monophosphates of the inositol could be detected.

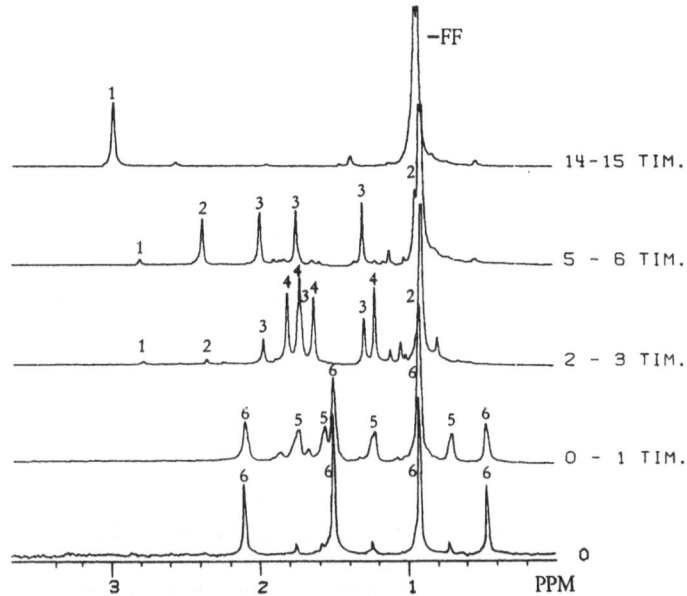

Fig. 1. ^{31}P-NMR spectra showing the hydrolysis of pure phytate (8 mg/ml) with commercial phytase (0 - 8 mg/ml) at pH 5 - 1 and 24 ^{0}C at the time intervals indicated. The numbered peaks represent resonances due to inositol hexa (6)-, pental (5)-, tetra (4)-, tri (3)-, di (2)- and mono (1)- phosphates and to inorganic phosphate (P_1).

When comparing a ferric ion method with the ^{31}P-NMR spectroscopy, the result showed that esters of inositol lower than the hexaphosphate and also inorganic phosphate were precipitated and measured with the ferric ion method. The ferric ion method seems to become less reliable when the hydrolysis proceeds.

Figure 2 shows the hydrolysis of phytate in a breaddough. The breakdown was slower in the dough than in the isolated system, probably due to the lower amount of phytase present in the dough, and the presence of other components which may have an inhibitory effect on the breakdown. The spectra were, however, the same as the spectra for the isolated system.

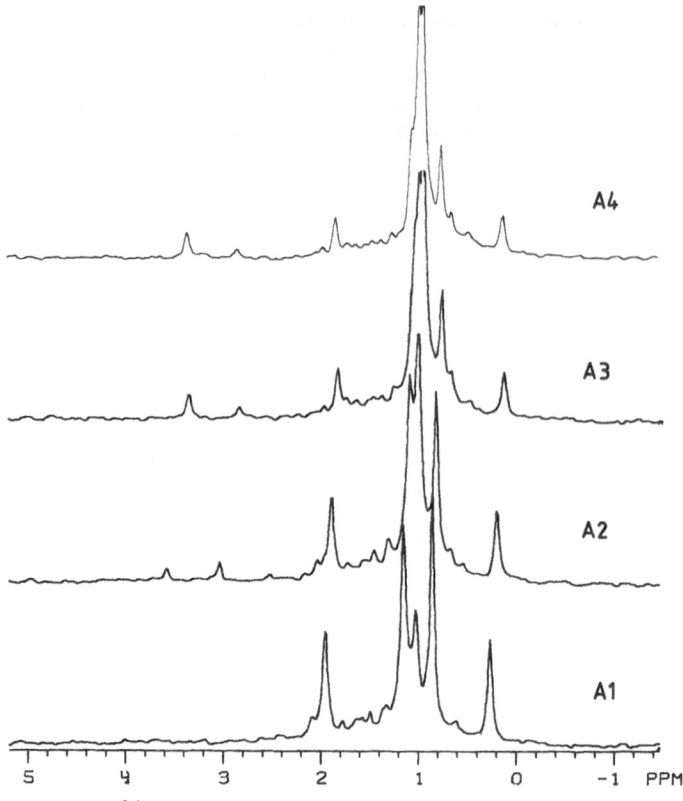

Fig. 2. ^{31}P-NMR spectra showing the hydrolysis of phytate during fermentation of bread dough at 37 ^{0}C. Samples were taken after 0-time (A1), 2 hours (A2), 8 hours (A3) and 8 hours plus additional 45 min baking at 200 ^{0}C (A4). The spectra were recorded at pH 5 and 23 ^{0}C. The full intensity of the resonance due to inorganic phosphate is not shown.

PHYTASE ACTIVITY IN OAT, RYE AND BARLEY

The principal function of the enzyme phytase (myo-inositol hexaphosphate phosphorylase) in grains is most likely to provide phosphate from phytate during germination. Phytase activity has been shown to increase during germination (Peers, 1953, Eskin and Wiebe, 1983).

Phytase from various sources has been studied most intensively, however, the one from wheat (Cosgrove, 1966). The phytase from barley has been shown to have a somewhat lower activity than the phytase from wheat (Bartnik and Szafranska, 1987). Rye has been claimed to be the cereal with the highest phytase activity (Nagai and Funahishi, 1962). Studies performed on phytase in oat are scarce and the results are conflicting, either indicating no (Hoff-Jørgensen et al., 1946) or low to medium activity (Bartnik and Szafranska, 1987, Lockart and Hurt, 1986).

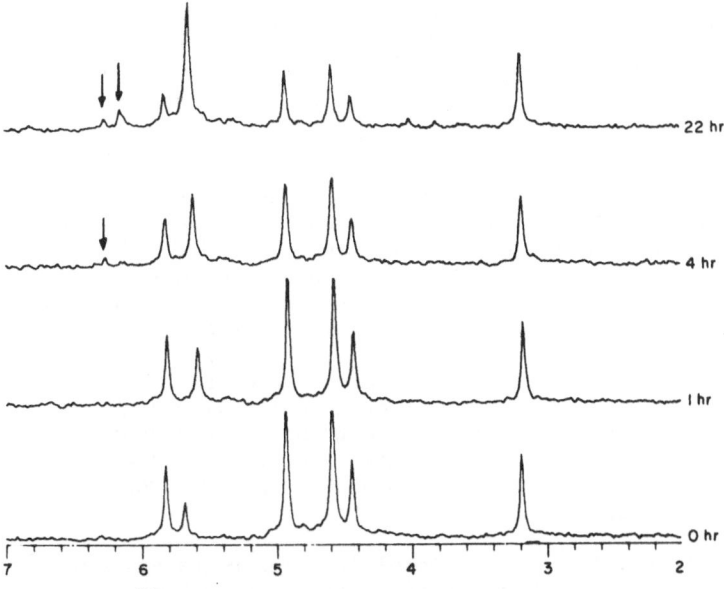

Fig. 3. ^{31}P-NMR spectra showing the hydrolysis of
phytate in a mixture of untreated oats and
water at 37 °C. Samples were taken after
the incubation times shown. The spectra
were recorded at pH 13 and 25 °C. The
arrows indicate lower inositol phosphates.

^{31}P-NMR-spectroscopy was used to study the breakdown of phytate in
oats (Frølich et al., 1987), rye (Ønning and Frølich, 1989) and barley
(Liljeberg and Frølich, 1989). The degradation of phytate will give an indication of the presence of the enzyme in the cereals.

The NMR-spectra for the three cereals were quite similar to those of
wheat, and for all the cereals around 90 % of the total phosphate was
present in the form of phytate.

The oat available on the marked is "stabilized" to inactivate lipase,
usually by heat-treatment at 96 - 100 °C at high moisture for 2 - 3
minutes. No significant hydrolysis of the phytate was observed during the
22 hours of incubation at 37 °C of the heat-treated oat mixture, whereas
around 65 % of the phytate was hydrolysed in the untreated oats during the
same period under the same conditions, as can be seen i fig 3. Addition of
wheat phytase to the heat-treated oat samples caused hydrolysis similar to
the one in the untreated oat.

After incubation for one hour in a similar system, but with whole
grain rye, 63 % of the original phytate was broken down. With an additional 18 hours of incubation all the phytate had diseappeared (Ønning and
Frølich, 1989). There is an increased breakdown of phytate present in the
rye with an addition of wheat phytase, indicating that the enzyme phytase
is not spesific for the different cereals (table 1).

An even faster breakdown of phytate is observed with barley where all
the phytate present is broken down after 4 hours of incubation at the same
conditions as for the other cereals. Only 25 % of the original phytate is
recovered after 1 hour incubation. The phytate disappeared even faster
with an extra addition of wheat phytase.

For all the cereals the speed of the breakdown is highest in the beginning. As breakdown products appear the rate of hydrolysis decrease with, probably due to product inhibition from inorganic phosphate.

MINERALS

For all the cereals most of the minerals and trace elements were recovered in the soluble fiber fraction, while only small amounts could be recovered in the insoluble fiber fraction. The only exception is iron where around 40 % of the total iron in wheat and 10 % of the iron in rye and barley was present in the insoluble fraction. In oat, however, nearly all the iron was recovered in the soluble fraction.

This could be explained by the presence of an insoluble component in wheat, not being present in the other cereals, with high chelating properties. Another explanation could be that the soluble fraction of the three other cereals have a considerably higher content of β-glucans than wheat, and that this fiber component has a higher affinity for iron than the insoluble fraction in wheat.

In wheat only 10 % of the total iron was released during breakdown of phytate, indicating association to other fiber components than phytic acid (Frølich et al., 1985).

To study the different binding patterns for the different minerals in the dietary fiber complex in wheat, rye and barley, a potentiometric method was used (Persson et al., 1987). The soluble fiber fraction of wheat interacted strongly with zinc, cadmium and copper (table 2). The interaction was strongly pH dependent, starting at pH 3.5 - 4.0 and generally completed at about pH 5.5.

At increasing pH-values, copper ions become extensively hydrolysed and at pH higher than 6, the hydroxycomplexes started to dominate over the fiber complexes. Zinc and cadmium ions were considerably less hydrolysed.

A reduction in the binding could be observed for cadmium and copper after phytase treatment up to pH 5. At higher pH no reduction in the binding was obtained by phytase treatment.

Soluble fiber also interacted strongly with the zinc ions, starting at pH 3.5, and being almost completed at pH 5. If the fraction was treated with phytase, the binding was substantially reduced (more than 50 %) which is in accordance to earlier studies on wheat (Frølich et al., 1985). In contrast to the findings for copper and cadmium in wheat, phytase treatment reduced zinc association substantially at all pH values tested.

Table 1. Phytic acid recovered after hydrolysis (% of original total phytic acid)

Time of hydro-lysis (hours)	Rye	Wheat	Barley	Oat
0	100	100	100	100
1	37	42	25	96
4	18	29	0	66
22	0	0	0	37

Table 2. Fraction of metal ions bound to soluble fiber components and/or phytate from wheat (adapted from Persson et al., 1987)

Fiber source	Metal	Bound metal (%) at different pH				
Soluble fiber from wholegrain wheat breaddough	Zn	0	20	28	35	66
	Cd	0	12	32	28	54
	Cu	0	30	56	75	10
Phytase treated	Cd	0	8	16	33	60
	Cu	0	9	30	75	10
Soluble fiber from bran	Zn	0	34	53	54	63
Phytase treated	Zn	0	14	17	18	19

The fraction of metal ions bound was closely proportional to the concentration of fiber at low pH. In wheat phytic acid seemed to be an important chelator for the three metals studied. At pH higher than 6, i.e. physiological pH, other components than phytic acid could be important for copper and cadmium. Phytic acid on the other hand is a strong chelator for zinc also at this pH.

Also for rye (table 3) and barley (table 4) the soluble fiber fraction interacted with zinc, cadmium and copper, being pH dependent for all the three minerals. At pH 6 the interaction with copper reached a maximum at around 50 % for the two cereals. At increasing pH values, copper ions become extensively hydrolysed, and hydroxycomplexes started to dominate over the fiber complexes. At pH 7, less than 10 % of the copper was associated to the fiber fraction for both cereals. When the soluble fraction from rye is treated with phytase to breakdown the phytate, less copper is associated to the soluble fiber fraction. The difference in association between the treated and not treated fraction, however, is minor compared to the one for wheat. For barley on the other hand, there is no effect of phytase treatment on the association of copper even when all the phytate is broken down.

Table 3. Fraction of metal ions bound to soluble fiber components from rye (adapted from Ønning/Frølich, 1989)

Fiber source	Metal	Bound metal (%) at different pH				
		3.0	4.0	5.0	6.0	7.0
Soluble fraction	Cu	0	10	33	52	9
	Cd	0	2	21	26	28
	Zn	0	4	16	21	28
Phytase treated	Zn	0	9	16	47	6
	Cd	0	3	15	24	41
	Cu	0	7	13	20	35

Table 4. Fraction of metal ions bound to soluble fiber components from barley (Liljeberg/Frølich, 1989)

| Fiber source | Metal | Bound metal (%) at different pH | | | | |
		3.0	4.0	5.0	6.0	7.0
Soluble fraction	Cu	0	13	32	48	9
	Cd	0	1	22	28	31
	Zn	0	14	27	30	36
Phytase treated	Zn	0	20	26	53	8
	Cd	0	4	20	28	42
	Cu	0	12	27	30	46

For cadmium there is an increase in the association of minerals to the soluble fibers for both rye and barley with 20 - 30 % of the total cadmium present in the cereals at pH 5. At higher pH there is only a slight increase in the association, as the chelation more or less seems to reach a plateau. When the phytate is broken down, there is only a minor tendency to decrease in the association of cadmium to the soluble fraction, indicating that the cadmium seems to be associated to other components in the soluble dietary fiber complex than phytic acid for both barley and rye.

The association of zinc to the soluble fiber fraction increases with increasing pH to around 30 % at pH 7 for both cereals. In contrast to wheat there is no reduction in the zinc associated to the soluble fraction when the phytate is broken down, indicating that phytic acid is not a chelator for zinc in neither rye nor barley as it is for wheat. The phytate in rye and barley do not seem to be a chelator of any significance for the three minerals studied, not even at pH lower than physiological pH. In contrast to wheat the soluble fiber fractions from rye and barley, contain components like tannins and β-glucans. These components are claimed to be potent chelators and might have a higher affinity for the minerals than phytate in rye and barley.

CONCLUSIONS

The results obtained from these studies indicate that different cereals have different chelating properties due to different fiber components present in the cereals studied.

For the four cereals (wheat, rye, barley and oats) studied, the minerals seem to be associated to the soluble fiber components, where also the phytate is recovered. Phytate seems to be a potent chelator for some of the minerals in wheat, but not in the other cereals studied, where most probably other fiber components seem to be of greater importance for the mineral association.

Unrefined cereal products are, however, good sources of minerals in the human diet, and in spite of a somewhat decreased percentage absorption for some minerals, a positive balance of the minerals seems to be retained, due to the considerably higher levels of minerals present in the high fiber products.

REFERENCES

Asada, K., Tanaka, K. and Kasai, Z., 1969, Formation of phytic acid in cereal grains, Ann. N.Y. Acad. Sci., 165:801

Asp, N.-G., Johansson, C.-G., Hallmer, H. and Siljeström, M., 1983, Rapid enzymatic assay of insoluble and soluble dietary fiber, J. Agric. Food Chem., 31:476.

Bartnik, M. and Szafranska, I., 1987, Changes in phytate content and phytate activity during the germination of some cereals, J. Cereal Sci., 5:23.

Eskin, N.A.M. and Wiebe, S., 1983, Changes in phytase activity and phytate during germination of two Fababean cultivars, J. Food Sci., 48:270.

Fretzdorff, B. and Weipert, D., 1986, Phytinsäure in getreide und getreide-erzeugnissen Mitteilung, Zeitschrift für Lebensmit.-Untersuch. u. Forsch., 182:287.

Frølich, W. and Asp. N.-G., 1985, Minerals and Phytate in the Analysis of Dietary Fiber from Cereals III, Cereal Chem., 62:238.

Frølich, W., Drakenberg, T. and Asp, N.-G., 1986, Enzymatic Degradation of Phytate (myo-inositol hexaphosphate) in Whole grain Flour Suspension and Dough. A comparison between ^{31}P-NMR spectroscopy and a Ferric Ion Method, J. Cereal Sci., 4:325.

Frølich, W. and Nyman, M., 1988, Minerals, Phytate and Dietary Fibre in Different Fractions of Oat-grain, J. Cereal Sci., 7:73.

Frølich, W., Wahlgren, M. and Drakenberg, T., 1988, Studies on Phytase Activity in Oats and Wheat using ^{31}P-NMR Spectroscopy, J. Cereal Sci., 8:47.

Gosgrove, D.J., 1980, "Inositol phosphates", Elsevier, Amsterdam.

Hoff-Jørgensen, E., Anderson, O. and Nielsen, G., 1946, The effect of phytic acid on the absorption of calcium and phosphorus, Biochem. J., 40:555.

Holt, R., 1955, Studies on dried peas. I. Determination of phytate phosphorus, J. Sci. Food Agric., 26:1461

Liljeberg, H. and Frølich, W., 1989, Mineralämnenes associering med kostfiberkomplex i kornmjöl, Master Diss., University of Oslo.

Lockhart, H.B. and Hurt, H.D., 1986, Nutrition of Oats, in "Oats, Chemistry and Technology", F.W. Webster, ed., AACC., St. Paul M.N.

Lolas, G.M., Palamidis, N. and Markakis, P., 1976, The phytic acid - total phosphorus relationship in barley, oats, soybeans and wheat, Cereal Chem., 53:867.

Maga, J.A., 1982, Phytate: Its chemistry, occurence, food interactions, nutritional significance and methods of analysis, J. Agric. Food Chem., 30:1.

Nagai, V. and Funahashi, S., 1962, Phytase (myo-inositolhexaphosphate phosphohydrolase) from Wheat Bran, Agr. Biol. Chem., 26:794.

Nyman, M., Siljeström, M., Pedersen, B., Bach Knudsen, K.E., Asp, N.-G., Johansson, C.-G. and Eggum, B.O., 1984, Dietary fiber content and composition in six cereals at different extraction rates, Cereal Chem., 61:14.

Peers, F.G., 1953, The phytase of wheat, Biochem., 53:102.

Persson, H., Nair, B.M., Frølich, W., Nyman, M. and Asp, N.-G., 1987, Binding of mineral elements by some dietary fibre components - in vitro (II), Food Chem., 26:139.

Reddy, N.R., Sathe, S.K. and Salunkhe, D.K., 1982, Phytates in legumes and cereals, Adv. Food Res., 28:1.

Önning, G. and Frølich, W., 1989, Mineralassociering til kostfiberkomplexet i råg, Master Diss., University of Oslo.

MECHANISMS OF ACTION ON DIETARY FIBRE ON SMALL INTESTINAL ABSORPTION AND
MOTILITY

Christine Edwards

Subdepartment of Human Gastrointestinal Physiology and
Nutrition
University of Sheffield
Sheffield, United Kingdom

INTRODUCTION

The physical form in which food is ingested and the physical
properties of the intestinal chyme produced may determine the rate of
intestinal digestion, absorption and transit. When food is eaten as
whole plant material, nutrients may be trapped within the cellular
structure of the plant. Processes such as cooking[1], grinding[2] and
chewing[3] will act to break down this cellular structure releasing the
nutrients and allowing greater access to human enzymes speeding up
digestion and absorption. Postprandial glycaemia may also be
influenced by interactions between carbohydrate and protein within the
plant matrix[4].

Soluble dietary fibres which form viscous solutions or gels in
the stomach and small intestine may also act to slow absorption and
transit by increasing the viscosity of luminal contents. This chapter
will discuss firstly the effects of an increased luminal viscosity on
intestinal absorption and motility and second the factors which
influence the luminal viscosity achieved at different sites in the
intestine when soluble fibres are ingested.

VISCOUS POLYSACCHARIDES SLOW ABSORPTION

It is well established that mixing viscous polysaccharides such
as guar gum with a carbohydrate meal reduces the postprandial
hyperglycaemia[5] suggesting an impairment of carbohydrate absorption.
Viscous polysaccharides, however, appear to slow the rate of
absorption of simple molecules rather than to reduce the amount
absorbed. Blood xylose levels after a xylose meal were reduced when
guar gum was included in the meal but xylose excretion by the kidneys
was unaffected[5]. Studies of ileostomy effluent in rats and in man
have shown that more fat is egested from the small intestine after
pectin ingestion[6,7] indicating a malabsorption of lipid and in the
distal ileum the reabsorption of bile acids may also be reduced[8].
The exact mechanisms of this inhibition of absorption have not been
fully established and are probably different for each molecule
involved.

New Developments in Dietary Fiber
Edited by I. Furda and C. J. Brine
Plenum Press, New York, 1990

Gastric Emptying

Increased viscosity of stomach contents may slow the emptying of liquid meals. This could delay small intestinal absorption by restricting the availability of nutrients to the small intestine[9]. However, gastric emptying is not delayed in all subjects in whom postprandial hyperglycaemia is reduced[10,11] and some viscous polysaccharides such as locust bean gum (1%) may even accelerate gastric emptying[11]. Indeed, there is no correlation between the change in gastric emptying rate induced by guar gum and the reduction in postprandial hyperglycaemia[10].

The effect of viscous polysaccharides on the gastric emptying of solid meals is less clear. The emptying of disruptable solids such as egg and bread may be slowed[12]. Gastric emptying of disruptable solids in a mixed meal occurs only after the liquid has emptied from the stomach[13]. If viscous polysaccharides slow the emptying of the liquid portion of a meal, therefore, this could result in a corresponding delay in the emptying of disruptible solids. In addition it is likely that the disruption of solids by gastric contractions is impaired by an increased luminal viscosity. On the other hand, viscous polysaccharides may accelerate the emptying of small undisruptible solids (3.2mm diam) by trapping and entraining them so that they empty with the liquid phase[14].

Small Intestinal Absorption

Increasing the luminal viscosity of small intestinal contents can reduce the absorption of glucose without any influence of gastric emptying. Studies by Blackburn and colleagues[10] showed that the presence of guar gum on glucose solutions infused into the jejunum of human subjects reduced the disappearance of glucose from the lumen. Similar observations have been made using tied and perfused intestinal loops in animals[15,16,17].

This inhibition of glucose movement can be mimicked with in vitro models using dialysis tubing to simulate the intestinal mucosa[18,19]. The reduction in glucose movement appears to be related directly to the viscosity of the complex polysaccharide solution used[5].

Molecules in the intestinal lumen move by simple diffusion or by the convective forces set up by intestinal contractions. Intestinal contractions cause turbulence in the luminal contents. This results in not only mixing of substrates and enzymes but also brings the nutrients from the bulk phase close to the absorptive epithelium[20]. The nutrients must then diffuse across the relatively unstirred layer of fluid lying adjacent to the epithelium[10] where they are absorbed.

Viscous polysaccharides have been shown to increase measurements of the apparent thickness of the unstirred layer[15,21]. The unstirred layer, however, is not an anatomical reality but is a functional concept which has been invented to explain the changes in absorption that occur under stirred or less stirred conditions[22]. An increase in the unstirred layer thickness could imply a decrease in the diffusion coefficient for the test solute across an unstirred layer of unchanged thickness or it could imply a reduction in luminal convection causing an actual increase in the unstirred zone.

Increasing luminal viscosity could reduce the movement of molecules, and hence absorption, by inhibiting both mixing and diffusion. It is difficult to study the effect of dietary fibre on diffusion and mixing separately. Studies which claim to show an effect of viscous polysaccharides on diffusion have usually measured the movement of glucose out of dialysis bags[18,19] or across diffusion chambers[23] under stirred conditions and have not eliminated the effect of convection[24]. To avoid the problems of measuring diffusion of a solute along a concentration gradient without mixing, we measured the conductivity (and hence ion mobility) of unstirred electrolyte solutions in the presence of increasing concentrations of guar gum[25]. Despite observing the expected changes in conductivity when the temperature or ionic strength of the solutions was increased, there was no change in the conductivity when the guar concentration was increased from 0 to 1%; a 500 fold increase in viscosity (Table 1). In contrast, the time taken for two electrolyte solutions of different concentration to mix under the influence of a constant rotary force was proportional to their viscosity[25]. We also designed a model to simulate the effects of intestinal contractions on absorption. This consisted of a dialysis tube filled with 10% glucose and 0.9% saline anchored at each end in a perspex trough of distilled water. Intestinal contractions were simulated using two pairs of paddles positioned at each end and driven by a motor so that each pair alternately occluded the dialysis tubing at a rate of 36 or 72 contractions per minute. When the contraction rate was increased with non-viscous solutions in the dialysis tubing the appearance of glucose in the external solution increased. This effect was inhibited by including 1% guar gum in the dialysis solution[25].

These in vitro studies suggest that the reduction in absorption of small molecules by viscous polysaccharides is probably caused by

Table 1. Mean conductivity (+ SEM) of saline solutions at 23°C and 37°C in the presence of increasing concentrations of guar gum

NaCl concentration (% w/v)		Conductivity ($\Omega^{-1}m^{-1}$) of solutions containing guar gum at:			
		0	0.5%	0.75%	1% w/v
0.45)	23°C	0.72(+0.01)	0.71(+0.02)	0.71(+0.02)	0.72(+0.02)
0.9)		1.45(+0.02)	1.4 (+0.06)	1.39(+0.04)	1.39(+0.04)
0.45)	37°C	0.89(+0.01)	0.86(+0.03)	0.87(+0.01)	0.88(+0.01)
0.9)		1.79(+0.03)	1.72(+0.09)	1.75(+0.00)	1.72(+0.04)

increased resistance to the convective effects of intestinal contractions rather than a decrease in diffusion coefficients. This may not be the case with large molecules or complexes such as micelles, however, where reductions in diffusion may play a greater role[26].

The effects of increased intraluminal viscosity may be augmented with some polysaccharides by the presence of certain residues such as uronic acid and sulphated groups. These may bind minerals[27,28] and hydrophobic interactions may cause the binding of bile acids and lipids at low pH[8,29].

In addition to limiting the access of digested nutrients to the epithelial surface, the reduction in intraluminal mixing would also reduce the interaction between food and enzymes and impair digestion of complex molecules. Though this may be compensated for by increased pancreatic enzyme production in longterm feeding of viscous polysaccharides[30,31]. These effects would tend to delay absorption reducing nutrient levels in the blood and resulting in a greater delivery of nutrients to more distal sites in the small intestine which may in turn induce morphological changes which indicate that the ileum adapts to play a greater role in the absorption of nutrients. The reduced access of nutrients to the epithelial surface and the change in site of absorption could change the pattern of hormonal release[32,33,34], and modulate neurohumoral reflexes responsible for gastrointestinal function.

Small Intestinal Propulsion and Transit

Ingestion of a viscous non-nutrient meal produces a highly active motor pattern of clusters of contractions which propagate over long distances in experimental animals and human volunteers[35,36]. This is surprising since viscous polysaccharides tend to delay small bowel transit. An in vitro preparation of the rat small intestine was therefore used to investigate the effect of increased luminal viscosity on the production and efficiency of small intestinal propulsion (Murray, Rumsey, Edwards and Read unpublished data). This model consisted of short ileal segments mounted in a Trendelenberg apparatus[37,38]. Repetitive and reproducible waves of peristaltic contractions were stimulated by application of a small hydrostatic load. When the luminal viscosity was increased, by incorporating increasing concentrations of guar gum in the luminal solution, the frequency of these peristaltic waves remained unchanged. The volume of luminal contents moved by each wave was reduced, however, and the duration of each wave was prolonged (Murray et al unpublished data).

This suggests that although an active motor pattern is stimulated by ingestion of viscous polysaccharide meals, the high viscosity of the luminal contents resist the propulsive actions of the intestinal contractions and thus luminal flow is reduced. The effect of guar gum on small bowel transit, however, has been studied only after ingestion of nutrient containing meals in contrast to the motility studies described above, viscous nutrient meals produce a more random motor pattern of contractions propagated over relatively short distances[35].

Another mechanism by which viscous polysaccharides may slow transit through the small intestine is related to their action in delaying small intestinal absorption. If this delay in absorption resulted in a greater amount of nutrients, especially lipids, in the distal ileum this may activate the ileal brake. Infusion of fat into the ielum reduces propagation of jejunal contractions[36] and slows small bowel transit[39]. This would have particular relevance to the transit of a second meal after one containing the viscous polysaccharide.

Delaying the transit of a carbohydrate meal through the small intestine would in theory increase glucose absorption unless the spread of the meal was restricted to a small portion of the jejunum providing only a small contact area for the nutrients. As studies which measured the effect of guar gum on the spread of a meal down the small intestine failed to show any difference between the guar and the control meals[40], it appears that although viscous polysaccharides slow transit the effects on reducing intraluminal mixing predominate and overall absorption is slowed. In addition the action of viscous polysaccharides on transit may be confined to the distal small intestine (see below)[41].

INTESTINAL SECRETIONS MAY REDUCE LUMINAL VISCOSITY

Since it is the luminal viscosity that determines the rate of intestinal absorption and flow, it is important to consider the factors which influence the viscosity achieved at each of the sites in the gastrointestinal tract which have particular physiological importance.

It is important to note that preparations of the same polysaccharide may have different viscosities at the same concentration[42]. The factors that influence viscosity are chemical structure, molecular size, concentration, shear rate and ionic environment. Viscosity increases with molecular weight and in some molecules such as carboxymethylcellulose the number and pattern of specific groups or side chains are also important.

The concentration and shear rate at which viscosity is measured are very important. The relationships between the concentration, shear rate and viscosity of four polysaccharide preparations are shown in the Figure[11]. The degree to which an increase in concentration or shear rate affect the viscosity is different for each polysaccharide tested. This is important when comparing polysaccharides since a polysaccharide with a higher viscosity at 1% concentration (Figure) such as guar gum compared with xanthan, may have a relatively lower viscosity than xanthan at 0.25% at low shear rates.

These relationships may well be critical in the gut where concentrations and shear rate are unknown. The effect of shear rate may also be important as the food passes along the gastrointestinal tract since the shear rate in the antrum and pylorus may be very different from that of the jejunum.

In addition to the effects of shear rate and concentration on the initial viscosity of a fibre, the effects of intestinal secretion on intraluminal viscosity must also be considered. A mixture of xanthan and locust bean gum (X/LBG; 1% w/v) is three times as viscous as the same concentration of guar gum. However, when fed to human

Table 2. Effect of acidification and reneutralisation on the viscosity of guar gum and a xanthan/locust bean gum mixture

			Viscosity of gum solutions			
Concentration	Initial 1%	Acidified 0.89%	Saline diluted control 0.89%	Reneutralised 0.68%	Saline-saline diluted control 0.68%	Acid-saline diluted control 0.68%
G	2396+51	1939+78*	1714+160	1209+51**	1227+38	1305+53
X/LBG	9922+524	1294+142+	4474+595	2138+68+***	1391+78	1009+53

Mean + SEM; n = 5; G is guar, X is xanthan, LBG is locust bean gum.

* = p<0.05)
) significantly different from saline control at same concentration
+ = p<0.001)

** = p<0.05)
) significantly different from acid-saline control.
*** = p<0.001)

volunteers in a glucose drink, these polysaccharides were equally potent in reducing postprandial hyperglycaemia. This apparent discrepency between the higher viscosity of X/LBG and its effects on blood glucose was explained when we simulated the effects of intestinal secretions on the viscous properties of the gums. The viscosity of X/LBG was reduced by dilutiong with saline, or by acidification to pH1 and reneutralisation, to a much greater extent than the viscosity of guar gum (Table 2) and the final viscosities of the two gums were very similar after these treatments (Table 2[11]). Thus preingestion viscosity measurements may be misleading if used to predict the action of viscous polysaccharides on postprandial glycaemia.

The effects of intraluminal dilution on the action of viscous polysaccharides is also evident when the transit of a viscous meal

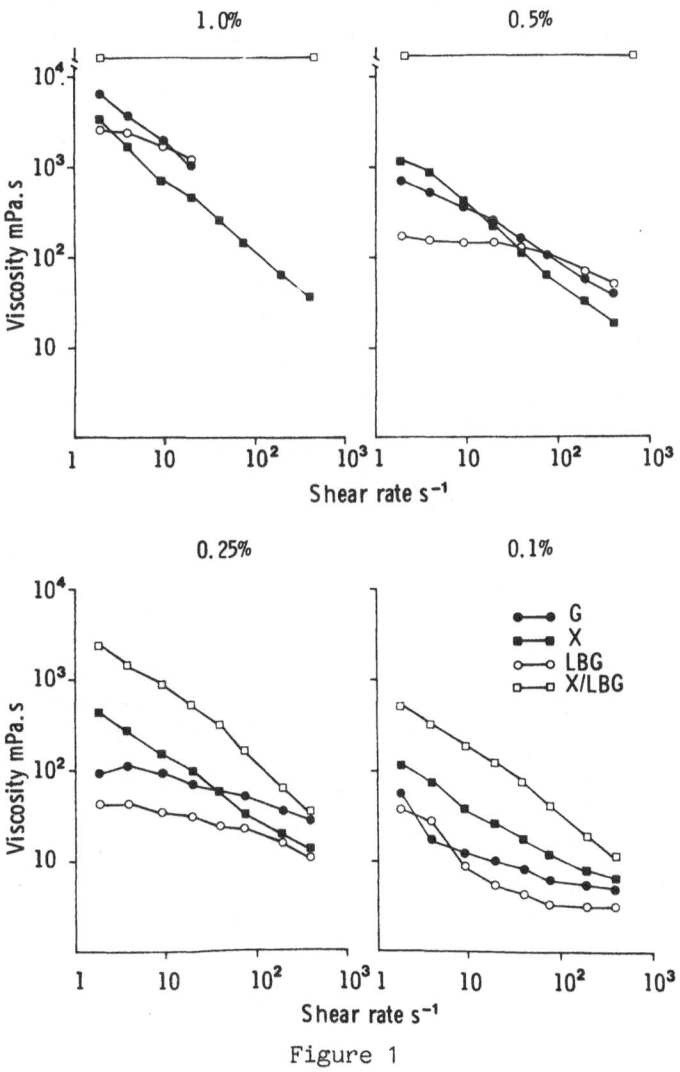

Figure 1

The effect of shear rate and concentrations on viscosity of viscous polysaccharides

was followed along the small intestine. Studies of small bowel transit time using the breath hydrogen technique measure only the overall mouth to caecum transit time. To study the transit along each section of the intestinal tract animal studies are necessary. Studies in rats have shown that the greatest effects of guar gum on transit were seen in the stomach and ileum[41]. Transit through the upper portion of the small intestion remains unchanged[41], probably because the viscosity· in the upper small intestine is reduced by digestive secretions, but by the time the remnants of the meal reach the ileum, most of this fluid has been reabsorbed and the intraluminal viscosity is recovered. The rate of caecal filling (or ileal emptying) is also reduced by viscous polysaccharides[43,41].

In conclusion, the physical form of food and the presence of viscous polysaccharides may have profound effects on small intestinal function. Viscous polysaccharides act mainly as antimotility agents reducing intraluminal mixing, and hence nutrient absorption, and also intraluminal flow along those portions of the gastrointestinal tract in which the viscosity is not lost by dilution with intestinal secretions.

REFERENCES

1. Collings, P., Williams, C., MacDonald, I. Effects of cooking on serum glucose and insulin responses to starch. Br Med J, 282; 1032 (1981)
2. O'Dea, K., Snow, P., Nestel, P. Rate of starch hydrolysis in vitro as a predictor of metabolic responses to complex carbohydrates in vivo. Am J Clin Nutr 34; 1991 (1981)
3. Read, N.W., Welch, I.McL., Austen, C.J., Barnish, C., Bartlett, C.E., Baxter, A.J., Brown, G., Comptom, M.E., Hume, K.E., Storie, I., Wording, J. Swallowing food without chewing; a simple way to reduce postprandial glycaemia. Br J Nutr 55; 43 (1986)
4. Jenkins, D.J.A., Thorne, M.J., Wolever, T.M.S., Jenkins, A.L., Rao, A.V., Thompson, L.A. The effect of starch-protein interaction in wheat on the glycaemic response and rate of in vitro digestion. Am J Clin Nutr 45; 946 (1987)
5. Jenkins, D.J.A., Wolever, T.M.S., Leeds, A.R., Gassull, M.A., Haisman, P., Dilawari, J., Goff, D.V., Metz, G.L., Albert, K.G.M.M. Dietary fibres, fibre analogues and glucose tolerance: importance of viscosity. Br Med J 1; 1392 (1978)
6. Isaksson, G., Asp, N-G., Ihse, I. Effect of dietary fibre on pancreatic enzyme activities of ileostomy evacuates and on excretion of fat and nitrogen in the rat. Scand J Gastro 18; 417 (1983)
7. Sandberg, A., Andersson, H., Hallgeren, B., Hasselblad, K., Isaksson, B. Experimental model for the in vivo determination of dietary fibre and its effect on the absorption of nutrients in the small intestine Br J Nutr 45; 283 (1981)
8. Vahouny, G.V. Dietary fibres and intestinal absorption of lipids, in: "Dietary Fiber in Health and Disease", G.V. Vahouny, D. Kritchevsky, eds., Plenum Press, New York (1982)
9. Holt, S., Heading, R.C., Carter, D.C., Prescott, L.F., Tothill, P. Effect of gell forming fibre on gastric emptying and absorption of glucose and paracetomol Lancet 1; 636 (1979)
10. Blackburn, N.A., Redfern, J.S., Jarjis, M., Holgate, A.M., Hanning, I., Scarpello, J.H.B., Johnson, I.T., Read, N.W. The mechanism of action of guar gum in improving glucose tolerance in man Clin Sci 66; 329 (1984)

11. Edwards, C.A., Blackburn, N.A., Craigen, L., Davison, P., Tomlin, J. Sugden, K., Johnson, I.T., Read, N.W. Viscosity of food gums determined in vitro related to their hypoglycaemic actions Am J Clin Nutr 46; 72 (1987)

12. Sandhu, K.S., El Samahi, M.M., Mena, I., Dooley, C.P., Valenzuela, J.E. Effect of pectin on gastric emptying and gastroduodenal motility in normal subjects. Gastroenterology 92; 486 (1987)

13. Houghton, L.A., Read, N.W., Heddle, R., Horowitz, M., Collins, P.J., Chatterton, B., Dent, J. The relationship of the motor activity of the antrum, pylorus and duodenum to gastric emptying of a solid/liquid mixed meal Gastroenterology 94; 1285 (1988)

14. Meyer, J.H., Elashoff, Y.G.J., Reedy, T., Dressman, J., Amidon, G. Effects of viscosity and fluid outflow on postcibal gastric emptying of solids Am J Physiol 250; G161 (1986)

15. Johnson, I.T., Gee, J.M. Effect of gel-forming food gums on the intestinal unstirred layer and sugar transport in vitro Gut 22; 398 (1981)

16. Elsenhans, B., Zenker, D., Caspary, W.F., Blume, R. Guaran effect on rat intestinal absorption. A perfusion study. Gastroenterology 86; 645 (1984)

17. Rainbird , A.L., Low, A.G., Zebrowsky, T. Effect of guar gum on glucose and water absorption from isolated jejunal loops in conscious growing pigs Br J Nutr 52: 4898 (1984)

18. Taylor, R.H., Wolever, T.M.S., Jenkins, D.J.A., Ghafari, H., Jenkins, M.J.A. Viscosity and glucose diffusion: potential for modification of absorption in the small intestine Gut 21; A452 (1980)

19. Jenkins, D.J.A. Slow release carbohydrate and the treatment of diabetes Proc Nutr Soc 40; 227 (1981)

20. Macagno, E.C., Christensen, J., Lee, C.I. Modelling the effect of wall movements on absorption in the intestine Am J Physiol 243; G541 (1982)

21. Flourie, B., Vidon, N., Florent, C.H., Bernier, J.H. Effect of pectin on jejunal glucose absorption and unstirred layer thickness in normal man Gut 25; 936 (1984)

22. Dietschey, J.M., Sallee, V.L., Wilson, F.A. Unstirred water layers and absorption across the intestinal mucosa Gastroenterology 61; 932 (1971)

23. Ebihara, K., Masuhara, R., Kirkyama, S. Major determinants of plasma glucose. Flattening activity of a water soluble dietary fibre. Effect of Konjac mannan on gastric emptying and intraluminal glucose diffusion Nutr Rep Internat 23; 1145 (1981)

24. Lucas, M.L. Estimation of sodium chloride diffusion coefficient in gastric mucins Dig Dis Sci 29; 336 (1984)

25. Edwards, C.A., Johnson, I.T., Read, N.W. Do viscous polysaccharides slow absorption by inhibiting diffusion or convection? Europ J Clin Nutr 42; 307 (1988)

26. Phillips, D.R. The effect of guar gum in solution on diffusion of cholesterol mixed micelles J Sci Food Agric 37; 548 (1986)

27. Sandstead, H.H., Munoz, J.M., Jacob, R.A., Klevay, L.M., Reck, S.J., Logan, G.M., Dintzis, F.R., Inglett, G.E., Shuey, W.C. Influence of dietary fibre on trace element balance Am J Clin Nutr 31; S180 (1978)

28. Reinhold, J.G., Faradji, B., Abadi, P., Ismail-Beigi, F. Decreased absorption of calcium, magnesium, zinc and phosphorous by humans due to increased fibre and phosphorus intake as wheat bread J Nutr 106; 493 (1976)

29. Eastwood, M.A., Hamilton, D. Studies on the absorption of bile salts to non-absorbed components in the diet Biochim Biophys Acta 152; 165 (1968)

30. Isaksson, G., Lilja, P., Lunquist, I., Ihse, I. Influence of dietary fibre on exocrine pancreatic function in the rat Digestion 27; 57 (1983)

31. Ikegamu, S., Tsuchihashi, N., Nagayama, S., Innami, S. Effect of viscous indigestible polysaccharides on pancreatic exocrine and biliary secretion in rats. Nutr Rep Internat 26; 239 (1982)

32. Jenkins, D.J.A., Bloom, S.R., Albuquerque, R.H., Leeds, A.R., Sarson, D.L., Metz, G.L., Albert, K.G.M.M. Pectin and complications after gastric surgery: normalisation of post-prandial glucose and endocrine responses Gut 21; 57 (1980)

33. Morgan, L.M., Goulder, T.J., Tsiolakis, D., Marks, V., Albert, K.R.M.M The effect of unabsorbable carbohydrate on gut hormones Diabetologia 17; 85 (1979)

34. Morgan, L.M., Tredger, J.A., Madden, A., Kwasowski, P., Marks, V. The effect of guar gum on carbohydrate fat and protein-stimulated gut hormone secretion. Modification of postprandial gastric inhibitory peptide and gastrin responses Br J Nutr 53; 567 (1985)

35. Schemann, M., Erhlein, H.J. Postprandial patterns of canine jejunal motility and transit of luminal contents Gastroenterology 90; 991 (1986)

36. Welch, I.McL., Worlding, J. The effect of ileal infusion of lipid on the motility pattern in humans after ingestion of a viscous, non-nutrient meal J Physiol 378 12P (1986)

37. Weems, W.A., Seygal, G.E. Fluid propulsion by cat intestinal segments under conditions requiring hydrostatic work Am J Physiol 240; G147 (1981)

38. Fisher, S., Murray, B.E., Rumsey, R.D.E. Factors affecting the volume expelled from isolated segment of rat small intestine during the peristaltic reflex J Physiol 378 15P (1986)

39. Read, N.W., MacFarlane, A., Kinsman, R., Bates, T., Blackhall, N.W., Farrar, G.B.G., Hall, J.C., Moss, G., Morris, A.P., O'Neill, B., Welch, I. Effect of infusion of nutrient solutions into the ileum on gastrointestinal transit and plasma levels of neurotensi and entroglucagon in man. Gastroenterology 86; 274 (1984)

40. Blackburn, N.A., Holgate, A.M., Read, N.W. Does guar gum improve post prandial hyperglycaemia in humans by reducing small intestinal contact area? Br J Nutr 52; 197 (1984)

41. Brown, N.J., Worlding, J., Rumsey, R.D.E., Read, N.W. The effect of guar gum on the distribution of a radiolabelled meal in the gastrointestinal ltract of the rat Br J Nutr 59; 223 (1988)

42. O'Connor, N., Tredger, J., Morgan, L. Viscosity differences between various guar gums Diabetologia 20; 612 (1981)

43. Spiller, R.C., Brown, M.L., Phillips, S.F. Emptying of the terminal ileum in intact humans. Influence of meal residue and ileal motility Gastroenterology 92; 724 (1987)

THE EFFECTS OF UNDIGESTIBLE FRUCTOOLIGOSACCHARIDES ON

INTESTINAL MICROFLORA AND VARIOUS PHYSIOLOGICAL FUNCTIONS

ON HUMAN HEALTH

Hidemasa Hidaka, Masao Hirayama,
Takahisa Tokunaga and Toshiaki Eida

Meiji Seika Kaisha, Ltd., Bio Science Laboratories, 5-3-1, Chiyoda, Sakado-shi, Saitama
350-02, Japan

INTRODUCTION

Fructooligosaccharides, containing 1-kestose (GF_2),
nystose (GF_3) and fructofuranosylnystose (GF_4), are found
undigestible and to be selectively utilized by beneficial intestinal bacteria, particularly by Bifidobacteria. An increased
population of beneficial intestinal bacteria would help to
improve constipation to reduce the production of putrefactive
substances.

A number of clinical studies has been done and the results
have shown that undigestible fructooligosaccharides are good
for human health to improve constipation or loose stool, to
decrease the production of putrefactive substances in large
intestine and to improve serum lipids in hyperlipidemia, and
reduce serum total cholesterol, triglycerides, blood glucose
and blood pressure.

These favorable results are almost the same as those of
dietary fiber functions.

CHEMICAL STRUCTURE AND PRODUCTION OF FRUCTOOLIGOSACCHARIDES

Fructooligosaccharides are composed of one moiety of
sucrose and several moieties of fructose as shown in Fig. 1.
They are widely existed in many kinds of plants[1] such as
onions, garlic, edible burdock, wheat, banana and so on (Table
1 and 2). Particularly, onion and burdock contain a large quantity of fructooligosaccharides. The fructooligosaccharides can
easily be prepared from sucrose by action of microbial or plant
enzymes.[2,3]

New Developments in Dietary Fiber
Edited by I. Furda and C. J. Brine
Plenum Press, New York, 1990

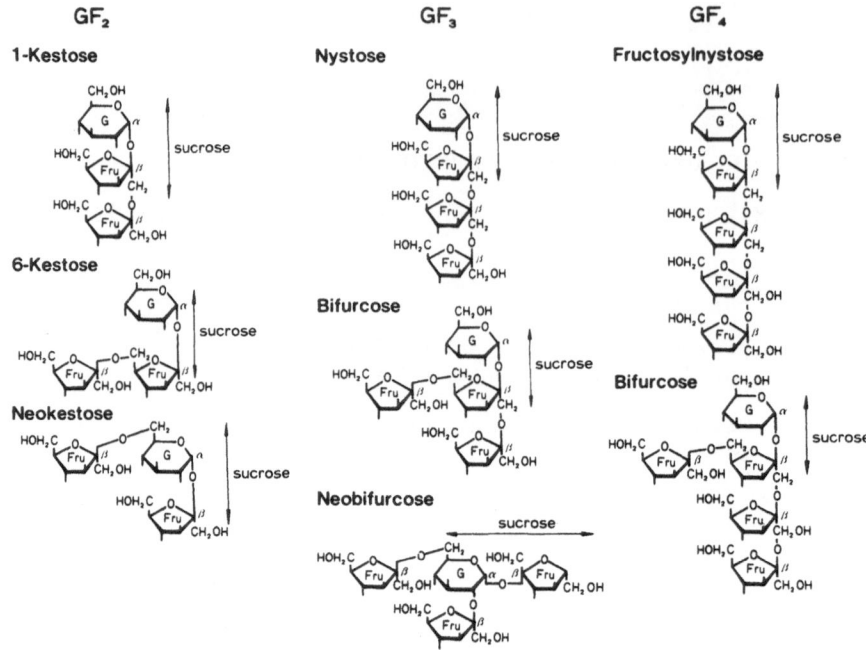

Fig. 1.   Chemical structure of fructooligosaccharides

Table 1.   Distribution of fructooligosaccharides in plants

Oligosaccharides		Distribution
GF$_2$	1-kestose	onion, edible burdock, rye, asparagus, chinese chive, jerusalem artichoke
	6-kestose	Gramineae plants
	neokestose	onion, banana, asparagus, sugar maple, Gramineae plants
GF$_3$	nystose	onion, edible burdock, asparagus,
	bifurcose	rye
	neobifurcose	oat
GF$_4$	fructosylnystose	onion, edible burdock, asparagus
	bifurcose	rye

Table 2. Amount of fructooligosaccharides in some edible plants

| plants | fructooligosaccharides (%) | | | moisture | total |
	in fresh material	in dry material	in total sugars	(%)	(%)
onion	2.8	25.0	29.7	89.0	9.3
welsh onion	0.2	1.9	3.6	91.5	4.4
garlic	1.0	2.2	3.9	57.1	24.3
burdock	3.6	16.7	22.0	78.5	16.4
rye	0.7	0.7	0.9	11.5	69.5
banana	0.3	1.3	1.6	75.5	19.2

Quantity of fructooligosaccharides was measured by HPLC; total sugars were determined by phenol-sulfuric acid method.

Enzymatic Preparation of Fructooligosaccharides

Enzyme. Fructooligosaccharides can be prepared from sucrose through the transfructosylating action of enzyme,[2,3] that is, β-fructofuranosidase (EC 3.2.1.26) and β-D-fructosyl-transferase (EC 2.4.1.9) from microorganisms and plants.

It is well known that β-fructofuranosidases commonly possess both transfructosylating (U_t) and hydrolyzing activity (U_h).[4] Estimation of the transfructosylating ability of different microorganisms has been done (Table 3). U_t and U_h were assayed by HPLC as the amounts of trisaccharides and fructose formed from sucrose. The U_t/U_h ratio indicates the relative strength of the transfructosylating activity of each strain, and the enzyme productivity was defined as the cell growth multiplied by the corresponding specific activity. For the efficient production of fructooligosaccharides, it is preferable to have both a high U_t/U_h ratio and high enzyme productivity. As judged from the data in Table 3, the cells of A. niger ATCC 20611 showed the highest values for both parameters. These data thus indicate that this strain is the most suitable for the preparation of fructooligosaccharides.

A fructooligosaccharide-producing β-fructofuranosidase was purified from cells of A. niger ATCC 20611. Studies on the enzyme[4] made clear the following enzymatic characteristics as follows. The molecular weight was 340,000 by gel filtration. The optimum pH of the enzyme was 5.0-6.0 and the optimum temperature was 50-60°C. The enzyme was inactivated by 1 mM Hg^{2+} and the Km value for sucrose was 0.29 M. The enzyme catalyzed an almost exclusively fructosyl transfer reaction in a 50% sucrose solution to produce a mixture of fructooligosaccharides and glucose, but both fructosyl transfer and hydrolytic action were observed in a 0.5% solution. The β-fructofuranosidase showed a high regiospecificity to transfer the fructosyl moiety for the 1-OH group of terminal fructofuranosides.

Preparation. Fructooligosaccharides could be more effectively prepared with a higher concentration of sucrose. Treatment of 50% (w/v) sucrose with the A. niger enzyme afforded a mixture of fructooligosaccharides (Fig. 2).

Table 3. Distribution of transfructosylating (U_t) and hydro-lyzing activity (U_h) in cells of various microorganisms[a]

No.	organisms[b]	incubation time (day)	cell growth (mg/ml)	specific activity (units/ml) U_t	specific activity (units/ml) U_h	U_t/U_h
1	A. niger ATCC 20611	1 3	143 346	0.271 0.266	0.019 0.022	14.2 12.2
2	A. niger NRRL 4337	1 3	139 275	0.040 0.023	0.020 0.013	2.1 1.8
3	A. niger ATCC 9612	1 3	114 264	0.124 0.016	0.016 0.005	8.0 3.1
4	A*. pullulans ATCC 20612	1 3	35 126	0.028 0.008	0.004 0.002	7.8 0.7
5	A. oryzae IAM-2609pp13	1 3	73 268	0.038 0.013	0.011 0.003	3.5 5.0
6	A. oryzae RIB-143	1 3	158 419	0.003 0.004	0.003 0.002	0.8 2.6
7	A. awamori K 6611	1 3	116 380	0.044 0.011	0.017 0.033	2.6 0.3
8	C. paradoxa FERM P6868	1 3	85 129	0.007 0.010	0.021 0.022	0.4 0.4
9	F. lini IAM-5008	1 3	216 274	0.001 0	0.003 0.003	0.2 0
10	P. nigricans IAM-7218	1 3	143 294	0.010 0	0.003 0	3.1 0
11	S. cerevisiae MIS 2-6 9021	1 3	45 78	0.007 0.005	0.166 0.130	0.04 0.04
12	A*. pullulans[c] ATCC 20612	3	80	0.133	0.014	9.9

a Cell growth was expressed as mg of cells per ml of cultured broth. Specific activity was defined as units per mg of cells. The U_t/U_h ratio shows the relative strength of the transfructosylating activity.

b A, Aspergillus; A*, Aureobasidium; C, Chalara; F, Fusarium; P, Penicillium; S, Saccharomyces.

c Optimum culture conditions for producing U_t of this strain.

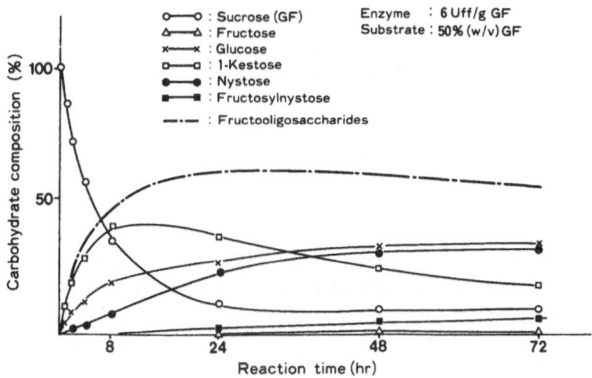

Fig. 2. Time course of the formation of fructooligosaccharides

Manufacturing Procedure of Fructooligosaccharides

Fructooligosaccharides production flow sheet is shown in Fig.3. The upper line shows the manufacturing procedure of the enzyme and the method of immobilizing the enzyme. On the middle

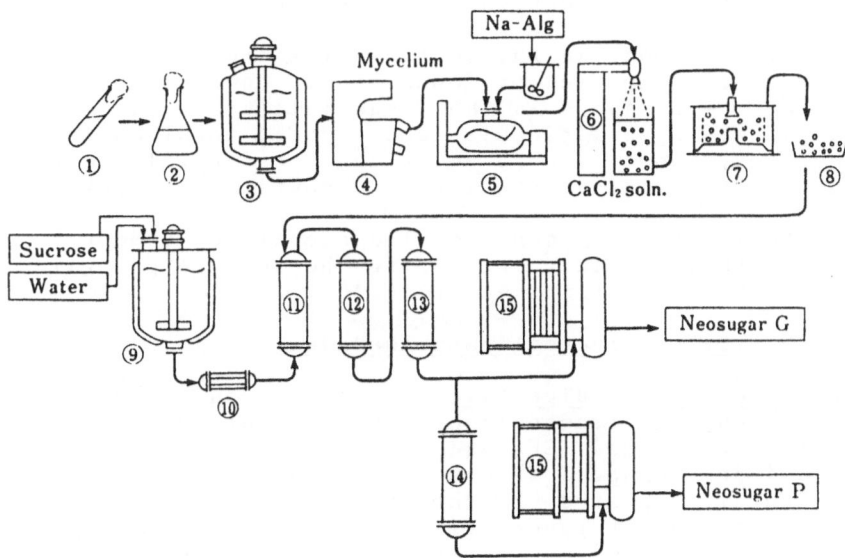

Fig. 3. Production of fructooligosaccharides (Neosugar) (1), slant culture; (2), seed culture; (3), production culture; (4), separator; (5), vacuum mixer; (6) gel forming chamber; (7), separator; (8), immobilized β-fructofuranosidase; (9), storage tank; (10), heater; (11), immobilized enzyme column; (12), active carbon column; (13), ion exchange column; (14), separating column; (15), concentrator.

Table 4. Composition of Neosugar G and P.

Trade name	mono-s G + F	sucrose GF	fructooligosaccha. GF2	GF3	GF4	total of FO
Neosugar G	35	10	25	25	5	55
Neosugar P		<5	40	45	10	>95

line, 50-60% of sucrose syrup run through the immobilized enzyme column, an active carbon column and an ion exchange column for transfructosylation and purification. Neosugar G is obtained from the reaction mixture by a purification process of decolourization and desalination, and Neosugar P is prepared from Neosugar G by removing monosaccharides and disaccharides. The sugar composition of Neosugar used in the present study is shown in Table 4. Neosugar G contained 35% of glucose and fructose, 10% of sucrose, and 55% of fructooligosaccharides, of which GF_2, GF_3, GF_4, were 25, 25 and 5%, respectively. Neosugar P contains above 95% of fructooligosaccharide, of which GF_2, GF_3 and GF_4 are 40, 45 and 10%, respectively. These products are commercially available in Japan.

EFFECTS OF FRUCTOOLIGOSACCHARIDES ON INTESTINAL MICROFLORA AND HUMAN HEALTH

Fructooligosaccharides, were found to be undigestible for humans and animals.[5] The selective utilization of fructooligosaccharides by intestinal bacteria led to a remarkable increase in bifidobacteria in the stool after intake of fructooligosaccharides and brought about various favorable phenomena and results.[6]

The improvement in the intestinal microflora was followed by relief of constipation or loose stool, decreased formation of putrefactive products in the large intestine and improved serum lipids in hyperlipidemia, and reduced total cholesterol, triglycerides, blood glucose and blood pressure. These favorable results in clinical studies were confirmed in several hospitals.[7,8,9]

Fructooligosaccharides are also useful for domestic animals for their rapidly increasing the body weight gain and increasing the feed efficiency rate.[10]

Undigestibility of Fructooligosaccharides

In vitro test by digestive enzyme.[5] In an attempt to investigate whether fructooligosaccharides are hydrolized by human salivary enzymes, GF_2 and GF_3 were incubated in vitro at 37°C for 24 hr. Comapred with sucrose and maltose, GF_2 and GF_3 were hardly digested. Further tests on GF_2 and GF_3 were performed using a rat pancreatic homogenate at 37°C for 120 min: again, GF_2 and GF_3 were hardly digested. These results indicate that fructooligosaccharides are not hydrolyzed by animal digestive enzymes such as salivary α-amylase and small intestinal digestive enzymes.

Sugar tolerance test. Undigestibility of fructooligosaccharides can be shown by sugar tolerance tests in healthy subjects. Fig. 4 shows the result of the test which shows that fructooligosaccharides do not increase blood glucose level nor

110

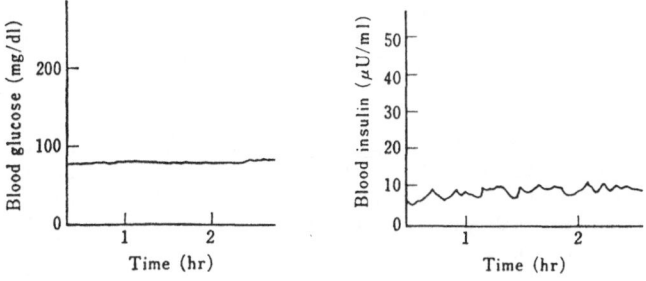

Fig. 4. Sugar tolerance test of fructooligosaccharides in a healthy subject. Left shows the blood glucose level and right shows the blood insulin level.

blood insulin level even after the intake of fructooligosaccharides in oral.

Utilization of Fructooligosaccharides by Intestinal Bacteria

Utilization of several sugars by intestinal bacteria in vitro. The results are shown in Table 5. Fructooligosaccharides were utilized by almost Bifidobacterium strains except bifidum but were not by Escherichia coli and Clostridium perfringens

Table 5. Utilization of several sugars by intestinal bacteria[a]

Bacterial species	No. of strains	Glucose	Lactulose	Fructooligo-saccharides	Bacterial species	No. of strains	Glucose	Lactulose	Fructooligo-saccharides
Bifidobacterium adolescentis	4	⧺	⧺	⧺	*Bacteroides melaninogenicus*	1	⧺	−	⧺
Bifidobacterium longum	3	⧺	⧺	⧺	*Fusobacterium varium*	2	⧺	−	−
Bifidobacterium breve	3	⧺	⧺	+	*Megamonas hypermegas*	2	⧺	⧺	⧺
Bifidobacterium infantis	2	⧺	⧺	⧺	*Mitsuokella multiacidus*	2	⧺	⧺	V
Bifidobacterium bifidum	2	⧺	⧺	−	*Escherichia coli*	2	⧺	⧺	−
Lactobacillus acidophilus	3	⧺	⧺	−	*Klebsiella pneumoniae*	1	⧺	−	⧺
Lactobacillus fermentum	4	⧺	⧺−	−	*Enterococcus faecalis*	1	⧺	+	+
Lactobacillus salivarius	2	⧺	⧺	+	*Enterococcus faecium*	1	⧺	⧺	+
Lactobacillus casei	1	⧺	⧺	−	*Streptococcus intermedius*	2	⧺	⧺	⧺
Lactobacillus plantarum	1	⧺	⧺	+	*Peptostreptococcus prevotii*	1	⧺	−	−
Eubacterium aerofaciens	1	⧺	⧺	+	*Peptostreptococcus parvulus*	1	⧺	−	⧺
Eubacterium limosum	1	⧺	−	−	*Clostridium perfringens*	4	⧺	⧺	−
Eubacterium lentum	1	−	−	−	*Clostridium difficile*	2	⧺	−	−
Propionibacterium acnes	1	⧺	−	−	*Clostridium paraputrificum*	2	⧺	+	−
Bacteroides fragilis	4	⧺	⧺	⧺	*Clostridium clostridiforme*	2	⧺	⧺	+
Bacteroides thetaiotaomicron	3	⧺	⧺	⧺	*Clostridium ramosum*	2	⧺	⧺	+
Bacteroides vulgatus	2	⧺	⧺	⧺	*Clostridium butyricum*	1	⧺	⧺	⧺
Bacteroides distasonis	1	⧺	⧺	⧺	*Veillonella dispar*	2	−	−	−
Bacteroides ovatus	1	⧺	⧺	⧺	*Megasphaera elsdenii*	1	−	−	−

a Judgement of bacterial growth: ⧺, same level of growth compared to glucose; +, weaker growth compared to glucose; −, no growth; V, variable (strains may be either + or −).

at all. In this experiment, spiecies of the Bacteroides fragilis group such as B. fragilis, B. thetaiotaomicron, B. vulgatus, B. distasonis and B. ovatus utilized all the examined sugars.

　　Changes in intestinal microflora. Fructooligosaccharides containing foods (equivalent of 8 g of fructooligosaccharides) were administered daily for two weeks to 23 senile inpatients ranging from 50 to 90 years old in an asylum for the aged. The number of bifidobacteria and total bacteria increased significantly from 4 to 7 days after administration of Neosugar (Table 6). The average number of bifidobacteria per gram of stool increased about 10 times from $10^{8.8}$ (frequency of occurrence, 87%) to $10^{9.7}$ (frequency of occurrence, 100%) after 14 days of

Table 6. Changes in intestinal microflora of senile patients by intake of Neosugar

Organisms	Days from the beginning					
	0 (23)[a]	4 (18)	8 (17)	11 (16)	14 (16)	22[b] (15)
Total counts	10.0 ± 0.3	10.3 ± 0.4*	10.6 ± 0.3**	10.5 ± 0.4**	10.3 ± 0.2*	10.2 ± 0.2
Bacteroidaceae	9.8 ± 0.4[c]	10.0 ± 0.4	10.1 ± 0.3*	10.0 ± 0.5	9.9 ± 0.4	9.9 ± 0.3
	(100)[d]	(100)	(100)	(100)	(100)	(100)
Bifidobacteria	8.8 ± 1.1	9.6 ± 0.7*	10.0 ± 0.6**	9.9 ± 0.6**	9.7 ± 0.5**	9.3 ± 0.7
	(89.5)	(94.4)	(94.1)	(87.5)	(100)	(100)
Enterobacteriaceae	8.0 ± 1.2	7.4 ± 1.3	7.6 ± 1.3	7.7 ± 1.2	7.5 ± 1.5	7.9 ± 1.5
	(100)	(100)	(100)	(100)	(100)	(100)
Streptococci	7.5 ± 1.0	7.4 ± 1.1	7.4 ± 1.2	7.9 ± 1.1	7.4 ± 1.2	7.3 ± 1.2
	(100)	(100)	(100)	(100)	(100)	(100)
Clostridia						
C. perfringens	5.2 ± 1.8	5.7 ± 2.4	5.4 ± 0.5	6.2 ± 1.6	5.0 ± 0.8	6.4 ± 1.1
	(47.4)	(16.7)	(23.5)	(12.5)	(31.3)	(33.3)
Other	7.2 ± 1.3	6.1 ± 1.1**	7.3 ± 1.5	7.2 ± 1.1	7.2 ± 1.4	6.9 ± 1.3
	(100)	(100)	(100)	(100)	(100)	(100)
Lactobacilli	6.9 ± 1.4	7.4 ± 1.6	7.6 ± 1.8	7.8 ± 1.5	7.9 ± 1.4*	7.5 ± 1.3
	(100)	(100)	(100)	(100)	(100)	(100)
Veillonellae	5.7 ± 1.5	5.3 ± 1.7	5.7 ± 1.6	5.9 ± 1.6	6.5 ± 0.9	6.4 ± 1.1
	(78.9)	(83.3)	(82.4)	(93.8)	(75.0)	(73.3)

[a]　Number of examined subjects.
[b]　8-day after discontinuing administration of Neosugar.
[c]　Mean$\pm$SD of log bacterial counts (when present).
[d]　Frequency of occurrence (%).
Significant difference from the counts of 0-day: * $p<0.05$; ** $p<0.01$.

administration of Neosugar. The total counts of intestinal bacteria were also increased from $10^{10.1}$ to $10^{10.3}$. These results suggest that bifidobacteria utilize fructooligosaccharides more rapidly than the spiecies of the Bacteroides fragilis group, which are the dominant bacteria in the human intestine.

　　Although there was no significant difference, lactobacilli showed a tendency to increase from $10^{7.0}$ to $10^{7.8}$, and the occurrence of Clostridium perfringens showed a tendency to decrease from 43.5% to 31.6%.

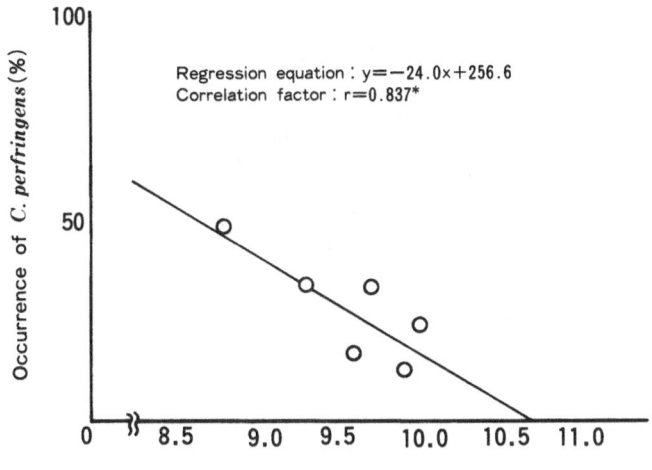

Fig. 5. Correlation between the average number of bifidobacteria and the occurrence of <u>Clostridium</u> <u>perfringens</u>
in senile patients.   * Significant correlation at <u>p</u><0.05.

Fig. 5 shows the correlation between the average count of
bifidobacteria and <u>Clostridium</u> <u>perfringens</u> in each examination
day. As a result, a significant negative correlation was recognized between them. This result suggests that a large number
of bifidobacteria may suppress the growth of <u>Clostridium</u>
<u>perfringens</u> in the human intestine by producing acetic and
lactic acids. In this experiment, 6 subjects had loose stool
before the test, but by taking fructooligosaccharides the
stool condition was gradually improved and became normal after
8 days of administration in all the subjects.
    On the other hand, we tried to administer fructooligo-
saccharides to patients with serious or moderate constipation. As a result, fructooligosaccharides were effective in
patients with constipation, and the number of bifidobacteria
also increased in those effective cases.[11]
    These facts suggest that the intake of fructooligosac-
charides normalizes the stool conditions of diarrhea or
moderate constipation by improving the balance of the intestinal microflora.

<u>Dose-response</u> <u>of</u> <u>fructooligosaccharides</u> <u>administration</u>
<u>on</u> <u>microflora.</u> We have already observed that intestinal bifido-
bacteria were selectively increased by taking 8 g of Neosugar
per day. To determine the minimum dose giving this effect,
we administrated 1, 2 and 4 g per day of fructooligosaccharides to 21 senile persons ranging from 54 to 88 years of
age. Fig. 6 shows the effect of the amount of fructooligosaccharides on increasing the intestinal bifidobacteria counts
in individual subjects. Intake of even 1 g of fructooligosaccharides per day increased the bifidobacteria population
although the rate of increase was slower as the dose was
reduced.

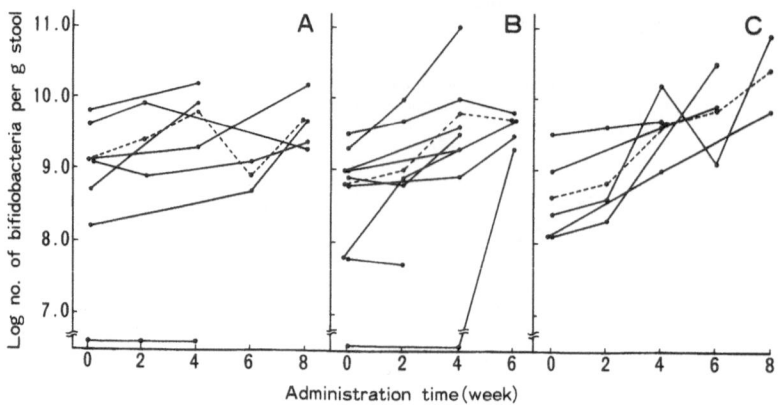

Fig. 6. Effect of the amount of administration of fructooligosaccharides on intestinal microflora. Dose of Neosugar:
A, 1 g per day; B, 2 g per day; C, 4 g per day.

Effect of fructooligosaccharides on the production of
volatile fatty acids (VFAs) and putrefactive substances. To
confirm the beneficial effects of fructooligosaccharides, they
were administered to healthy adults every day for 2 months.
Fig. 7 shows the changes in the intestinal microflora, VFAs
and putrefactive products in a healthy subject. In the case
of Fig. 7, whose microflora had been relatively poor in
bifidobacteria before the administration, the number of
bifidobacteria increased from $10^{7.0}$ to $10^{10.0}$ per gram of
stool, and their ratio to the total counts also increased from
0.9% to 71.3% by administration of Neosugar (Fig. 7-A). The
amount of VFAs in this subject increased (Fig. 7-B), while the
putrefactive products decreased (Fig. 7-C), reflecting the
changes in the microflora.

Effects of Fructooligosaccharides on Blood Glucose and Serum
Lipids in Diabetic Subjects[8]

Daily intake of 8.0 g per day of fructooligosaccharides
for fourteen days significantly reduced mean fasting blood
glucose levels by 15 mg/dl, mean serum total cholesterol levels
by 19 mg/dl and LDL-cholesterol levels by 17 mg/dl in diabetic
subjects (n=10) (Fig. 8A-8D) who were given 5.0 g per day of
sucrose showed no significant changes. The levels of serum
HDL-cholesterol, triglycerides or free fatty acids were not
significantly affected either by fructooligosaccharides nor
sucrose. These results indicate that the daily intake of
fructooligosaccharides ameliorates the derangements of carbohydrate and lipid metabolism in diabetic subjects.

Available Energy of Fructooligosaccharides[12]

Utilization of fructooligosaccharides, was investigated by
a radiorespirometric study and anaerobic incubation of [U-^{14}C]
fructooligosaccharides with feces. About 49% and 55% of the
administered radioactivity were detected in expired $^{14}CO_2$
after 24 and 48 h, respectively. In the anaerobic incubation,
the saccharides were catabolized to $^{14}CO_2$ (9.6%), microbial
cell constituents(10.4%), and ^{14}C-volatile fatty acid (acetic

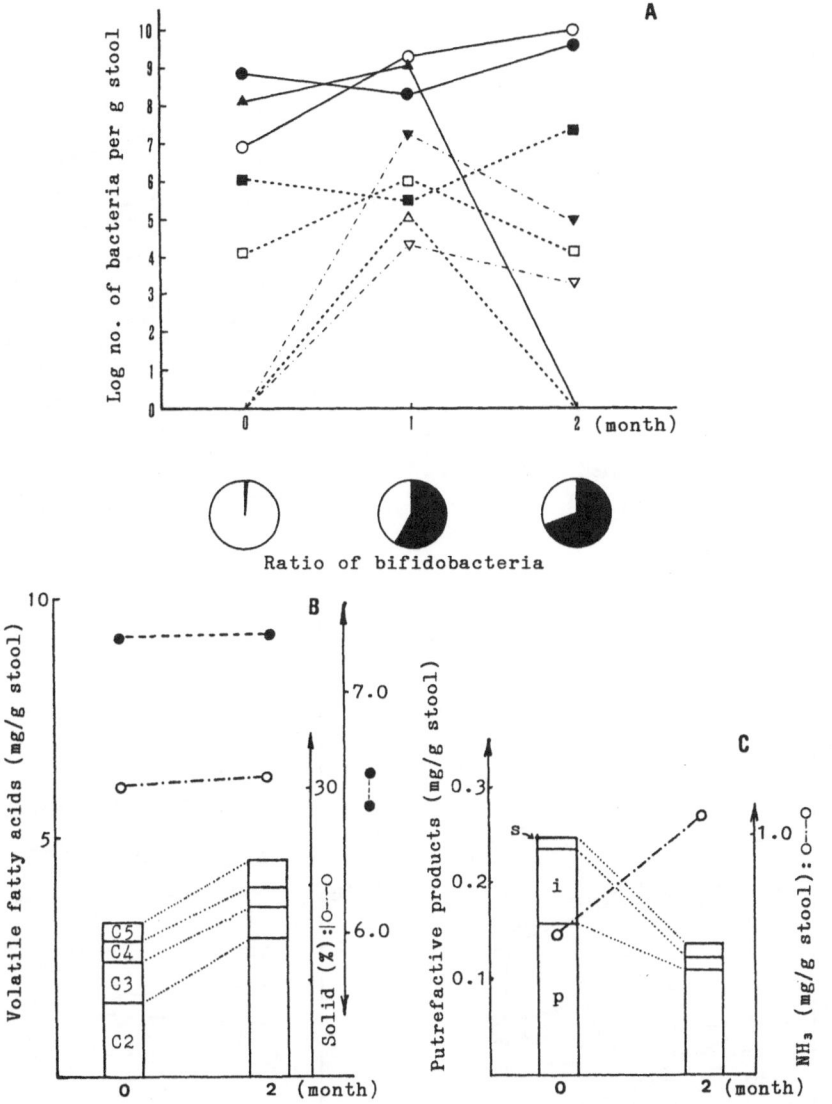

Fig. 7. Effect of fructooligosaccharides on the intestinal microflora and fecal components in a subject who had a low number of bifidobacteria. (A), change in the intestinal microflora; (B), change in the VFAs, pH and solid in stool; (C), change in the putrefactive products and ammonia in stool. Symbols: in (A) ●, bacteroidaceae; o, bifidobacteria; ■, enterobacteriaceae; □, streptococci; ▲, eubacteria; △, lactobacilli; ▼, *Clostridium perfringens*; ▽, veillonellae; : in (B) C2, acetic acid; C3, propionic acid; C4, n and iso-butyric acid; C5, n and iso-valeric acid: in (C) s, skatole; i, indole; c, p-cresol; p, phenol.

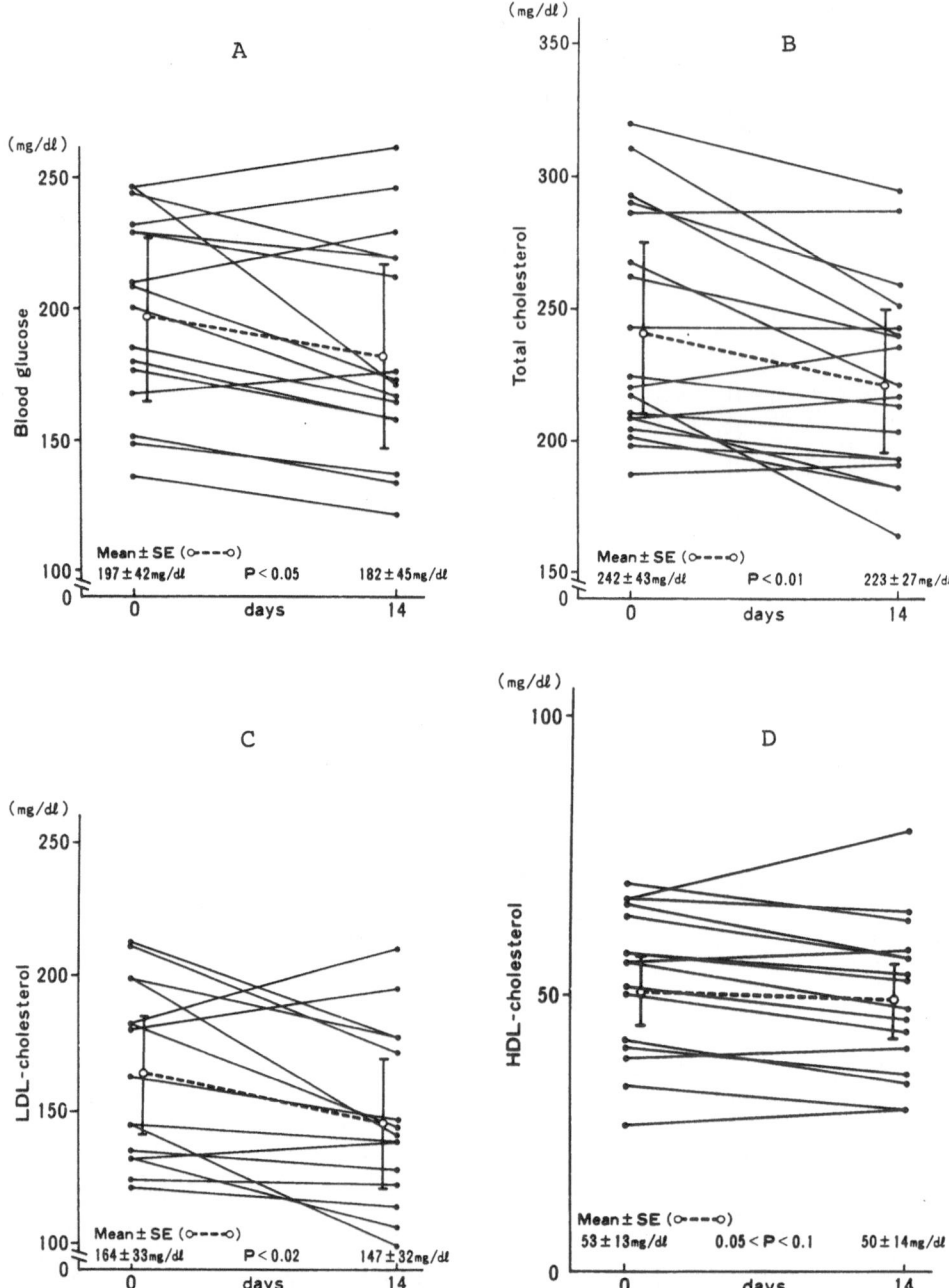

Fig. 8. Effect of fructooligosaccharides intake in diabetic subjects. Eight grams of the saccharides were taken for 14 days. Mean　SE was indicated as　o----o. (8A), fasting blood glucose; (8B), serum total cholesterol levels; (8C), serum LDL-cholesterol levels; (8D), serum HDL-cholesterol levels.

acid, 24.1%: propionic acid, 20.2%; butyric acid, 11.4% and valeric acid, 2.2%).

Complemental interpretation of two studies led to the elucidation of catabolic pathway in which the $[^{14}CO_2]$fructo-oligosacchariodes were fermented by intestinal bacteria into $^{14}CO_2$ and $[^{14}C]$VFAs which were absorbed and utilized in man to give the respiratory $^{14}CO_2$. Therefore, caloric utilization of the saccharides could be quantified by estimating the amount of the $[^{14}C]$VFAs which were absorbed from the colon, and the available energy would be estimated to be between 1.5 and 2.0 kcal/g.

REFERENCES

1. P. L. Wistler and C. L. Smart, "Polysaccharide Chemistry," Academic Press, New York, p. 276 (1952).
2. J. Edelman, The formation of oligosaccharides by enzymic transglycosylation, in :"Advances in Enzymology," Vol. XVII, F. F. Nord, ed., Interscience Publishers, Inc., New York, p. 189 (1956).
3. H. Hidaka, M. Hirayama and N. Sumi, _Agric. Biol. Chem.,_ 52: 1181 (1988).
4. M. Hirayama, N. Sumi and H. Hidaka, _Agric. Biol. Chem.,_ 53: 667 (1989).
5. T. Oku, T. Tokunaga and N. Hosoya, _J. Nutr.,_ 114: 1574 (1984).
6. H. Hidaka, T. Eida, T. Takizawa, T. Tokunaga and Y. Tashiro, _Bifidobacteria Microflora,_ 5: 37 (1986).
7. Y. Hata, T. Hara, T. Oikawa, M. Yamamoto, N. Hirose, T. Nagashima, N. Torihama, K. Nakajima, A. Watabe and M. Yamashita, _Geriatric Medicine,_ 21: 156 (1983) (in Japanese).
8. K. Yamashita, K. Kawai and M. Itakura, _Nutr. Res._ 4: 961 (1984).
9. T. Sano, M. Ishikawa, Y. Nozawa, K. Hoshi and K. Someya, Application of fructooligosaccharides for diabetic subjects. The effect of fructooligosaccharides on blood glucose, "Reports of Neosugar Conference (II)," N. Hosoya ed., Meiji Seika Kaisha, Ltd., Tokyo, p. 79 (1985) (in Japanese).
10. T. Fukuyasu and T. Oshida, Use of fructooligosaccharides in piglets, "Reports of Neosugar Conference (III)," N. Hosoya ed., Meiji Seika Kaisha, Ltd., Tokyo, P. 71 (1986) (in Japanese).
11. S. Kameoka, H. Nogata, H. Yoshitoshi and K. Hamano, _Jap. J. Clin. Nutr.,_ 68: 823 (1986)(in Japanese).
12. N. Hosoya, B. Dhorrannintra and H. Hidaka, _J. Clin. Biochem. Nutr.,_ 5: 67 (1988).

PHYSICOCHEMICAL PROPERTIES AND PHYSIOLOGICAL EFFECTS

OF THE (1→3)(1→4)-β-D-GLUCAN FROM OATS

Peter J. Wood

Food Research Centre
Agriculture Canada
Ottawa, Ontario K1A 0C6

INTRODUCTION

Oats have a a strong traditional reputation as a nutritious cereal. Reports of the benefits of oatmeal in management of diabetes, and clinical evidence for the value of oatmeal in lowering serum cholesterol appeared as long ago as 1913 and 1963 respectively (Allen, 1913; deGroot 1963). Today's feverish commercial activity in oat products is a consequence of the newly developed concern amongst consumers of the relationship between diet and health, and more particularly epidemiologically based evidence that the normal diet of the Western developed societies is deficient in dietary fiber (Burkitt and Trowell, 1975). Animal and clinical studies indicating the specific value of soluble dietary fiber in regulation of glucose metabolism and reduction of serum cholesterol levels have particularly stimulated public interest in oat bran (Anderson and Chen, 1979; Roehrig, 1988).

NATURE OF OAT BRAN AND β-GLUCAN

J.W. Anderson was first to recognise the potential value of oat bran, perhaps reasoning that like wheat, the bran should be richer in the desired active fiber component (Anderson and Chen, 1979). Oat bran is indeed enriched in dietary fiber, but two factors make oat bran quite different from wheat. Firstly, unlike wheat bran, oat bran is a good source of soluble dietary fiber. The second related reason is that, in a sense, oat bran is less pure than wheat bran, containing more endospermic contamination. The endospermic cell walls are the main source of the soluble fiber, oat gum, of which the major component is (1→3)-(1→4)-β-D-glucan, or β-glucan. In particular, the β-glucan rich subaleurone endosperm cell walls are thick, resistant to milling attrition and therefore produce coarse particles which may be fractionated to give β-glucan enriched products. This is not, however, a traditional milling process as with wheat, and differences in morphology and component distribution in oats make production of a bran that is comparable to wheat bran difficult and perhaps not even desirable. Important differences in microchemical organization between oats, wheat and barley which relate

Contribution No. 836 of the Food Research Centre, Agriculture Canada, Ottawa, Ontario, Canada K1A 0C6

New Developments in Dietary Fiber
Edited by I. Furda and C. J. Brine
Plenum Press, New York, 1990

both to milling characteristics and the value of the cereal as a β-glucan source have been demonstrated using histochemical methods (Fulcher
and Wood, 1983). Wheat is a poor source of β-glucan. Barley endosperm
cell walls however are rich in this polysaccharide, and some cultivars
have a β-glucan content in the whole kernel as high as oat brans (Aman
and Graham, 1987).

Dietary fiber and β-glucan contents of commercial rolled oats and
oat bran and some experimental oat brans are shown in Table 1. Clearly
oats are a potentially good source of dietary fiber (DF), particularly
soluble DF, and the experimental products are exceptionally rich in this
component. This presents opportunities not readily available with other
fiber sources. The purified soluble fiber, or oat gum, is relatively
easily prepared (Wood et al., 1978, 1989) although large scale pilot
plant production is expensive (≈$5K/Kg). Technically, however, there
are no difficulties in working with both a palatable bran, rich in the
native fiber, or a purified extract of the bran. A particular advantage
in studies of cereal β-glucan is that it is the only dietary fiber for
which there is a specific method of analysis commercially available.
Indeed there are two methods, one enzymic (McLeary and Glennie Holmes,
1985) and one based on calcofluor fluorescence (Joergensen, 1988).

Structurally, oat β-glucan is very similar to the more extensively
studied barley β-glucan. The polysaccharide is unbranched and contains
about 70% 4-linked and 30% 3-linked β-D-glucopyranosyl units. The 3-
linked units mostly occur singly whereas the 4-linked units mostly occur
in groups of two or three. Details of fine structure, obtained by analysis of fragments released by enzyme action, reveal differences in
structure between oat and barley β-glucan, with oats having relatively
more of the groups of three consecutive 4-linked units (Wood et al.,
1989a).

As part of studies of oats as an ingredient source, large quantities of oat brans and oat gum were produced under a contract at the POS

Table 1. Dietary fiber and β-glucan content (% dry weight basis) of
some oat brans

| Material | Dietary Fiber | | β-Glucan[a] |
	Soluble	Total	
Commercial Rolled Oats	nd[b]	≈ 5[c]	3.8 - 4.2[d]
Commercial Bran	nd[b]	≈13[c]	7 - 10[d]
POS Pilot Plant: Bran 1	15.6[e]	33.5[e]	14.5
Bran 2	18.8[e]	40.1[e]	16.6
FRC Bran[f]	19.4[e]	54.9[e]	18.6

[a] Determined by method of McCleary and Glennie-Holmes (1985)
[b] nd, not determined
[c] Manufacturer's specifications, as is basis
[d] Typical range
[e] Determined by R. Mongeau using method of Mongeau and Brassard
  (1986)
[f] Prepared by D. Paton according to method of Burrows et al (1984)

Pilot Plant in Saskatoon (Wood et al., 1989). The availability of a purified polysaccharide preparation allowed the question of whether oat β-glucan was the major active ingredient in reducing serum cholesterol, or modifying glycemic response, to be pursued. This article focuses on studies carried out at the Ottawa Civic Hospital on the effect of oat gum and bran on glucoregulation. Because in the past similar studies have been done with guar gum, this gum was included for comparison.

VISCOSITY

Viscosity has been reported as important in determining effectiveness of soluble fibers in lowering glycemic response (Jenkins et al., 1978). Some rheological characteristics of oat and guar gum have therefore been compared (Wood et al., 1990).

Both oat and guar gum produced pseudoplastic solutions above certain concentrations (Wood, 1986). Accordingly, meaningful comparisons of viscosity must be made at known concentration and shear rates. Alternatively the shear stress (σ) - shear rate (τ) curves may be analysed for a particular concentration according to the equation

$$\sigma = c\tau^n$$

and the shear rate index (n), an indicator of pseudoplasticity, and the plastic viscosity coefficient (c), theoretical viscosity at low shear rate (1 s^{-1}), compared. The data shown (Table 2) was obtained with 1 and 2.9% w/v solutions prepared on a solids basis. Similar results were obtained when solutions were prepared on the basis of polysaccharide content (Wood et al., 1990). The 2.9% solutions were products as used in nutritional studies and contained aspartame, flavour and coloring.

At 1% (w/v) guar gum showed distinctly higher viscosity than oat gum at relatively low shear rates. Apparent viscosity was 1.6 times

Table 2. Viscosity characteristics of pilot plant and bench prepared oat gum and guar gum prepared at 1% (w/v) solids, and of gum gels prepared at 2.9% (w/v) solids as used in feeding study (average of duplicates)

Sample	Apparent Viscosity mPa.s at 30 s^{-1}	n[a]	c[b]
Pilot Plant Oat gum	391	0.60	14.6
Guar Gum	632	0.43	41.9
Bench Oat Gum[c]	1475	0.31	139
Oat Gum Gel	6822	0.43	470
Guar Gum Gel	8309	0.22	1217

[a] shear rate maxima 400-430 s^{-1}
[b] A theoretical value of the apparent viscosity ($\times 10^{-2}$) in mPa.s at 1 s^{-1}
[c] Restricted availability of sample caused uncertainty in concentration. Shear stress - shear rate curves were therefore displaced leading to difference between duplicate viscosity values of ≈25%

that of oat gum at 30 s⁻¹ shear rate and plastic viscosity coefficient 3 times that of oat gum. At higher shear rates the difference was less apparent. More relevant to the physiological effects is perhaps the data obtained with the gel products as fed. Although differences exist, the relative magnitudes of these are less. At 30 s⁻¹ shear rate guar was just 1.2 times the viscosity of oat gum and the ratio between plastic viscosity coefficients was 2.6.

The viscosity of oat gum (81% β-glucan) isolated in the laboratory (bench gum) was also determined. For reasons that are not entirely clear (Wood et al., 1989) significant losses in viscosity occurred in pilot plant gum relative to bench gum. Thus, the plastic viscosity coefficient was ≈10 times and viscosity at 30 s⁻¹ shear rate ≈4 times that of pilot plant oat gum.

EFFECT OF OAT BRAN AND GUM ON POSTPRANDIAL BLOOD GLUCOSE AND INSULIN IN HEALTHY SUBJECTS

Wood et al. (1990) studied the effect of oat bran and oat gum on postprandial glycemia in healthy subjects. The bran used contained 14.5% β-glucan (Bran 1, Table 1). The oat gum was made from a bran prepared by air classification of defatted oat flour (Wood et al., 1989) and contained approximately 80% β-glucan, 7.8% starch, 7.2% protein, 2.9% pentosan and 1.6% ash.

Guar gum was obtained commercially. Analysis (acid hydrolysis and HPLC) showed 82% galactomannan with a mannose to galactose ratio of 1.5. Protein content was ≈4% and ash 0.5% (manufacturer's data).

Table 3. Glycemic index values and areas under the two hour blood glucose curves of glucose (Gluc), glucose plus oat (OG) and guar gum gels (GG), cream of wheat (CW), cream of wheat plus oat gum (CW + OG) and oat bran (OB) meals [Wood et al., 1990. Reproduced with permission from J. Agric. Food Chem.]

Meal	Mean Area Under Curve (± SE)[a]	Glycemic Index[b]
Gluc	158 ± 29	100
OG[c]	65* ± 29	41
GG[c]	82* ± 31	52
CW	144 ± 17	100
CW + OG	87** ± 17	60
OB	90** ± 15	62

* Significantly different from glucose control (p < 0.05)
** Significantly different from cream of wheat control (p < 0.010)
a Standard error
b Relative to glucose (=100) for oat gum and guar gum, and relative to cream of wheat (=100) for cream of wheat + oat gum and oat bran
c Data shown omits one subject (outlier): inclusion of this subject gives a GI of 62 for oat gum and 64 for guar gum but areas under curves were not significantly different

In the first experiment the effects of oat gum and guar gum on plasma glucose and insulin after a 50g glucose load were compared, using the model developed by Jenkins et al. (1978) in which the glucose is consumed in the presence or absence of 14.5 g of gum in 500 mL of water. To make this more palatable, coloring, artificial sweetener and flavour were added to produce products somewhat like gel puddings.

Fasting and postprandial blood glucose and insulin were determined over three hours using nine healthy subjects. Peak blood glucose, was similarly reduced (p<0.01) by both oat and guar gums with significant reductions (p<0.05) between 20 and 60 min. Plasma insulin levels were similary reduced (Braaten et al., 1988a; Wood et al., 1990).

In a second experiment, ten healthy subjects consumed oat bran, cream of wheat with oat gum and cream of wheat alone as control. Meals were in the form of porridge and were supplemented with white bread to contain approximately 60 g of available carbohydrate.

Peak blood glucose again occurred at 30-40 min. The responses for the oat gum and bran meals were almost identical and significantly (p<0.01) lower than cream of wheat alone from 20-40 min. Insulin responses were similarly reduced (Braaten et al., 1988b; Wood et al., 1990).

The results are summarised in Table 3 in which the glycemic indices (Jenkins et al., 1981) relative, as appropriate, to either glucose or cream of wheat are compared. Oat and guar gum had comparable glycemic indices of 41% and 52% respectively, and oat gum as added to cream of wheat had essentially the same glycemic index as oat bran ($\approx$60%).

THE EFFECT OF OAT GUM AND OTHER FIBERS ON GASTROINTESTINAL FUNCTION AND GLYCEMIA IN THE RAT

The effect of oat and guar gum and cellulose on gastrointestinal function and glycemia in the rat was compared with a fiber free diet (Vachon et al., 1988, Bégin et al., 1989).

Table. 4 Effect of oat and guar gum on postprandial insulin (µU/mL) in the rat (from Vachon et al., 1988)

Diet	Concentration of fiber in diet (%)	Time from meal (min)					
		0	30	60	90	150	210
Fiber Free	0	32	95	76	67	54	59
Cellulose	20	27	83	65	52	43	43
Oat Gum	2.5	40	74*	51*	54	49	51
	5.0	38	67**	54*	52	53	59
	7.5	25	56**	40**	40**	39	34*
Guar Gum	2.5	38	76*	63	53	52	46
	5.0	30	62**	43**	36**	35	34*
	7.5	20	51**	40*	36**	33*	35*

* p<0.05; ** p<0.01: compared to fiber free group

Table 5. Effect of oat and guar gum (5% of diet) and cellulose (20% of diet) on rate of dry matter disappearance from rat intestine (from Bégin et al., 1989)

Diet	Time from meal (min)			
	45	90	210	360
		%a		
Fiber Free	19	59	89	96
Cellulose	16**	39**	81*	98
Oat Gum	14	36**	84	92
Guar Gum	12*	35**	79**	93

* p<0.05; ** p<0.01: compared to fiber free group
a Dry matter content expressed as percent of intake

Both oat and guar gum significantly reduced postprandial insulin in a dose related fashion (Table 4). There was no effect on blood glucose, possibly because a small meal size was required for the voluntary test meal design used.

For all meals, gastric emptying was complete by 210 min and guar and pilot plant oat gum similarly and significantly reduced gastric emptying at 45 and 90 min. Cellulose also reduced gastric emptying at these times, and its effect at 45 min was significantly greater than the two soluble fibers (Bégin et al., 1989).

Small intestinal transit of dry matter, was also significantly altered by the three fibers (Bégin et al., 1989). The rate of disappearance of dry matter, expressed as a percentage of intake, is shown in Table 5. In these studies cellulose, oat and guar gum all delayed gastric emptying and intestinal transit, but only the soluble fibers significantly decreased postprandial insulinemia.

ROLE OF VISCOSITY OF OAT GUM

Jenkins and colleagues showed a correlation between viscosity of gums and their ability to reduce postprandial glucose (Jenkins et al., 1978). Dialysis rates were also reduced by the viscous gums. In the studies of Wood et al., (1990), although pilot plant oat gum was of lower viscosity than guar gum, it was equally effective as guar in reducing postprandial glucose. The rates of dialysis of glucose from the oat and guar gum gel meals were also similar (Wood et al., 1990) and markedly less than from glucose alone. Viscosity differences with oat and guar gums was not therefore an important factor at the levels used.

Oat gum prepared in the laboratory (bench gum) was more viscous than the pilot plant oat gum or commercial guar gum. The native cell wall polysaccharide as found in the bran might therefore be expected to develop the same or greater viscosity than the bench oat gum. Indeed, the β-glucan in crude extracts of oat is of higher molecular weight than that of isolated bench or pilot plant gum (Wood, unpublished data). However there was no difference in the hypoglycemic efficacy of oat bran and isolated oat gum. Although this might be coincidence, arising from multiple factors influencing the hypoglycemic nature of oat bran, it again suggests that under the conditions of study the differences in viscosity of the soluble fiber are not important. Edwards et al. (1988) found no significant difference in hypoglycemic response to gums and gum mixtures of widely differing viscosities. However, the viscosity variations of these systems were highly sensitive to pH and ionic strength. This is not the case with the essentially uncharged galactomannan of guar, and β-glucan of oats.

It may be that the physiological response to viscosity, as changed either by concentration or physicochemical characteristics of the polysaccharide, reaches a plateau. Such an effect has been described by Johnson and Gee (1982). Both guar gum and o-(carboxymethyl)cellulose increased the resistance in vitro to diffusion, or thickness of the unstirred layer, of rat jejunum sections. The effect increased to a plateau at a viscosity of ≈100 mPa s^{-1} (shear rate 50 s^{-1}). Our viscosity data suggests that bench oat gum would achieve this limit at about half the concentration of the pilot plant gum. Recently, Lund et al. (1989) showed that pilot plant oat gum (at 0.85% w/v) increased the thickness of the unstirred water layer of the rat jejunum similarly to guar gum.

Although gum gels delayed in vitro diffusion and in vivo absorption of glucose (Wood et al., 1990), real meals and the situation in the human gastrointestinal tract are considerably more complex than this. Assessment of the process should start with a consideration of the structure of the starting material (food) and the effects of cooking.

A recent study (Yiu et al., 1988) has shown that the β-glucan solubilized by cooking rolled oats represents only a small fraction of the total present in the groat and this amount is significantly influenced by whether the oats were brought to the boil gradually or added to boiling water. Even after 10 minutes of cooking the amount of solubilized β-glucan might be as little as 10% of the total. Intact, β-glucan rich endospermic cell walls remain after cooking and these structures might inhibit digestion. Microscopy highlighted the heterogeneous nature of the system, in which the role of β-glucan is controlled by microstructure, dispersion and interaction with other food components. A precise understanding of the role of viscosity in slowing absorption may require knowledge of rates of solubilization and effects of interaction with other food components.

SUMMARY

In summary, despite the many uncertainties regarding mode of action, these results have established that the soluble dietary fibre from oats, mainly (1→3)(1→4)-β-D-glucan, is physiologically active in a fashion similar to guar gum. Extraction processes reduce the molecular size of the β-glucan relative to the native cell wall polymer, and the product was of lower viscosity than a guar gum used for comparison, but this did not appear to affect either in vivo or in vitro activity under the conditions used. There remains a need to develop a clearer picture of the relationship, if any, of viscosity to activity, through dose response studies with products of different viscosity.

REFERENCES

Allen, F.M., 1913, "Studies Concerning Glycosuria and Diabetes", W.M. Leonard, Boston, MS.
Aman, O. and Graham, H., 1987, Analysis of total and insoluble mixed-linked (1→3)(1→4)-β-D-glucans in barley and oats, J. Agric. Food Chem. 35:704.

Anderson, J.W. and Chen, W-J.L., 1979, Plant fiber. Carbohydrate and lipid metabolism, <u>Amer</u>. <u>J</u>. <u>Clin</u>. <u>Nutr</u>. 32:346.

Bégin, F., Vachon, C., Jones, J.D., Wood, P.J. and Savoie, L., 1989, Effect of dietary fibers on glycemia and on gastrointestinal function in rats, <u>Can</u>. <u>J</u>. <u>Physiol</u>. <u>Pharmacol</u>., in press.

Braaten, J.T., Wood, P.J., Scott, F.W., Brulé, D., Riedel, D. and Poste, L.M., 1988a, High beta glucan oat gum, a soluble fiber which lowers glucose and insulin after an oral glucose load: comparison with guar gum, <u>Fed</u>. <u>Amer</u>. <u>Soc</u>. <u>Expt</u>. <u>Biol</u>. 72nd Annual Meeting, p. 198, No. 5257.

Braaten, J.T., Wood, P.J., Scott, F.W., Riedel, D., Brulé, D., Collins, M. and Poste, L.M., 1988b, Effect of oat gum and oat bran on glucoregulation in diabetic and non-diabetic individuals, <u>Diabetes</u>, 37 (Suppl.1):214A.

Burkitt, D.P. and Trowell, H.C., 1975, "Refined Carbohydrate Food and Disease: Some Implications of Dietary Fibre", Academic Press, London.

Burrows, V.D. Fulcher, R.G. and Paton, D., 1984, Processing aqueous treated cereals, <u>U.S</u>. <u>Patent</u> 4,345,429.

Edwards, C.A., Blackburn, N.A., Craigen, L., Davison, P., Tomlin, J., Sugden, K., Johnson, I.T. and Read, N.W., 1987, Viscosity of food gums determined in vitro related to their hypoglycemic actions, <u>Amer</u>. <u>J</u>. <u>Clin</u>. <u>Nutr</u>., 46:72.

deGroot, A.P., Luyken, R. and Pikaar, N.A., 1963, Cholesterol lowering effect of rolled oats, <u>Lancet</u> 2:303.

Farr, D.T., 1987, Consumer attitudes and the supermarket, <u>Cereal</u> <u>Foods</u> <u>World</u>, 32:413.

Fulcher, R.G. and Wood, P.J., 1983, Identification of cereal carbohydrates by fluorescence microscopy, <u>in</u>: "New Frontiers in Food Microstructure", D.B. Bechtel, ed., Amer. Assoc. Cereal Chem., St. Paul, MN.

Jenkins, D.J.A., Wolever, T.M.S., Leeds, A.R., Gassull, M.A., Haisman, P., Dilawari, J., Goff, D.V., Metz, G.L. and Alberti, K.G.M.M., 1978, Dietary fibres, fibre analogues, and glucose tolerance: importance of viscosity, <u>Brit</u>. <u>Med</u>. <u>J</u>., 1:1392.

Jenkins, D.J.A., Wolever, T.M.S., Taylor, R.H., Barker, H., Fielden, H., Baldwin, J.M., Bowling, A.C., Newman, H.C., Jenkins, A.L. and Goff, D.V., 1981, Glycemic index of foods: a physiological basis for carbohydrate exchange, <u>Amer</u>. <u>J</u>. <u>Clin</u>. <u>Nutr</u>., 34:362.

Joergensen, K.G., 1988, Quantification of high molecular weight (1→3)-(1→4)-β-D-glucan using calcofluor complex formation and flow injection analysis, Analytical principal and its standardisation, <u>Carlsberg</u> <u>Res</u>. <u>Commun</u>., 53:277.

Johnson, I.T. and Gee, J.M., 1982, Influence of viscous incubation media on the resistance to diffusion of the intestinal unstirred water layer in vitro, <u>Pflügers</u> <u>Archiv</u>., 393:139.

Lund, E.K., Gee, J.M., Brown, J.C., Wood, P.J. and Johnson, I.T., 1989, Effect of oat gum on the physicochemical properties of the gastrointestinal contents and on the uptake of D-galactose and cholesterol by rat small intestine in vitro, <u>Brit</u>. <u>J</u>. <u>Nutr</u>., 62:91.

McCleary, B.V. and Glennie-Holmes, M., 1985, Enzymic quantification of (1→3)(1→4)-β-D-glucan in barley and malt, <u>J</u>. <u>Inst</u>. <u>Brew</u>., 91:285

Mongeau, R. and Brassard, R., 1986, A rapid method for the determination of soluble and insoluble dietary fibre: comparison with AOAC total dietary fibre procedure and Englyst's method, <u>J</u>. <u>Food</u> <u>Sci</u>., 51:1333.

Roehrig, K., 1988, The physiological effects of dietary fiber - a review, <u>Food</u> <u>Hydrocolloids</u>, 2:1.

Vachon, C., Jones, J.D., Wood, P.J. and Savoie, L., 1988, Concentration effect of soluble dietary fibers on postprandial glucose and insulin in the rat, <u>Can</u>. <u>J</u>. <u>Physiol</u>. <u>Pharmacol</u>., 66:801.

Wood, P.J., 1986, Oat β-glucan: structure, location and properties, in: "Oats: Chemistry and Technology", F.H. Webster, ed., Amer. Assoc. Cereal Chem, St. Paul, MN.

Wood, P.J., Siddiqui, I.R. and Paton, D., 1978, Extraction of high viscosity gum from oats, Cereal Chem., 55:1038.

Wood, P.J., Weisz, J., Fedec, P. and Burrows, V.D., 1989, Large scale preparation and properties of oat fractions enriched in (1→3)-(1→4)-β-D-glucan, Cereal Chem., 66:97.

Wood, P.J., Anderson, J.W., Braaten, J.T., Cave, N.A., Scott, F.W. and Vachon, C., 1989a, Physiological effects of β-D-glucan rich fractions from oats, Cereal Foods World, 34:878.

Wood, P.J., Braaten, J.T., Scott, F.W., Riedel, D. and Poste, L.M., 1990, Comparisons of viscous properties of oat and guar gum and the effects of these and oat bran on glycemic index, J. Agric. Food Chem., accepted for publication.

Yiu, S.H., Wood, P.J. and Weisz, J., 1987, Effects of cooking on starch and β-glucan of rolled oats, Cereal Chem., 64:373.

FIBER AND PHYSIOLOGICAL AND POTENTIALLY THERAPEUTIC EFFECTS OF SLOWING

CARBOHYDRATE ABSORPTION

David JA Jenkins, Alexandra L Jenkins, Thomas MS Wolever and
Vladimir Vuksan

Department of Nutritional Sciences, University of Toronto

Furio Brighenti and Guilio Testolin

Department of Science and Food Technology, University of Milan

INTRODUCTION

The dietary recommendations of diabetes associations, heart founda-
tions, and cancer agencies encourage the use of higher carbohydrate
intakes derived from minimally processed or high fiber foods (1-5). A
characteristic of such foods is often that they are more slowly absorbed
than many refined and highly processed foods. The assumption that slow
absorption of nutrients is of benefit is central to the original fiber
hypothesis (6) and has been described as a new therapeutic principle (7).
Fibers that have proved especially useful include the soluble fibers.
Early on purified soluble fibers such as guar and pectin were found to be
particularly useful in lowering cholesterol in patients at high risk of
heart disease (8,9). This paved the way for current interest in high
soluble fiber foods such as beans, oats (e.g. oat bran), barley, etc., all
of which have been shown to have these effects (10). Our own current
interest has been in the use of traditional starchy foods, e.g. beans,
dried peas, lentils, barley, etc., or starchy foods processed in traditional ways, e.g. pumpernickel bread (whole kernel rye), bulgur (cracked
parboiled wheat), pasta, etc. These foods reduce serum cholesterol in
patients at risk of heart disease and also provide better blood glucose
control in patients with diabetes. There is also evidence that they may
be of benefit in liver disease and in kidney disease because of their
effects on amino acid and nitrogen metabolism. A growing body of evidence
suggests that many traditionally processed foods, especially those still
eaten in parts of the world where heart disease, diabetes and colonic
diseases are rare, may owe much to the fact that their carbohydrate,
possibly due to fiber content, is more slowly absorbed. Other means by
which these same effects can be produced include the use of specific
digestive enzyme inhibitors and increased feeding frequency.

The effects of dietary fiber will therefore be discussed as part of a
general approach to slowing carbohydrate absorption which includes the
following: 1) Use of soluble fiber supplements and soluble fiber foods;
2) Lowering the glycemic index of diets by appropriate food selection; 3)
Alpha-glucosidase inhibition; 4) Nibbling (as opposed to gorging).

The argument which will be pursued is that theoretically, these manipulations would be accompanied by lower blood glucose and insulin responses, the latter being important also as possible stimulus to hepatic lipogenesis and lipid synthesis in arterial wall.

Soluble Fiber

This has attracted much attention both from the standpoint of blood lipid control, but also in terms of diabetes control and possibly reduction in blood urea levels in chronic renal failure. Consumption of both soluble fiber supplements (e.g. guar, pectin, locust bean gum) and fiber rich foods (e.g. barley, oats, legumes, beans, peas, and lentils) have been associated with improved blood glucose control in patients with diabetes (11-14) and lower blood lipids in normal (14), diabetic (12) and hyperlipidemic individuals (8). In keeping with the objective of controlling metabolic events by slowing the rate of absorption, soluble fiber and high fiber foods such as legumes reduce the rate of glucose uptake in vivo (15) and slow carbohydrate digestion (16), glucose diffusion (15), and amino acid uptake (17) in vitro.

Problems with the use of fiber have arisen from administration of ineffective insoluble fiber types or, in the case of supplements, failure to allow for adequate mixing with the food. Of greater concern with the use of supplements, as opposed to high fiber foods, in the effectiveness in the long term and the degree to which compliance can be maintained.

Some of the lessening of effectiveness with time may be the result of adaption of the colonic microflora with increased acetate generation and absorption. The acetate may contribute to cholesterol synthesis and so offset some of the metabolic advantage gained by slowing small intestinal absorption and reducing insulin levels.

Low Glycemic Index Foods

This is another strategy to slow the absorption of carbohydrate foods. Many of the high fiber foods which have been used successfully to lower serum cholesterol or enhance glycemic control have in fact had low glycemic indices, i.e. produce a low glycemic response relative to that predicted from the food carbohydrate content. The separation of useful high fiber and low glycemic index foods may therefore be somewhat artificial. Nevertheless there is a need to attempt to be numerate in assessing the physiological effect of foods. Therefore direct ranking of foods by a physiological system according to the degree to which equicarbohydrate amounts raise the blood glucose has also been proposed as an indicator of a slower absorption rate (18). In this respect, it has been demonstrated that the glycemic index (GI) is related to the rate of in vitro digestion (16). It was hoped that this classification would encompass the many aspects of foods which alter their absorption rate. These factors include the nature of the starch, amylose versus the more readily digested amylopectin; fiber; and the antinutrients, phytates, lectins, tannins and saponins (19). The idea is not new, but the reawakening of interest in formal studies is recent and work in both type I (20) and type II diabetes (11) has shown promise. Lower day glucose profiles and reduced urinary C-peptide excretion have been noted in normal volunteers (10), together with reduced total and LDL cholesterol and apo-B lipoprotein levels. These changes have also been associated with lower serum fructosamine levels. In both insulin-dependent (IDDM)(20) and noninsulin-dependent diabetics (NIDDM)(11), improvement in HbA_{1c} has been noted on low GI diets. Not all groups have found a relationship between the calculated GI and the glycemic impact of the meal; however, where large difference in the GIs of

individual carbohydrate meal components are selected, differences have
also been seen in the glucose and insulin responses to the meal. Further
exploration of the factors determining the glycemic responses to mixed
meals will enhance the utility of existing tables. Nevertheless, the
concept of selecting foods with slower rates of absorption on the basis of
glycemic index fits well with the overall principle of controlling
metabolism by gut-related mechanisms.

Enzyme Inhibition

The concept of slow absorption has been taken up by the pharmaceutical
industry as an approach to the treatment of metabolic disease especially
diabetes. Their approach has been related to the development and use of
the alpha-glucosidase inhibitors. Early studies with kidney bean and
wheat derived alpha-amylase inhibitors demonstrated that the rate of
starch absorption could be reduced and postprandial glucose and insulin
levels flattened (21). The bacterially derived inhibitors (e.g. acarbose)
have a broader spectrum of activity with anti-amylolytic, anti-sucrase,
and anti-maltase activity (22). Their use has been associated with
reduced gut hormone responses (23) and flattened postprandial profiles of
glucose and insulin (24) and in the longer term reduced HbA_{1c} and serum
VLDL triglyceride levels (25). Concern has been raised that high monosaccharide intake and low starch and disaccharide consumption may reduce the
effect. Again the long-term effects remain to be documented. However
initial successes have resulted in the instigation of major longer term
trials of this form of therapy.

Meal Frequency

Eating huge amounts at once has always been associated with various
aspects of ill health. In the last century Charles Dickens described it
most graphically in one of his novels, "Martin Chuzzlewit":

> "It was a numerous company - eighteen or
> twenty perhaps, ... All the knives
> and forks were working at a rate that was
> quite alarming; very few words were
> spoken; and everybody seemed to eat his
> utmost in self-defense, as if a famine
> were expected to set in before breakfast
> to-morrow morning, and it had become high
> time to assert the first law of nature.
> The poultry, . . . disappeared as rapidly
> as if every bird had had the use of its
> wings, and had flown in desperation down a
> human throat. The oysters, stewed and
> pickled, leaped from their capacious
> reservoirs, and slid by scores into the
> mouths of the assembly. The sharpest
> pickles vanished; whole cucumbers at once,
> like sugar-plums; and no man winked his
> eye. Great heaps of indigestible matter
> melted away as ice before the sun. It was
> a solemn and an awful thing to see.
> Dyspeptic individuals bolted their food in
> wedges; feeding, not themselves, but
> broods of nightmares, who were continually
> standing at livery within them. Spare
> men, with lank and rigid cheeks, came out
> unsatisfied from the destruction of heavy

dishes, and glared with watchful eyes upon the pastry.
What Mrs. Pawkins felt each day at dinnertime is
hidden from all human knowledge. But she had one
comfort. It was very soon over."

Over the last 30 years or more the effects of gorging have been contrasted
with those of nibbling through scientific investigation and demonstrated
that "nibbling" as opposed to "gorging" resulted in lower serum
cholesterol, phospholipid, and triglyceride levels, as well as improved
glucose tolerance (26). In preliminary studies, we have confirmed the
hypocholesterolemic effect of nibbling and demonstrated reduction in Apo-B
lipoprotein levels.

In respect to carbohydrate metabolism, this concept had itself been
anteceded by the Staub-Traugott phenomenon, where one carbohydrate meal
was found to improve the response to the next in proportion to the time
interval between the meals. The shorter the interval, the better the
subsequent tolerance.

In this context, our own studies have demonstrated that 50 g glucose
taken as a bolus drink produced a significantly greater insulin response
area than is seen when the same amount of glucose is sipped at an even
rate over 3.5 h. Furthermore, subsequent glucose disposal at 4 h following a 5 g IVGTT was improved as judged by a significant increase in
glucose disappearance rate (i.e. a reduction in K_g) at the end of the
sipping compared with the bolus experiment. Insulin levels during the
IVGTTs were similar but over the preceding 4 h period there had been an
increase in urinary catecholamine excretion and reduced glucagon levels
were also seen during the post-sipping IVGTT. As further evidence of
increased insulin sensitivity, branched chain amino acid levels were
reduced despite the similar insulin levels. Such studies may be seen as
supporting the concept of slow absorption as a means of favourably
influencing carbohydrate and lipid metabolism to achieve the goals of
diabetes management. In this context we have also demonstrated beneficial
effects on glucose and insulin levels in type II diabetes who were given
more frequent meals of common foods over a one day period as compared to
three meals daily.

CONCLUSION

There are a number of approaches by which carbohydrate absorption may
be slowed, these include soluble fiber or low glycemic index diets, enzyme
inhibitors and increased meal frequency. All these approaches may reduce
the need for insulin secretion which may have additional benefits
especially in view of the implication of raised insulin levels and
cardiovascular disease (27). The effects of slow absorption in enhancing
SCFA synthesis is also apparent. The exact role of SCFA in altering
cholesterol synthesis in man remains to be defined.

REFERENCES

1. American Diabetes Association Policy Statement, Nutritional
       recommendations and principles for individuals with
       diabetes mellitus, Diabetes Care 10:126-132 (1986).
2. Nutrition Sub-Committee of the British Diabetic Association's Medical
       Advisory Committee, Dietary recommendations for diabetics
       for the 1980's - a policy statement by the British
       Diabetic Association, Hum Nutr App Nutr 36A:378-394
       (1982).

3. Canadian Diabetes Association, Guidelines for the nutritional management of diabetes mellitus, <u>J Can Dietet Assoc</u> 42:110-118 (1980).

4. AHA Committee Report, Rationale of the diet-heart statement of the American Heart Association, <u>Circulation</u> 65:839A-854A (1982).

5. Committee on Diet, Nutrition, and Cancer, Diet, nutrition and cancer, National Research Council-National Academy of Sciences, Washington D.C. (1982).

6. D. P. Burkitt and H. C. Trowell, eds., "Refined carbohydrate foods and disease", Academic Press, New York (1975).

7. W. Creutzfeldt, Introduction, <u>in</u>: "Delaying Absorption as a Therapeutic principle in Metabolic Diseases", W. Creutzfeldt and U. R. Fosch, eds., Thieme-Stratton, New York (1983).

8. T. A. Miettinen and S. Tarpila, Effect of pectin on serum cholesterol, fecal bile acids and biliary lipids in normolipidemic and hyperlipidemic individuals, <u>Clin Chim Acta</u> 79:471-474 (1977).

9. D. J. A. Jenkins, D. Reynolds, B. Slavin, A. R. Leeds, A. L. Jenkins and E. M. Jepson, Dietary fiber and blood lipids: treament of hypercholesterolemia with guar crispbread, <u>Am J Clin Nutr</u> 33:575-581 (1980).

10. D. J. A. Jenkins, T. M. S. Wolever, G. R. Collier, A. Ocana, A. V. Rao, G. Buckley, Y. Lam, A. Mayer and L. U. Thompson, The metabolic effects of a low glycemic index diet, <u>Am J Clin Nutr</u> 46:968-975 (1987).

11. D. J. A. Jenkins, T. M. S. Wolever, G. Buckley, K. Y. Lam, S. Giudici, J. Kalmusky, A. L. Jenkins, R. L. Patten, J. Bird, G. S. Wong and R. J. Josse, Low glycemic index starchy foods in the diabetic diet, <u>Am J Clin Nutr</u> 48:248-254 (1988).

12. J. W. Anderson and K. Ward, Long-term effects of high carbohydrate, high fiber diets on glucose and lipid metabolism: a preliminary report on patients with diabetes, <u>Diabetes Care</u> 1:77-82 (1978).

13. J. W. Anderson, L. Story, B. Sieling and W. L. Chen, Hypocholesterolemic effects of high-fibre diets rich in water-soluble plant fibres, <u>J Can Diet Assoc</u> 45:140-149 (1984).

14. W. J. L. Chen and J. W. Anderson, Hypocholesterolemic effects of soluble fibers, <u>in</u>: "Dietary Fiber: Basic and Clinical Aspects", G. V. Vahouny and D. Kritchevsky, eds., Plenum Press, New York (1986).

15. N. A. Blackburn, J. S. Redfern, J. Jarjis, A. M. Holgate, I. Hanning, J. H. B. Scarpello, I. T. Johnson and N. W. Read, The mechanism of action of guar gum in improving glucose tolerance in man, <u>Clin Sci</u> 66:329-336 (1984).

16. D. J. A. Jenkins, H. Ghafari, T. M. S. Wolever, R. H. Taylor, H. M. Barker, H. Fielden, A. L. Jenkins and A. C. Bowling, Relationship between the rate of digestion of foods and postprandial glycaemina, <u>Diabetologia</u> 22:450-455 (1982).

17. B. Elsenhaus, U. Sufke, R. Blume and W. F. Caspary, the influence of carbohydrate gelling agents on rat intestinal transport of monosacchardies and neutral amino acids in vitro, <u>Clin Sci</u> 59:373-380 (1980).

18. D. J. A. Jenkins, T. M. S. Wolever and A. L. Jenkins, Starchy foods and glycemic index, <u>Diabetes</u> 11:149-159 (1988).

19. M. J. Thorne, L. U. Thompson and D. J. A. Jenkins, Factors affecting starch digestibility and the glycemic response with special reference to legumes, <u>Am J Clin Nutr</u> 38:481-488 (1983).

20. A. M. Fontvielle, M. Acosta, S. W. Riskalla, F. Bomet, P. David, M. Letanouse, G. Tchobroutsky and G. Slama, A moderate switch from high to low glycemic-index foods for 3 weeks improves the metabolic ontrol of type I (IDDM) diabetes, <u>Diabetes Nutr Metabol</u> 1:139-143 (1988).

21. W. Puls and V. Kreup, Influence of an alpha-amylase inhibitor (BAY d 7791) on blood glucose, serum insulin and NEFA in starch loading tests in rats, dogs and man, <u>Diabetologia</u> 9:97 (1973).

22. D. J. A. Jenkins, R. H. Taylor, D. V. Goff, H. Fielden, J. J. Hisiewicz, D. L. Sarson, S. R. Bloom and K. G. M. M. Alberti, Scope and specificity of acarbose in slowing carbohydrate absorption in man, <u>Diabetes</u> 30:951-954 (1981).

23. U. R. Solsch, R. Ebert and W. Creutzfeldt, Response of serum levels of gastric inhibitory polypeptide and insulin to sucrose ingestion during long-term application of acarbose, <u>Scand J Gastroenterol</u> 16:629-632 (1981).

24. R. J. Walton, G. A. Now, I. T. Sherif and K. G. M. M. Alberti, Improved metabolic profiles in insulin treated patients given an alpha-glucosidehydrolase inhibitor, <u>Br Med J</u> 1:220-221 (1979).

25. I. Hillebrand and K. Boehma, Clinical studies on Acarbose during 5 years, <u>in</u>: "Proceedings of the First International Symposium on Acarbose, Montreux, October 1981", W. Creutzfeldt, ed.,Excerpta Medica, Amsterdam (1982).

26. G. Gwinup, R. C. Byron, W. H. Roush, F. a. Druger and G. J. Hamere, Effect of nibbling versus gorging on serum lipids in man, <u>Am J Clin Nutr</u> 13:209-213 (1963).

27. P. Ducimetiere, E. Eschwege, L. Papoz, J. L. Richard, J. R. Claude and G. Rosselin, Relationship of plasma insulin levels to the incidence of myocardial infarction and heart disease mortality in a middle aged population, <u>Diabetalogia</u> 19:205-210 (1980).

PHYSIOLOGICAL IMPLICATIONS OF WHEAT AND OAT DIETARY FIBER

K.E. Bach Knudsen, Inge Hansen, B. Borg Jensen and
Karin Østergård

NATIONAL INSTITUTE OF ANIMAL SCIENCE
ANIMAL PHYSIOLOGY AND BIOCHEMISTRY
P.O. BOX 39
DK-8830 TJELE, DENMARK

INTRODUCTION

The physiological implications of various types of dietary fiber (DF) are highly correlated to the chemical and structural composition of cell wall materials (CWM) (e.g. Anderson and Chen, 1979; Eastwood and Kay, 1979; Cummings and Branch 1982; Selvendran, 1984; Wisker et al. 1985). These factors in turn determine the physical and physicochemical properties of DF of which the water holding capacity, viscosity and cationic exchange capacity have attracted most interest (Van Soest and Robertson, 1976; Eastwood and Kay, 1979). Wheat bran has a high proportion of insoluble lignified CWM (Selvendran, 1984) and behaves more or less like a inert marker in the gastrointestinal (GI) tract. Wheat bran has little or no effect on digestion and absorption of nutrients in the small intestine and is highly resistant to microbial degradation in the large intestine of monogastrics including man (Stephen and Cummings, 1980). Consequently wheat bran, due to its physical presence, is one of the most efficient DF sources in increasing fecal bulk and in decreasing mouth-to-anus transit time (Cummings et al. 1978; Spiller et al. 1986).

In contrast to the highly insoluble wheat bran, soluble DF such as oat bran and oat products derived from the oat endosperm may increase the viscosity of the intraluminal content within the GI tract, so delaying gastric emptying, increasing mouth-to-caecum transit time and reducing the rate of nutrient absorption from the small intestine (Jenkins et al. 1978; Holt et al. 1979). This in turn may have consequences for the carbohydrate and lipid metabolism (Anderson and Chen, 1979; Chen and Anderson, 1982). Oat bran in particular has demonstrated significant hypocholesterolemic effects (Chen and Anderson, 1982). Like other soluble DF sources oat fibers are likely to be readily fermented by colonic microorganisms and therefore to exhibit only marginal effect on fecal bulk and mouth-to-anus transit time (Cummings et al. 1978).

Recently we have studied the physiological implications of wheat and oat dietary fiber; here we will present the results in order to discuss interactions between the chemical composition of CWM and their physiological effects. Pigs cannulated at the end of the small intestine were used as the experimental model in the study.

New Developments in Dietary Fiber
Edited by I. Furda and C. J. Brine
Plenum Press, New York, 1990

MATERIAL AND METHODS

The following fractions of wheat were prepared by a dry milling process: wheat flour, wheat bran and two fractions rich in aluerone and peri-
carp/testa cell walls. In the following these latter fractions are referred to as aleurone and pericarp/testa. The oat products used were: rolled
oats and oat bran.

The experimental diets were prepared from: wheat flour (WF), wheat
flour + aleurone (WFA), wheat flour + pericarp/testa (WFPT), wheat flour +
wheat bran (WFWB), wheat flour + oat bran (WFOB), rolled oats, (RO) and
rolled oats + oat bran (ROOB)(Table 1). The diets were tested in two series, using diet WF as the low DF control (diet WF1 and WF2, Tabel 1). The
low DF control provided 54 g DF/d, diet WFA, WFPT, WFWB and WFOB, 95-102 g
DF/d, diet RO, 130 g DF/d and diet ROOB 173 g DF/d. Adjustment for dietary
protein was made with casein. To each diet was added 4 g/kg dry matter
Cr_2O_3 as a marker.

As experimental model was used 40-50 kg pigs, cannulated at the end
of the small intestine. Surgery was performed at 30-35 kg and the cannula
was placed in the ileum approximately 15 cm anterior to the ileo-caecal
junction. After a 7-d adaptation period, feces was quantitatively collected on d 7-11 and ileal digesta on d 12-14. The whole procedure was repeated with the same pigs at d 15 to d 28. Feces was collected twice daily,
frozen and stored at -20°C, while the ileal digesta was collected on ice
for a total period of 12 hours; two times 2 hours on each of the three
days, frozen immediately after collection and stored at -20°C. After
finishing the balance periods, the pigs were fed an additional 6 d (Series
1) and 14 d (Series 2). On d 39, the pigs in Series 2 were surgically fitted with catheters in the peripheral circulation (jugular vein). The following day, blood samples were drawn from -30 min to 120 min after the
morning meal. On day 34 (Series 1) and d 42 (Series 2), the pigs were fed
the morning ration and the pigs killed 4 hours post the morning feeding.
Immidiately after killing, the GI tract was removed and separated by liga-
tures into 12 sections. These consisted of the cranial and caudal halves
of the stomach (S_1, S_2) three equal segments of the small intestine (SI_1,
SI_2, SI_3), the caecum (Ce) and six segments of the colon (C_1, C_2, C_3, C_4,
C_5, C_6). The latter consisted of the proximal, two ascending, two descending and the distal segments. The total content of each GI segment was collected for the determination of dry matter, pH, adenine nucleotides, and
short chain fatty acids.

Protein (Nx6.25), low-molecular weight (LMW) sugars, fructans and
starch were analysed essentially as described by Bach Knudsen et al.
(1987), and total mixed linked β(1->3,1-<4)-D-glucans (ß-glucans) by the
fluorometrically method of Jørgensen and Aastrup (1987). Total, insoluble,
and soluble nonstarch polysaccharides (T-NSP, I-NSP, and S-NSP) of DF were
determined as alditol acetates by gas-liqued chromatography (GLC) for neutral sugars, and by a decarboxylation method for uronic acids using a modification of the Theander and Englyst procedures (Theander and Åman 1979;
Englyst et al., 1982; Theander and Westerlund, 1986).
Content of cellulose in NSP was calculated as:
$$Cellulose = NSP_{glucose} - \beta-glucans$$
arabinoxylans (AX) as:
$$AX = (arabinose + xylose + uronic acids)$$
and S-NSP as:
$$S-NSP = T-NSP_{sugars} - I-NSP_{sugars}$$
Klason lignin was measured gravimetrically as the residue resistant
to 12 M H_2SO_4 (Theander and Westerlund, 1986) and Cr_2O_3 by the methods of
Schürch et al., (1950).

Plasma levels of glucose was quantified by the method of Trinder
(1969) and immunoreactive insulin in plasma in accordance to Tindal et al.
(1978).

Tabel 1. Composition of the experimental diets (g/kg dry matter)

Diet:	WF1	WFA	WFPT	WFWB	WF2	WFOB	RO	ROOB
Wheat flour	794	675	744	740	794	705	-	-
Wheat aleurone	-	174	-	-	-	-	-	-
Wheat pericarp/testa	-	-	72	-	-	-	-	-
Wheat bran	-	-	-	82	-	-	-	-
Rolled oats	-	-	-	-	-	-	892	794
Oat bran	-	-	-	-	-	154	-	151
Casein	122	79	105	100	122	66	70	21
Soy oil	46	39	43	42	46	41	-	-
Vitamin/mineral mixture	34	29	32	32	34	30	34	30
Cr2O3 (marker)	4	4	4	4	4	4	4	4
Chemical composition of the diets								
LMW-sugars	11	12	11	10	11	11	12	13
Fructans	10	12	11	12	10	10	1	1
Starch	654	648	628	629	652	659	600	593
NSP	30	48	54	54	34	47	80	96
S-NSP	nm	nm	nm	nm	20	28	42	52
I-NSP	nm	nm	nm	nm	14	19	38	44
β-glucans	3	5	4	4	3	12	38	47
Klason lignin	4	7	8	8	9	10	13	13
Dietary fiber	34	55	62	62	43	57	93	109

LMW-Sugars = low-molecular weight sugars.
NSP = nonstarch polysaccharides.
S-NSP = soluble nonstarch polysaccharides.
I-NSP = insoluble nonstarch polysaccharides.
nm = not measured.

Table 2. Polysaccharides and Klason lignin content of the wheat and oat fractions (g/kg dry matter)

	Wheat fractions				Rolled oats	Oat bran
	Flour	Aleurone	Pericarp/testa	Bran		
Starch	829	514	160	168	605	424
NSP:						
Cellulose	5	39	144	98	14	8
β-glucans	4 (4)[1]	9 (8)	9 (11)	14 (12)	42 (38)	98 (77)
AX	22 (11)	118 (25)	296 (36)	287 (46)	29 (8)	43 (11)
Arabinose	8 (4)	41 (6)	123 (10)	94 (11)	11 (4)	15 (4)
Xylose	13 (6)	66 (14)	143 (18)	168 (26)	13 (2)	21 (4)
Uronic acid	1 (<1)	11 (5)	28 (8)	25 (9)	5 (2)	7 (3)
Total NSP	34 (17)	174 (38)	465 (52)	414 (63)	91 (47)	155 (91)
Klason lignin	1	26	92	86	15	30
Dietary fiber	35	200	557	500	106	185

NSP = nonstarch polysaccharides.
AX = arabinoxylans.
1) Values in brackets are soluble NSP

138

The concentration of adenine nucleotides in digesta content was estimated by the luciferin-luciferace method (McElroy, 1947; Wolstrup and Jensen, 1976), and short chain fatty acids (SCFA) by a gas-liquid chromatographic method as described by Eggum et al. (1982).

The content of polysaccharide residues was calculated as anhydrosugers, and all digestibilities were calculated relative to the Cr_2O_3 content. The data from the balance experiment were examined by a one-way analysis of variance model and the data from the GI tract by a two-way analysis of variance model (Snedecor and Cochran, 1973).

RESULTS AND DISCUSSION

The wheat and oat fractions used in the current experiment varied significantly in chemical and structural composition. As expected from the composition of purified wheat CWM (Mares and Stone, 1973; Ring and Selvendran, 1980; Bacic and Stone 1981a,b) the main cell wall polysaccharides in the wheat fractions were AX (64-69%) and cellulose (15-31%)(Tabel 2). The composition of the oat products comprised β-glucans (46-63%) and AX (28-32%). The wheat bran and pericarp/testa fractions had the highest content of NSP (414 and 465 g/kg dry matter) and Klason lignin of 86 and 92 g/kg dry matter. The lowest solubility of NSP was found in the lignified wheat fractions (11%) while the highest solubility was found in oat bran where more than 60% of NSP was soluble. The high solubility of oat NSP is primarily due to high content of soluble β-glucans in contrast to the high extracted wheat flour where soluble AX is responsible for the relative high solubility of NSP.

Starch in the wheat based diets was almost completely digested in the small intestine (97.9-99.5%), while the digestibility of oat starch was significantly lower (96.6-96.8%) (Table 3). More marked differences

Table 3. Digestion of starch and nonstarch polysaccharides in the small intestine of pigs

Diet	Recovery, % of intake		
	Starch	NSP	β-glucans
WF1	0.6[B]	97[A]	12[B]
WFA	0.7[B]	104[A]	23[AB]
WFPT	0.5[B]	104[A]	22[AB]
WFWB	1.3[B]	90[AB]	36[A]
WF2	2.1[AB]	82[AB]	21[AB]
WFOB	1.4[B]	64[B]	25[AB]
RO	3.2[A]	64[B]	27[AB]
ROOB	3.4[A]	66[B]	36[A]

A-B: Groups in the same column with the same superscript do not differ significantly (P<0.05).
NSP = nonstarch polysaccharides.

between the two cereals were seen in the digestion of NSP. For wheat the recovery of NSP at this site was nearly complete (82-104%), while there was a substantial loss in the small intestine of oat NSP. Similar results were obtained in other studies using pigs and man (Sandberg et al. 1981;

Table 4. Digestion of β-glucans at various sites of the gastrointestinal
tract of pigs

Site	Recovery, % af intake*	SE
Stomach	103	11
SI_1	101	21
SI_2	106	20
SI_3	89	23
Terminal ileum**	29	16
Caecum	3	1
C_1	1	1

*Means of diet WFOB, RO and ROOB.
**Samples collected from ileum cannula.
SI_1, SI_2, SI_3 = proximal, middle and distal thirds of the small intestine.
C_1 = proximal colon.

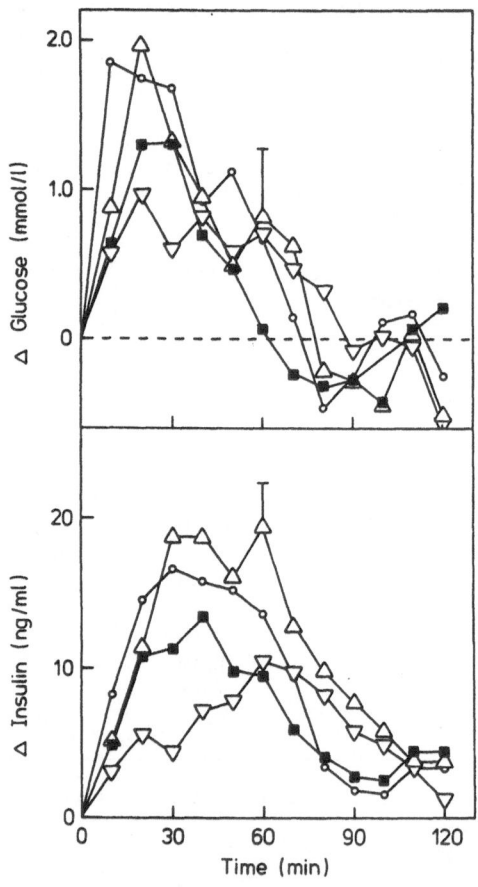

Figure 1. Postprandial glucose and insulin responses in pigs. Symbols corresponding to the four diets are: —△diet WF2; —■diet WFOB; —○diet RO
and —▽diet ROOB. Standard errors of means are represented by vertical
bars.

Sandberg et al. 1982; Millard and Chesson, 1984; Englyst and Cummings, 1985; Graham et al. 1986). When feeding a cereal based diet to pigs 80% of NSP was recovered in the ileal digesta while NSP recovery increased to 89% when one third of a basal diet was substituted with wheat bran (Graham et al. 1986). In man (ileostomy patients) Sandberg et al. (1981) recovered 80% of NSP from wheat bran while a total recovery of NSP (95-115%) was obtained by Englyst and Cummings (1985) when studying rolled oats, white bread and cornflakes. The NSP recovery in the latter study is higher than found for a similar diet in this study (rolled oats) or when feeding other soluble DF sources in form of beet pulp, swede (Brassica napus L.) and pectins (Sandberg et al. 1982; Millard and Chesson, 1984; Graham et al. 1986). For these diets NSP recovery varied from 56-66% in pigs to 70% in man.

In the present study a substantial degradation of β-glucans in the small intestine was found as only 12-36% escaped digestion in the upper digestive tract (Tabel 3). More detailed studies (Tabel 4) showed that the β-glucans mainly were digested in the distal segment of the small intestine indicating that the β-glucans thereby may retain its viscosity in the upper part of the small intestine. The high solubility of β-glucans (Aspinall and Carpenter, 1984; Wood, 1986) certainly made the β-glucans an easily degradable substrate for the bacteria which in the distal segment of the small intestine (ileum) reached considerable levels (Figure 2).

The importance to man of the low digestibility of β-glucans in the stomach and small intestine lies in the fact that the viscosity (Wood, 1986) of β-glucans may affect the rate of starch digestion and glucose absorption. As judged from the flattened response curve for insulin, oat bran seems to affect the rate of glucose absorption in the small intestine (Figure 1). This was also noted by Kirby et al.(1981)in a study with non-diabetic and diabetic patients. The oral glucose tolerance test of eight hypercholesterolemics were significantly improved by the oat bran as compared with the control diet. In the study of Kirby et al. (1981) one of the patients was a diabetic receiving 20 units of insulin per day. He was able to reduce his insulin requirements progressively to zero after 10 days on the oat bran diet while maintaining his fasting glucose concentration within and acceptable range (Gould et al. 1980). The study of Wood et al. (1989) further demonstrated that the β-glucans in oat bran are responsible for the reduction in postprandial glucose response. When consumed by healthy subjects after an overnigh fast, oat gum (80% β-glucans), incorporated into a cream of wheat meal to simulate oat bran porridge, and an equivalent oat bran meal, significantly and similarly reduced peak blood glucose levels relative to the control (cream of wheat). The glycemic index relative to the control of the oat gum meal was 60% compared to 62% for the oat bran meal.

In contrast to oat bran, rolled oats did not affect the rate of glucose absorption from the small intestine as rolled oats resulted in approximately the same increment in the postprandial glucose and insulin response as the control diet. This was the case although the intake of S--NSP (β-glucans) was approximately three times higher for the rolled oats diet compared to the wheat flour diet (91 g/d vs 33 g/d). For comparison addition of oat bran to the wheat flour or rolled oats diets only increased S-NSP by 19 g/d. Another factor which may be considered is starch malabsorption (Jenkins et al. 1987). However, there was no difference in the ileum loss between diet RO and ROOB, in spite of the latter diet resulting in a significantly flatter insulin response curve than the former diet (Figure 1). Thus the regulatory effect of the various oat fractions on the rate of delivering glucose to the enterocyte is complex. One can only speculate if the size differences of the cell walls of the aleurone and subaleurone layers relative to the central endosperm (Fulcher, 1986) may cause differences in the physicochemical properties of β-glucans from the different layers.

Extensive microbial degradation of NSP and starch residues took place in the large intestine (Table 5). For the nonlignified CWM of wheat flour (diet WF 1 and WF 2) and oats (diet WFOB, RO and ROOB) only from 8% to 17% survived breakdown during passage of the entire gut. CWM of aleurone and in particular pericarp/testa were more resistant to microbial degradation. For these latter CWM the recovery of NSP increased to 33% and 50% for diet WFA and WFPT. When wheat bran was added to the diet recovery was 38% i.e. significantly lower than for diet WFPT and slightly, but nonsignificantly higher than in diet WFA.

Table 5. Fecal digestibility of nonstarch polysaccharides in pigs

| Diet | Recovery, % of intake | | | |
	NSP	Cellulose	β-glucans	AX
WF1	17[C]	40[C]	t	15[CD]
WFA	33[B]	53[B]	t	32[B]
WFPT	50[A]	76[A]	t	50[A]
WFWB	38[B]	56[B]	t	38[B]
WF2	13[CD]	37[C]	t	10[D]
WFOB	9[D]	17[D]	t	10[D]
RO	10[D]	22[D]	t	18[C]
ROOB	8[D]	17[D]	t	16[CD]

A-D: Groups in the same column with the same superscript do not differ significantly (P<0.05).
NSP = nonstarch polysaccharides.
AX = arabinoxylans.
t = trace.

The differences in relative breakdown of the various CWM of wheat are probably due to the way in which the polysaccharides are organized and linked to other macromolecules within the different types of CWM. In pericarp/testa (lignified cell walls) the cellulose microfibrils are dispersed in AX polysaccharides and lignin while in endosperm and aleurone cell walls lignin is not present (Mares and Stone, 1973; Bacic and Stone, 1981a,b; Selvendran, 1983). The AX present in pericarp/testa are acidic with linkages to other macromolecules (e.g. lignin or proteins, or both) while in endosperm neutral AX are found. Moreover approximately one third of the endosperm AX are soluble in water in contrast to the acid AX from pericarp/testa which are indispensable in water (Ring and Selvendran, 1980; Selvendran, 1983). Thus the structural features of the two types of CWM explain, firstly, the higher breakdown of cellulose and AX in endosperm relative to pericarp/testa. Secondly, the organization of the polysaccharides and its crosslinking to other macromolecules make the pericarp/testa cell walls much more rigid than that of the endosperm. Therefore, during microbial fermentation cellulose and AX sugar monomers are released from the cell walls of pericarp/testa with approximately the same rate, while the more loose organization of endosperm cell walls allows AX polysaccharides to be released with a higher rate than that of cellulose. A further demonstation of the importance of the chemical and structural compostition of CWM for the digestibility of polysaccharides is seen when comparing the digestibility of oat AX with the analogous polysaccharide deriving from wheat flour. In agreement with the lower solubility of

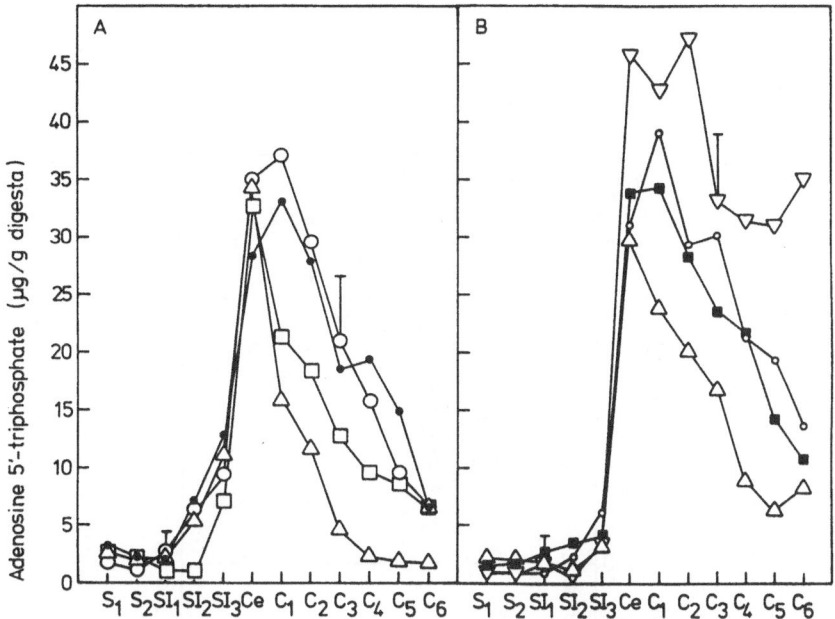

Figure 2. Adenosine5'-triphosphate concentration (ug/g digesta) of the
gastrointestinal content of the pigs. Symbols corresponding to the four
diets in Series 1 (A) are: —△diet WF1; —○diet WFA; —□diet WFPT; —●
diet WFWB and in Series 2 (B): —△diet WF2; —■diet WFOB; —○ diet RO
and —▽diet ROOB. Standard errors of means for values in stomach and
small intestine, respectively, are represented by vertical bars.

oat AX this polysaccharide was degraded to a lower extent than that of
wheat flour.

The microflora in the large intestine metabolize both dietary and
endogenous residues and play an important role in the normal process of
digestion in monogastrics including man (Mason and Just, 1976; Mason 1980;
Graham et al. 1986; Cummings and Englyst, 1987). Carbohydrates (NSP,
starch, and LMW-sugars) and other nonabsorbed residues passing the small
intestine are their choice of substrate (Cummings and Englyst, 1987). As
revealed from the present digestibility trial the amount of carbohydrates
fermented in the large intestine varies in respons to the amount of DF in
the diet and the digestibility. In this study, carbohydrates fermented in
the large intestine varied from 50 g/d (diet WF1) to 151 g/d (diet ROOB).
This has great impact on the microbial activity in this GI compartment;
with diet WF1 there was a rapid decline in microbial activity in the colon
most likely because the bacteria ran out of fermentable substrates, while
with diet ROOB the microbial activity was kept high throughout the colon
(Figure 2).

The importance of fermentation to man and other monogastrics, however, lies mainly in the type of products formed and their fate in the
body. The major end products of fermentation, SCFA (acetic, propionic, and
butyric acids) were the main anion in the caecum and colon where they,
independently of the dietary composition, were found in concentrations of
60-140 mmol/l in the caecum and the two upper segments of colon (C_1-C_2)
and from 35 mmol/l to 60 mmol/l in the lower portions of colon (C_3-C_6).

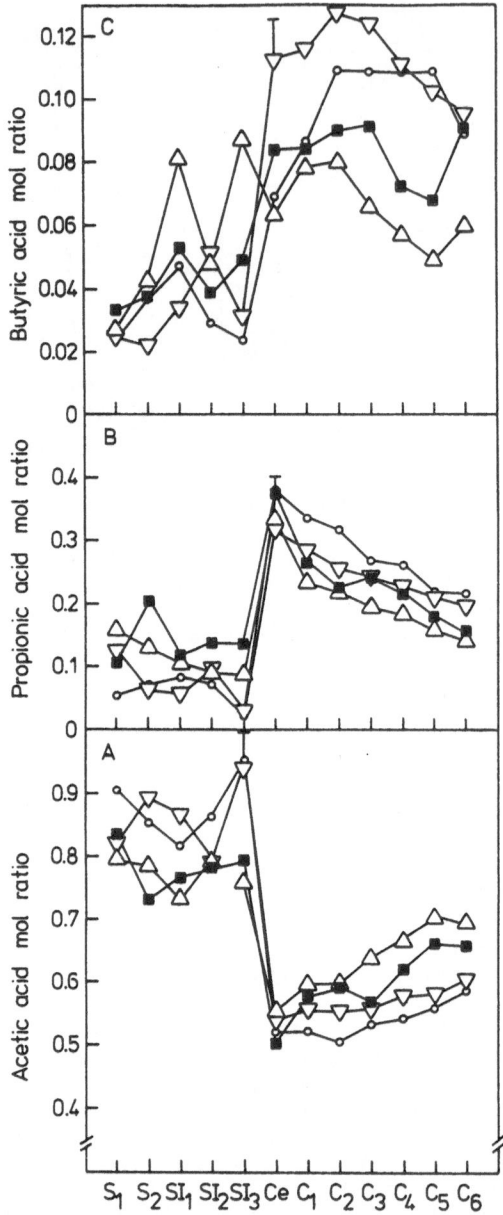

Figure 3. Acetic (A), propionic (B), and butyric acids (C) mol ratio of the short chain fatty acids in the gastrointestinal content of pigs fed the diets in Series 2. Symbols corresponding to the four diets are: ——△ diet WF2; ——■diet WFOB; ——○diet RO and ——▽diet ROOB. Standard errors of means are represented by vertical bars.

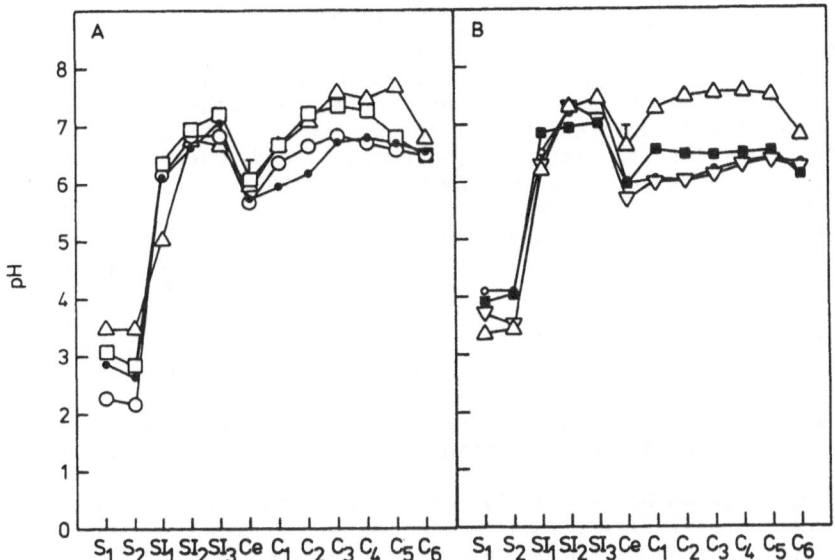

Figure 4. pH in gastrointestinal contents of pigs. Symbols corresponding to the four diets in Series 1 (A) are: —△diet WF1; —○diet WFA; —□ diet WFPT; —●diet WFWB and in Series 2 (B): —△diet WF2; —■diet WFOB; —○diet RO and —▽diet ROOB. Standard errors of means are represented by vertical bars.

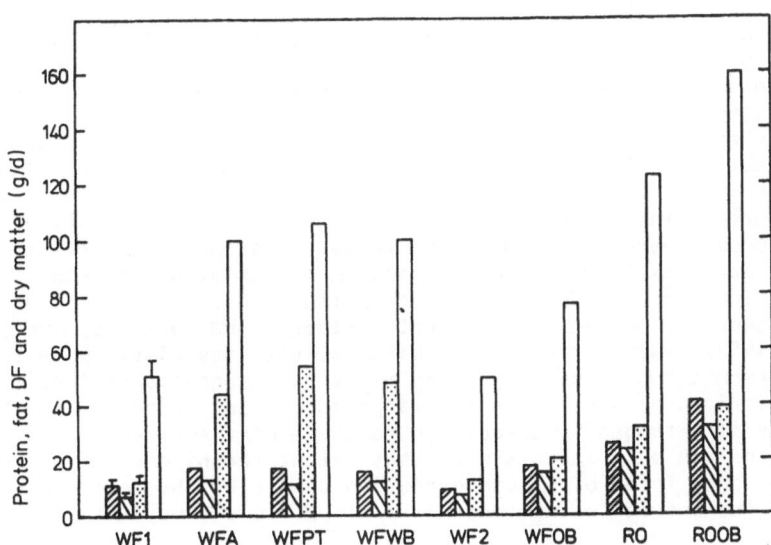

Figure 5. Fecal excreation (g/d) of protein (▨), fat (◨), dietary fiber (▥) and dry matter (□). Standard error of means are represented by vertical bars.

Assuming that the equation for converting carbohydrate into SCFA given for man (Miller and Wolin, 1979) is appropiate even for pigs, the amount of fermented carbohydrates in theory lead to a production of approximately 500 mmol SCFA/d for diet WF1 to 1500 mmol SCFA/d for diet ROOB. The SCFA produced are rapidly absorbed from the gut lumen of monogastrics (McNiel et al. 1978; Argenzio and Whipp, 1979), stimulate sodium and water homeostasis (Argenzio and Whipp, 1979) and play an important role for the energy supply of the colonic epithelial cells. Particularly butyric acid seems to be important for the metabolism, structure, and function of epithelial cells lining the large intestine (Ryan et al. 1979; Cummings and Brance, 1982; Sakata and Yajima, 1984) where it is the preferred fuel over glucose (Roediger, 1980). Interestingly enough, DF addition and oat DF in particular not only causes an increase in butyric acid concentration in accordance with the overall SCFA, but the type of DF source also affects the ratio of butyric acid relative to the other SCFA's (Figure 3). Hence, while the total production of SCFA increased from approximately 500 mmol/d for diet WF1 to 1500 mmol/d for diet ROOB; butyric acid production increased from 35 mmol/d to 170 mmol/l for the two diets. Certainly this is one of the most important factors to consider when discussing the mitogenic response in large intestine of certain DF sources (Lupton et al. 1988).

An association between a high DF intake and a low fecal pH was noted by Walker et al. (1979). The pH in the caecum and colon decreased as a consequence of fermentation of carbohydrates; the effect, however, was strongly influenced by source and amount of residues reaching the large intestine (Figure 4). Hence with the diets providing most fermentable substrates (diet RO and ROOB) the pH was kept in the range of 6.0 to 6.5 in all segments of caecum and colon while pH increased from approximately 6.0 in caecum to 7.5 to 8.0 in the lower segments of colon when feeding the DF depleted wheat flour diet. The luminal pH may influence the growth and metabolism of colonic epithelial cells (Lupton et al. 1988) and the degradation of bile acids. Under acid conditions the 7α-dehydroxylation of primary bile acids is inhibited possibly through a direct effect on the enzyme or the poor solubility of these compounds in acid conditions (Cummings and Branch, 1982).

When discussing the gastrointestinal implications of various DF sources the digestibility of the polysaccharides and the other major dietary constituents are important factor to consider. In agreement with other studies (Stephan and Cummings, 1980; Spiller et al. 1986) lignified DF sources (wheat bran and pericarp/testa) are only degraded to a small extent resulting in an increase of fecal bulk mainly by virtue of its physical presence (Figure 5). In colon these DF sources caused a dilution of the content. In contrast oat DF and to some extent cell walls from wheat aleurone were degraded in the large intestine. This stimulation of microbial activity leeds to an increase in microbial biomass in the fecal material. The same was found with cell walls from other soluble DF soruces: cabagge, pectins, and carrots (Stephan and Cummings, 1980; Nyman and Asp, 1982). The increased microbial activity associated with the higher fermentation of carbohydrates might have significant implications for the nitrogen metabolism in the large intestine. Certainly the higher fecal loss of nitrogen is associated to the higher excretion of microbial biomass. One can only speculate to what extent the higher fat excretion is due to a higher fecal bile acid excretion caused by the fermentation of carbohydrates and following lower luminal pH. In a study with rolled oats to man fecal fat excretion increased by 47% and faecal bile acids by 35% (Judd and Truswell, 1981) compared with the control group.

CONCLUSION

In conclusion, the physiological implications ·of wheat and oat DF are different. Oat bran, because of the high content of soluble DF in form of β-glucans seems to affect the rate of glucose absorption from the small intestine. The same was not seen with rolled oats. Wheat DF from lignified cell walls was almost completely recovered in feces, diluted colonic content and stimulated fecal output by virtue of its physical presence in feces. In contrast to that, oat DF and nonlignified wheat cell walls were digested in the large intestine to a high degree. As a consequence, these DF sources have a much more pronounced effect on the microbial fermentation in the large intestine. The increase in fecal bulk was in this case primarily caused by a higher excretion of nitrogen and fat.

ACKNOWLEDGEMENTS

This work was supported by the Danish Agricultural and Veterinary Research Council, The Danish Technical Research Council and The Carlsberg Foundation. The authors are indebted to Dr. Lars Munck, Carlsberg Research Laboratory, Denmark for placing milling facility at our disposal and to Dr. David Hurt, The Quaker Oats Company, USA for providing us with the oat bran.

REFERENCES

Anderson, J. W., and Chen, W. J. L., 1979, Plant fiber. Carbohydrate and lipid metabolism, Am. J. Clin. Nutr., 32:346.
Argenzio, R. A., and Whipp, S. C., 1979, Interrelationship of sodium, chloride, bicarbonate and acetate transport by the colon of the pig, J. Physiol, 295:365.
Aspinall, G. O., and Carpenter, R. C., 1984, Structural investigations on the non-starchy polysaccharides of oat bran, Carbohydrate Polymers, 4:271.
Bacic, A., and Stone, B. A., 1981a, Isolation and ultrastructure of aleurone cell walls from wheat and barley, Aust. J. Plant Physiol., 8:853.
Bacic, A., and Stone, B. A., 1981b, Chemistry and organization of aleurone cell wall components from wheat and barley, Aust. J. Plant Physiol., 8:475.
Bach Knudsen, K. E., Åman, P., and Eggum, B. O., 1987, Nutritive value of danish-grown barley varieties, I, Carbohydrates and other major constituents, J. Cereal Sci., 6:173.
Chen, W. J. L., and Anderson, J. W., 1986, Hypocholesterolemic effects of soluble fibers, in: "DIETARY FIBER Basic and Clinical Aspects," G. V. Vahouny and D. Kritchevsky, eds., Plenum Press, New York.
Cummings, J. H., and Branch, W. J., 1982, Postulated mechanisms whereby fiber may protect against large bowel cancer, in: "Dietary Fiber in Health and Disease,", G. V. Vahouny and D. Kritchevsky, eds., Plenum Press, New York.
Cummings, J. H., and Englyst, H. N., 1987, Fermentation in the human large intestine and the available substrates, Am. J. Clin. Nutr., 45:1243.
Cummings, J. H., Southgate, D. A. T., Branch, W. J., Houston, H., Jenkins, D. J. A., and James, W. P. T., 1978, Colonic response to dietary fibre from carrot, cabbage, apple, bran and guar gum, Lancet, 1:5.
Eastwood, M. A., and Kay, R. M., 1979, An hypothesis for the action of dietary fiber along the gastrointestinal tract, Am. J. Clin. Nutr., 32:364.

Eggum, B. O., Andersen, J. O., and Rothenberg, S., 1982, The effect of
    dietary fibre level and microbial activity in the digestive tract
    on fat metabolism in rats and pigs, Acta. Agric. Scand., 32:145.

Englyst, H. N., and Cummings, J. H., 1985, Digestion of the polysacchari-
    des of some cereal foods in the human small intestine, Am. J.
    Clin. Nutr., 42:778.

Englyst, H. N., Wiggins, H. S., and Cummings, J. H., 1982, Determination
    of non-starch polysaccharides in plant foods by gas-liquid chroma-
    tography of constituent sugars as alditol acetates, Analyst,
    107:307.

Fulcher, R. G., 1986, Morphological and chemical organization of the oat
    kernel, in: "OATS Chemistry and Technology," F. H. Webster ed.,
    American Association of Cereal Chemists, St. Paul, Minnesota.

Graham, H., Hesselman, K. and Åman, P., 1986, The Influence of wheat bran
    and sugar-beet pulp on the digestibility of dietary components in
    a cereal-based pig diet, J. Nutr., 116:242.

Gould, M.R., Anderson, J.W., and O'Mahony, S., 1980, Biofunctional proper-
    ties of oats, in: "Cereals for Food and Beverages," G. G. Inglett
    and L. Munck, eds., Academic Press, New York.

Holt, S., Heading, R. C., Carter, D. C., Prescott, L. F., and Tothill, P.,
    1979, Effect of gel fibre on gastric emptying and absorption of
    glucose and paracectamol, Lancet, 1:636.

Jenkins, D. J. A., Wolever, T. M. S., Leeds, A. R., Gassull, M. A., Hais-
    man, P., Dilawari, J., Goff, D. V., Metz, G. L., and Alberti, K.
    G. M. M., 1978, Dietary fibres, fibre analogues and glucose tole-
    rance: Importance of viscosity. Br. Med. J., 1:1392.

Jenkins, D. J. A., Jenkins, A. L., Wolever, T. M. S., Collier, G. R., Rao,
    A. V., and Thompson, L. U., 1987, Starchy foods and fiber: reduced
    rate of digestion and improved carbohydrate metabolism, Scand. J.
    Gastro., 22, Suppl. 129:132.

Judd, P. A., and Truswell, S. A., 1981, The effect of rolled oats on blood
    lipids and fecal steroid excretion in man, Am. J. Clin. Nutr.,
    34:2061.

Jørgensen, K. G., and Aastrup, S., 1987, Determination of β-glucan in bar-
    ley, malt, wort and beer, in: "Modern Methods of Plant Analysis,
    Vol 7," H. F. Linskens and J. F. Jackson, eds., Springer-Verlag,
    Berlin.

Kirby, R. W., Anderson, J. W., Sieling, B., Rees, E. D., Chen, W. J. L.,
    Miller, R. E., and Kay, R. M., 1981, Oat-bran intake selectively
    lowers serum low-density lipoprotein cholesterol concentrations of
    hypercholesterolemic men, Am. J. Clin. Nutr., 34:824.

Lupton, J. R., Coder, D. M., and Jacobs, L. R., 1988, Long-term effects of
    fermentable fibres on rat colonic pH and epithelial cell cycle, J.
    Nutr., 118:840.

Mares, D. J., and Stone, B. A., 1973, Studies on wheat endosperm. I.
    Chemical composition and ultrastructure of the cell walls, Aust.
    J. Biol. Sci., 26:793.

Mason, V. C., 1980, Role of the large intestine in the processes of dige-
    stion and absorption in the pig, in: "Current Concept of Digestion
    and Absorption in Pigs," A. G. Low and I. G. Partridge, eds.,
    Technical Bulletin 3, The National Institute for Research in
    Dairying, Reading, England, The Hannah Research Institute, SJK,
    Scotland.

Mason, V. C., and Just, A., 1976, Bacterial activity in the hind-gut of
    pigs. 1. Its influence on the apparent digestibility of dietary
    energy and fat. Z. Tierphysiol., Tierernährg. u. Futtermittelkde.,
    36:301.

McElroy, W.D., 1947, The energy source for bioluminescence in an isolated
    system, Proc. Nat. Acad. Sci. USA, 33:342.

McNiel, N. I., Cummings, J. H., and James, W. P. T., 1978, Short chain fatty acid absorption by the human large intestine, Gut, 19:819.

Millard, P., and Chesson, A., 1984, Modification to swede (Brassica napus L.) anterior to the terminal ileum of pigs: some implications for the analysis of dietary fibre, Br. J. Nutr., 52:583.

Miller, T. L., Wolin, M. J., 1979, Fermentation by saccharolytic intestinal bacteria, Am. J. Clin. Nutr., 32:164.

Nyman, M., and Asp, N.-G., 1982, Fermentation of dietary fibre components in the rat intestinal tract, Br. J. Nutr., 47:357.

Ring, S. G., and Selvendran, R. R., 1980, Isolation and analysis of cell wall meterial from beeswing wheat bran (Triticum aestivum), Phytochemistry 19:1723.

Roediger, W. E. W., 1980, Role of anaerobic bacteria in the metabolic welfare of colonic mucosa in man, Gut, 21:793.

Sakata, T., and Yajima, T., 1984, Influence of short chain fatty acids on the epithelial cell division of the digestive tract, Q. J. Exp. Physiol., 69:639.

Sandberg, A.-S., Anderson, H., Hallgren, B., Hasselblad, K., Isaksson, B., and Hultén, L., 1981, Experimental model for in vivo determination of dietary fibre and its effect on the absorption of nutrients in the small intestine, Br. J. Nutr., 45:283.

Sandberg, A.-S., Ahderinne, R., Andersson, H., Hallgren, B., and Hultén, L., 1983, The effect of citrus pectin on the absorption of nutrients in the small intestine, Human Nutr.: Clin. Nutr., 37C:171.

Schürch, A., LLoyd, L. E., and Crapton, E. W., 1950, The use of chromic oxide as an index for determining the digestibility of a diet, J. Nutr., 50:629.

Selvendran, R. R., 1983, The chemistry of plant cell walls, in: "DIETARY FIBRE," G. G. Brich and K. J. Parker eds., Applied Science Publishers, London.

Selvendran, R. R., 1984, The plant cell wall as a source of dietary fibre: chemistry and structure, Am. J. Clin. Nutr., 39:320.

Snedecor, G. W., and Cochran, W. G., 1973, Statistical Methods, Iowa State University Press, Ames, Iowa.

Spiller, G. A., Story, J. A., Wong, L. G., Nunes, J. D., Alton, M., Petro, M. S., Furumoto, E. J., Whittam, J. H., and Scala, J., 1986, Effect of increasing levels of hard wheat fiber on fecal weight, minerals and steroids and gastrointestinal transit time in healthy young women, J. Nutr., 116:778.

Stephen, A. M., and Cummings, J. H., 1980, The microbial contribution to human faecal mass, J. Med. Microbiol, 13:45.

Theander, O., and Åman, P., 1979, Studies on dietary fibre. 1. Analysis and chemical characterization of water-soluble and water-insoluble dietary fibres, Sw. J. Agric. Res., 9:97.

Theander, O., and Westerlund, E. A., 1986, Studies on dietary fiber. 3. Improved procedures for analysis of dietary fiber, J. Agric. Food Chem., 34:330.

Tindal, J.S., Kanggs, G. S., Hart, I. C., and Blanke, S. A., 1978, Release of growthhormone in lactating and non-lactating goats in relation to behavior, stages of sleep, electroencephalograms, environmental stimuli and levels of prolactin, insulin, glucose and free fatty acids in circulation, J. Endocr., 76:333.

Trinder, P., 1969, Determination of glucose in blood using glucose oxidase with an alternative oxygen acceptor, Ann. Clin. Biochem., 6:24.

Van Soest, P. J., and Robertson, J. B., 1976, Chemical and physical properties of dietary fibre, in: "Dietary fiber," W. W. Hawkins, ed. Proc. of the Miles Symposium, Ontario, Canada.

Walker, A. R. P., Walker, B. F., and Segal, I., 1979, Fecal pH value and its modification by dietary means in South African black and white school children, S. A. Med. J., 55:495.

Wisker, E., Feldheim, W., Pomeranz, Y., and Meuser, F., 1985, Dietary
    fiber in cereals, in: "Advances in Cereal Science and Technology,
    Vol VII," Y. Pomeranz, ed., American Association of Cereal
    Chemists, St. Paul, Minnesota.
Wolstrup, J., and Jensen, K., 1976, Adenosine triphosphate in bovine rumen
    during maximum nutrient supply and starvation, J. Appl. Bact.,
    41:243.
Wood, P., 1986, Oat β-glucan: Structure, location, and properties, in:
    "OATS chemistry and Technology," F. H. Webster, ed. American
    Association of Cereal Chemistry, St. Paul, Minnesota.
Wood, P., 1989, Physicochemical properties and physiological effect of the
    (1->3)(1->4)-β-D-glucan of oats, in: "Dietary Fiber-New Develop-
    ments: Physiological Effects and Physiochemical Properties," I.
    Furda and C. J. Brine, eds., Plenum Press, New York.

POLYSACCHARIDE UTILIZATION BY HUMAN COLONIC BACTERIA

Abigail A. Salyers

Department of Microbiology
University of Illinois
Urbana, Illinois

POLYSACCHARIDES WHICH REACH THE COLON

A complex and constantly changing mixture of polysaccharides
enters the human colon every day. Most of these polysaccharides are
plant cell wall polysaccharides. Although plant cell wall polysaccharides can be degraded to a limited extent by exposure to acid in
the stomach, they are not digested at all by human small intestinal
enzymes and thus reach the colon virtually intact. Also included in
the mixture that enters the colon are host polysaccharides such as
glycoprotein mucins secreted by goblet cells and mucopolysaccharides
released during sloughing of intestinal mucosal cells. Since the rate
of mucin production and mucosal cell turnover increases as the amount
of fiber in the diet increases, the amount of host polysaccharide
entering the colon is not constant but varies with the composition of
the diet (Vahouny and Cassidy, 1986).

It is now well established that some of the starch in the diet
reaches the colon (see, for example, Englyst and Cummings, 1985).
Retrograded starch, which forms when starch is heated and cooled, is
relatively resistant to digestion by small intestinal amylases. Also,
starch in plant cells which remain intact in fragments of vegetable
matter would also escape digestion in the small intestine. Micro-
biologists have long suspected that appreciable amounts of starch
enter the colon because there are some major groups of bacteria in the
colon which cannot ferment any polysaccharide except starch (Salyers
et al., 1977; Salyers et al., 1978). It is unlikely that these bacteria could persist in such high numbers if their only source of carbohydrate was simple sugars lost by other colonic species which can
ferment plant or host polysaccharides.

A fairly accurate estimate of the amount of plant cell wall polysaccharide which reaches the colon can be obtained from an analysis of
fiber in the diet. An estimate of the amount of starch which reaches
the colon could also be made based on information about the effects of
various types of treatments on the starches in foods and measurements
of the amount of starch which is resistant to amylase digestion.
However, in the case of host glycoproteins and mucopolysaccharides,
such an estimate cannot be made easily. It would be possible to

estimate the amount of host polysaccharide entering the colon by measuring the concentration in ileal fluid of sugars such a hexosamines, fucose and sialic acid which are found mainly in host products, although there are technical difficulties associated with quantitation of hexosamines and sialic acids. Also, samples of ileal fluid are difficult to obtain. So far, only one attempt has been made to measure host products in ileal fluid (Vercellotti et al., 1977). The results of this study indicated that polysaccharides containing fucose, hexosamines and sialic acids were present in concentrations comparable to those of the plant cell wall polysaccharides. However, only soluble polysaccharides were analyzed in this study, so the amount of plant cell wall material was probably underestimated.

Information about the total amounts of the various polysaccharides entering the colon is of limited utility unless the physical states of the polysaccharides are also known. Factors such as solubility or binding to protein and lignin are at least as important as structure in determining whether a polysaccharide will be fermented by colonic bacteria. The critical importance of solubility is illustrated by cellulose digestion in the human colon. Nutritional studies have shown that the cellulose in cabbage is extensively degraded in the colon (Van Soest, 1978). Thus there are clearly bacteria in the colon which can ferment the hydrated celluloses in vegetables, although these microorganisms have not yet been identified. However, microcrystalline cellulose, a particularly insoluble form of cellulose, passes through the colons of most people without being degraded (Ehle et al., 1982). The importance of binding to nonpolysaccharide substances can be seen from studies of polysaccharide digestion by bacteria in the rumen of cattle. These studies have shown that binding of lignin to polysaccharides limits polysaccharide digestion. Accordingly, small amounts of a soluble polysaccharide which is not complexed to other materials could well have a greater impact on the colonic microflora than much larger amounts of a polysaccharide which is insoluble or in a sterically hindered complex. This could have practical significance in the case of sweeteners such as the fructo-oligosaccharides or in the case of food additives such as guar gum, both of which are soluble and rapidly digested by colonic bacteria.

MICROBIAL ECOLOGY OF THE COLON

Given the variety of complex carbohydrates entering the colon, it is not surprising that the microbial population of the colon is quite complex and consists primarily of carbohydrate fermenters (Moore et al, 1977). The genus Bacteroides, which accounts for 20% of all colon isolates, contains most of the strains which can ferment plant cell wall polysaccharides and host polysaccharides (Salyers et al., 1977). Some strains of Eubacterium, Bifidobacterium and Peptostreptococcus can also ferment plant cell wall polysaccharides (Salyers et al., 1978). However, the majority of colonic strains cannot ferment either plant cell wall polysaccharides or host polysaccharides, and it is not clear where they are obtaining the carbohydrate they need to maintain their numbers. As mentioned above, most of these bacteria can ferment starch, and enough starch may reach the colon to support them. Another possibility is that some of them may be able to ferment polysaccharides which have not been yet been tested in bacterial fermentation studies because they are not commercially available. Still another possibility is that consortia of these bacteria may be able to degrade polysaccharides which cannot be degraded by a single species.

Polysaccharide-degrading bacteria are unquestionably an important metabolic group in the colonic ecosystem, and as such they have received most of the attention. However, it is well to keep in mind that there are other metabolic groups, many of which subsist on the end products of the polysaccharide fermenters. Some colonic bacteria utilize fermentation products such as succinate, lactate or short chain fatty acids. Relatively little is known about these bacteria. Two other groups of end product utilizers have been better studied: sulfate reducers and methanogens. Many people have sulfate reducing bacteria in their colons. People with high levels of sulfate reducers generally have low levels of methanogens and vice versa, presumably because these two types of bacteria compete for the same source of reducing power (Gibson et al., 1988a; Gibson et al., 1988b). It has also been shown that breakdown of sulfated mucins is linked to sulfate reduction by sulfate reducing bacteria (Gibson et al., 1988c).

Interactions between different metabolic groups in the colonic ecosystem could have practical significance. For example, a sudden increase in the fiber content of the host's diet would be expected to lead to an increase in the numbers and activity of the polysaccharide-fermenters. This could result in a sudden accumulation of fermentation products, including gases, because the end product utilizers have not yet had time to adjust to the change. Some of the problems with flatulence and cramps experienced by people who suddenly switch to a high fiber diet could result from such a transient uncoupling of two linked metabolic groups. If so, a more gradual change in diet should prevent undesirable symptoms. Unfortunately, too little is known about bacterial interactions in the colon to formulate such hypotheses with confidence.

POLYSACCHARIDE UTILIZING COLONIC BACTERIA

Most of the work on polysaccharide utilizing colon bacteria has focused on Bacteroides, because Bacteroides are easy to cultivate and are the only colon anaerobes which can be manipulated genetically. Also, surveys of colonic isolates have shown that the most versatile carbohydrate fermenters are members of this genus. However, other genera, especially Eubacterium, could play an important role in polysaccharide breakdown. Although some work has been done on bile acid modification by Eubacterium species (Hylemon and Glass, 1983), virtually nothing is known about its mechanisms of polysaccharide breakdown. Since Bacteroides strains are Gram negative and Eubacterium strains are Gram positive, the strategies used by these two different types of bacteria for polysaccharide breakdown may not be the same.

Studies of polysaccharide utilization by colonic Bacteroides species have shown that these organisms rely on intracellular rather than extracellular enzymes for breaking down polysaccharides. Instead of secreting the enzymes into the medium, as is generally done by Gram positive bacteria, Bacteroides appear to use outer membrane receptors to bind a polysaccharide prior to bringing it through the outer membrane and into the periplasmic space where the degradative enzymes are located (McCarthy and Salyers, 1988). The monosaccharides and oligomers resulting from polysaccharidase digestion of the polysaccharide are then transported through the cytoplasmic membrane into the cell. This is an ecologically sensible strategy because few of the products of enzymatic breakdown are lost to competing bacteria. It is

clearly a very efficient system because Bacteroides can grow as
rapidly on many polysaccharides as they do on the constituent monosaccharides.

The polysaccharide utilization systems of Bacteroides are quite
complex. The starch utilization system of Bacteroides thetaiotaomicron provides an example of this complexity. B. thetaiotaomicron can utilize amylose, amylopectin and pullulan. Genes
encoding the starch degrading enzymes are all under the same
regulatory control and are induced by maltose as well as the polysaccharides themselves (Anderson and Salyers, 1989a; Anderson and
Salyers, 1989b). None of the starch-degrading enzymes is extracellular. Instead, the enzymes are found in the periplasm or
cytoplasm. Thus in addition to the degradative enzymes, there are
binding sites on the cell surface for amylose, amylopectin and
pullulan. The pullulan receptor appears to be separate from the
receptor for amylose and amylopectin. Genes for proteins involved in
polysaccharide binding are under the same regulatory control as the
genes encoding the polysaccharidases. Mutations that eliminate genes
encoding binding proteins abolish starch utilization. In fact all but
one type of mutant found during screens for starch-minus mutants were
mutants that were deficient in starch binding (Anderson and Salyers,
1989b). This indicates that starch binding to the cell surface is a
critical step in starch utilization.

Investigations of the starch degrading enzymes have identified
several different enzymes which are associated with growth on starch
or pullulan. One is a pullulanase which cleaves pullulan into malto-
triose units. This enzyme appears not to be essential for growth on
pullulan on starch because disruption of this gene in Bacteroides does
not affect the ability of the bacteria to grow on pullulan (Smith and
Salyers, 1989). In addition to this pullulanase, two other starch-
regulated enzymes have been purified and characterized. One is a
polysaccharidase which breaks the α-(1,4) linkages both in amylose and
in pullulan. The second is an α-glucosidase which is active on
maltose but not on starch (Smith and Salyers, unpublished data).
There is also some genetic and biochemical evidence for another
maltase which is a phosphohydrolase rather than a hydrolase (Anderson
and Salyers, unpublished data). Finally, there is presumably an
enzyme which cleaves the α-(1,6) linkage in amylopectin, but this
enzyme has not been characterized. Nothing is known about the cytoplasmic membrane proteins which mediate transport of products of the
periplasmic polysaccharidases into the cell. Based on information
obtained to date, it is clear that the starch utilization system consists of at least ten genes.

The same strains of B. thetaiotaomicron which ferment starch and
pullulan also ferment a variety of other polysaccharides, including
mucopolysaccharides, pectin and arabinogalactan. In all cases, the
polysaccharide utilization systems appear to be as complex as the
starch utilization system. As with the starch utilization, all are
regulated by the associated polysaccharide (McCarthy and Salyers,
1988).

CONTRIBUTION OF POLYSACCHARIDE UTILIZATION GENES TO SURVIVAL OF
BACTEROIDES IN VIVO

Since B. thetaiotaomicron is one of the more numerous species in
the human colon, the polysaccharide utilization systems found in this

species presumably represent a successful adaptation to the colonic environment. The fact that multiple polysaccharide utilization systems have been maintained during evolution indicates that there must have been some selection favoring each of them. This would be the case if B. thetaiotaomicron normally utilizes a variety of polysaccharides in the colon rather than relying primarily on one or two major sources. In an attempt to determine what types of polysaccharides might be important sources of carbon and energy in vivo and what contribution various polysaccharide utilization genes make to survival of B. thetaiotaomicron in its natural habitat, we have been testing polysaccharide utilization mutants of B. thetaiotaomicron for the ability to compete with wild type bacteria for colonization of the intestinal tracts of germfree mice.

Some of the mutations which affect plant polysaccharide utilization had no detectable effect on ability of the bacteria to compete equally with the wild type in vivo. Included in this group were several mutants which were unable to utilize starch and a mutant which was unable to utilize polygalacturonic acid (Salyers and Pajeau, in press). These results could be explained by the fact that starch and pectin are probably present in low concentrations in the intestines of mice fed a chow diet. Support for this explanation comes from preliminary experiments in which a defined diet, with starch and cellulose as the only polysaccharides, was fed to animals colonized with a mixture of wild type and one of the starch-minus mutants. As long as the animals were fed the chow diet, the ratio of mutant to wild type remained constant. But when the animals were switched to the defined diet, the wild type rapidly outcompeted the mutant.

To assess the importance of host products as a source of carbohydrate in vivo, mutants which were deficient in the ability to use host polysaccharides were also tested. These included several mutants unable to use mucopolysaccharides and a fucose-minus mutant which would presumably be deficient in the ability to utilize glycoprotein mucins (Salyers et al., 1988; Salyers and Pajeau, in press). The fucose-minus mutant and one of the mucopolysaccharide-minus mutants were able to compete equally with the wild type. This result seemed to indicate that host products are not an important source of polysaccharide in mice fed a chow diet. However, three other mucopolysaccharide-minus mutants were outcompeted by the wild type. Also outcompeted by the wild type was a mutant which was unable to utilize hexosamines and which should thus have been deficient in the ability to utilize both mucopolysaccharides and glycoprotein mucins. Work is currently underway to determine why some mutants which are unable to utilize host products are outcompeted in vivo whereas others are not. The results may lead us to a more sophisticated understanding of the factors that are important for colonization of the intestinal tract by Bacteroides. Clearly, more than the simple inability to ferment a polysaccharide is involved.

Two other mutants which were rapidly outcompeted by the wild type were a mutant which could not ferment galactose and a mutant which could not ferment uronic acids (Salyers and Pajeau, in press). These mutants were deficient in the ability to utilize more than one type of polysaccharide. The finding that mutations of this type were deleterious in vivo is consistent with the hypothesis that Bacteroides normally utilize many different polysaccharides in the intestinal tract of an animal fed a complex diet, and do not rely on any single polysaccharide as the major source of carbohydrate. Thus it is necessary to eliminate more than one polysaccharide utilization pathway before an effect becomes apparent. Further experiments are underway to test this hypothesis.

DO CHANGES IN THE POLYSACCHARIDE CONTENT OF THE DIET PRODUCE CHANGES IN THE COLONIC MICROFLORA?

Polysaccharide fermenters such as <u>Bacteroides</u> can change their activities in response to changes in polysaccharide composition of the host's diet because their polysaccharide utilization genes are regulated. Induction of these genes is rapid compared to the generation time of bacteria in the colon. From this, one could predict that although changes in the diet would be likely to cause changes in bacterial activities, changes in the relative concentrations of the different species need not occur because the bacteria adapt quickly to the new carbohydrate mixture. Of course, a change in the <u>total number</u> of bacteria would be expected to occur if the total amount of dietary carbohydrate increased. Such an increase is, in fact, observed when people shift from a low fiber to a high fiber diet.

The question of whether changes also occur in the <u>species composition</u> of this population is still unanswered. The difficulty of enumerating even the major species of colonic bacteria by conventional methods has discouraged most researchers from undertaking a study of the effect of changes in diet on the species composition of the colonic microflora. Despite the difficulties, a few attempts have been made to carry out such studies, but the results have been somewhat inconclusive. Each study finds some statistically significant effect, but the effect is not the same from one study to another. Moreover, large individual-to-individual variations are observed in the composition of the microflora. Because of this, only major changes in the flora would be scored as statistically significant. Thus, although it is probably safe to conclude that major changes in the composition of the microflora do not occur after a change in diet, the possibility that some changes do occur cannot be ruled out.

Although it is possible that the species composition of the colonic microflora of one individual differs substantially from that of another individual eating the same diet, it is also possible that these variations reflect the shortcomings of the techniques used for enumeration. There are three major shortcomings of the classical enumeration approach. First, it is not known whether plating efficiencies are the same for all of the major colonic species. Thus, enumeration by plating and counting colonies may be inherently biased in favor of the more easily cultivated species. Second, species identification, unless it is done with DNA probes, is somewhat subjective. Finally, there may still be some the major groups of colon bacteria which have not been cultivated.

Recent technical advances have made it possible to take a new approach to this problem (Stahl, 1988). Probes consisting of segments of 16S rRNA could be constructed for all of the known species or metabolic groups. Such probes, labeled with a fluorescent compound, could then be used to enumerate bacteria directly in smears of fecal material by <u>in situ</u> hybridization. If there are bacteria which are not currently being cultivated, they would be detected by such an approach because they are seen in the smears but do not hybridize to any known probe. If 16S rRNA could be obtained from these organisms, a tentative classification of the organism could be made based on the 16S rRNA sequence. 16S rRNA probes might also give information about growth rates because the number of ribosomes, and thus the strength of the fluorescent signal, increases with the growth rate of the bacterium.

The initial work on such a project would be technically demanding. But once a complete set of probes has been generated, it would be feasible for almost any laboratory to do routine enumerations of the major species in fecal specimens, because the techniques involved are quite simple and no bacteriological expertise is necessary. No group has yet undertaken the approach described above, but such work is now technically feasible.

ACKNOWLEDGMENTS

This work was supported by grant number AI 17876 from the Allergy and Infectious Diseases Institute of the National Institutes of Health.

REFERENCES

Anderson, K. and Salyers, A. A., 1989a, Biochemical evidence that starch breakdown by Bacteroides thetaiotaomicron involves outer membrane starch binding sites and periplasmic starch degrading enzymes, J. Bacteriol., 171: 3192.

Anderson, K., and Salyers, A. A., 1989b, Genetic evidence that outer membrane binding of starch is required for starch utilization by Bacteroides thetaiotaomicron, J. Bacteriol., 171: 3199.

Ehle, F. R., Robertson, J. B., and VanSoest, P. J., 1982, Influence of dietary fiber on fermentation in the human large intestine, J. Nutr., 112: 158.

Englyst, H. N., and Cummings, J. S., 1985, Digestion of the polysaccharides in some cereal foods in the human small intestine, Am. J. Clin. Nutr., 42: 778.

Gibson, G. R., Cummings, J. H., and Macfarlane, G. T., 1988a, Competition for hydrogen between sulphate-reducing bacteria and methanogenic bacteria from the human large intestine, J. Appl. Bacteriol., 65:241.

Gibson, G. R., Macfarlane, G. T., and Cummings, J. H., 1988b, Occurrence of sulphate-reducing bacteria in human feces and the relationship of dissimilatory sulphate reduction to methanogenesis in the large gut, J. Appl. Bacteriol., 65: 103.

Gibson, G. R., Cummings, J. H., and Macfarlane, G. T., 1988c, Use of a three-stage continuous culture system to study the effect of mucin on dissimilatory sulfate reduction and methanogenesis by mixed populations of human gut bacteria, Appl. Environ. Microbiol., 54: 2750.

Hylemon, P. B., and Glass, T. L., 1983, Biotransformation of bile acids and cholesterol by the intestinal microflora, in: Human Intestinal Microflora in Health and Disease, D. Hentges, ed., Academic Press, New York, pp. 189-214.

McCarthy, R. E., and Salyers, A. A., 1988, Effect of dietary fiber utilization on the colonic microflora, in: "The Role of the Gut Flora in Toxicity and Cancer," I. Rowland, ed., Academic Press, London, pp. 295-313.

Moore, W. E. C., Cato, E. P., and Holdeman, L. V., 1977, Some current concepts in intestinal bacteriology, Amer. J. Clin. Nutr., 31: S33.

Salyers, A. A., and Pajeau, M., Competitiveness of different polysaccharide utilization mutants of Bacteroides thetaiotaomicron in the intestinal tracts of germfree mice, Appl. Environ. Microbiol., in press.

Salyers, A. A., Pajeau, M. P., McCarthy, R. M., 1988, Importance of mucopolysaccharides as substrates for Bacteroides thetaiotaomicron growing in the intestinal tracts of germfree mice, Appl. Environ. Microbiol., 54: 1970.

Salyers, A. A., Vercellotti, J. R., West, S. E., and Wilkins, T. D., 1977, Fermentation of mucin and plant polysaccharides by strains of Bacteroides from the human colon, Appl. Environ. Microbiol., 33: 319.

Salyers, A. A., West, S. E., Vercellotti, J. R., and Wilkins, T. D., 1978, Fermentation of mucin and plant polysaccharides by anaerobic bacteria from the human colon, Appl. Environ. Microbiol., 34: 529.

Smith, K., and Salyers, A. A., 1989, Characterization of a cell-associated Bacteroides pullulanase and determination of its contribution to pullulan utilization, J. Bacteriol., 171: 2116.

Stahl, D. A., 1988, Phylogenetically based studies of microbial ecosystem perturbation, in "Biotechnology for Crop Protection," P. A. Hedin, J. L. Menn and R. M. Hollingworth, eds., American Chemical Society Press, pp. 373-390.

Vahouny, G. V., and Cassidy, M. M., 1986, Dietary fiber and intestinal adaptation, in: "Dietary Fiber: Basic and Clinical Aspects," G. V. Vahouny, D. Kritchevsky, eds., Plenum, New York, pp. 181-210.

Van Soest, P. J., 1978, Dietary fibers: their definition and nutritional properties, Amer. J. Clin. Nutr., 31: 512.

Vercellotti, J. R., Salyers, A. A., Bullard, W. S., and Wilkins, T. D., 1977, Breakdown of mucin and plant polysaccharides in the human colon. Canad. J. Biochem., 55: 1190.

EFFECT OF TYPES OF DIETARY FIBER ON FECAL MUTAGENS AND BACTERIAL ENZYMES

IN RELATION TO COLON CANCER

Bandaru S. Reddy

Division of Nutrition and Endocrinology
American Health Foundation, Valhalla, NY

Large bowel cancer affects men and women in Western coutries, including United States and Canada, with high frequency (1,2). Since Wynder et al. (3) first provided an evidence for an association between colon cancer and consumption of a more westernized diets including more fat in Japan, there have been several epidemiologic and animal model studies to test this hypothesis. The hypothesis that a diet high in fiber may protect against colon cancer was first proposed by Burkitt and Trowell (4) who observed that African blacks consuming high fibrous foods had lower death rates due to colon cancer than their white counterparts eating a low fiber diet. Subsequent studies demonstrated that, in certain populations consuming diets high in total fat, the intake of diets high in total fiber, fibrous foods, and certain whole grain foods has been associated with a reduced risk for colon cancer (5,6). A majority of international studies provided evidence for protective effect of dietary fiber and fiber-containing foods (7). Recent intracountry comparisons of dietary fiber and colon cancer mortality rates strongly supported the hypothesis that dietary fiber protects against colon cancer (8). Colon cancer incidence data from Scandinavian countries show a gradient of risk, with the lowest risk in rural areas in northern Finland (Kuopio and Parikkala), increasing to urban Helsinki, and highest in Copenhagen, Denmark (9,10). Finnish population consume diets high in fat, mainly saturated fat; but the low incidence of colon cancer in rural Finland may be explained in part by their consumption of diets high in fiber mainly from whole grain cereals and bread (10,11).

Case-control studies on the relationship between the dietary fiber and colon cancer provided conflicting results (12). Out of 19 case-control studies to assess the role of fiber and fiber-containing foods, 3 studies reported no protective effect, 2 found an increased risk, and 13 studies reported a protective effect of fiber-containing foods and vegetables (12). Several studies in laboratory animal models have demonstrated that dietary cellulose and wheat bran inhibit chemically-induced colon carcinogenesis (13-15), whereas corn bran and oat bran have no such protective effect (15,16), suggesting that the protective effect of dietary fiber depends on the nature and source of fiber in the diet. The discrepancy in human case-control studies on the relationship between dietary fiber and colon cancer risk might, at least in part, be explained on the basis that the comparison of fiber intake between the cases and controls was based on either total dietary fiber intake, vegetable and fruit intake, or fiber-containing foods rather than the types of fiber.

Several studies demonstrated that the concentrations of colonic (fecal) secondary biles acids, particularly deoxycholic acid and lithocholic acid, which have been shown to promote colon carcinogenesis (17) were found to be higher in popula-

New Developments in Dietary Fiber
Edited by I. Furda and C. J. Brine
Plenum Press, New York, 1990

159

tions who are at high risk for colon cancer and consuming high-fat/low-fiber diets than in populations who are at low-risk and consuming low-fat/low-fiber, high-fat/high-fiber fiber diets (6,18). With regard to genotoxic carcinogens in the colon, several mutagens (presumptive carcinogens) have been isolated from the stools of high-risk individuals consuming diets high in fat and low in fiber (19-22). Although there is no direct evidence that these mutagens may play a role in the etiology of human colon cancer, mutagens present in fried foods and also identified in the colonic contents, have been shown to induce colon and mammary tumors in laboratory animals (23-26). The protective effect of dietary fiber in colon carcinogenesis may be due to adsorption, dilution, and/or metabolism of co-carcinogens, promoters, and yet-to-be identified carcinogens by the components of the fiber. Different types of non-nutritive fiber could bind the tumorigenic compounds, affect the enterohepatic circulation of tumorigenic compounds, dilute potential carcinogens and co-carcinogens by its bulking effect and alter the metabolic activity of gut microflora. Thus, the humans consuming relatively large amounts of certain fibers would have a greater protection against carcinogenic and co-carcinogenic compounds than than do individuals consuming lesser quantities of these fibers.

In view of an inverse relationship between dietary fiber and colon cancer risk, and the potential significance of fecal mutagens, secondary bile acids and bacterial enzymes in the pathogenesis of colon cancer, a series of studies on the effect of types of fiber on these fecal constituents was studied in healthy subjects.

EFFECTS OF DIETARY FIBER ON FECAL MUTAGENS

In this study, the effect of dietary wheat bran plus rye fiber on fecal mutagenic activity was studied in healthy men and women from Helsinki, Finland who were consuming high-fat/moderately low-fiber diets and excreting high levels of fecal mutagens (27). The participants of the study were interviewed by a nutritionist and diet histories were recorded. A 3-day dietary record was obtained from each study subject. Twenty four hour stool samples were collected from each study subject for 2 days. They were then given about 10g of supplemental fiber (whole wheat and rye) in the form of bread with the 3 main meals. On the average, they consumed about 6 slices of bread/day. After 4 weeks on high-fiber regimen, 24-hour stool samples for 2 days and 3-day diet records were collected from the subjects. Nutrient and total dietary fiber analysis were performed using food tables. Stool samples collected from each participant before and at the end of the intervention period were processed for mutagen assay as described (27).

Nutrient analysis using 3-day food records collected before and at the end of the intervention period is summarized in Table 1. As might expected, the intakes of total calories, fat, and protein were similar during both the periods and the fiber intake was increased by about 10g during the intervention period. The daily fresh stool output during the control and fiber periods was 151 and 228g, respectively and the fecal pH was 6.9 and 6.7.

Figures 1-2 summarize the effect of dietary fiber on fecal mutagenicity. The mutagenic activity in tester strains (Salmonella typhimurium) TA 100 and TA 98 with and without microsomal (S9) activation was inhibited during the fiber period compared to control diet period.

The mechanism by which dietary fiber affects the mutagens observed in this study may not be ascribed to a simple dilution due to a high fecal bulk since the mean fecal mutagenic activity was decreased by 2.5 to 4.5 fold and the stool bulk was increased by about 35%. Dietary fiber affects the gut microflora by modifying both the metabolic activities and the composition of colonic microflora which, in turn, alters the breakdown products of dietary fiber as well as dietary fat and protein in the gut. The amount and type of breakdown products produced by the microflora in the gut may depend partly on the intake of certain dietary fibers and other nutrients. Therefore, it is possible that the intake of dietary fiber modifies the formation of certain types of mutagens in the colon and/or inhibits the activity of mutagens in the colon.

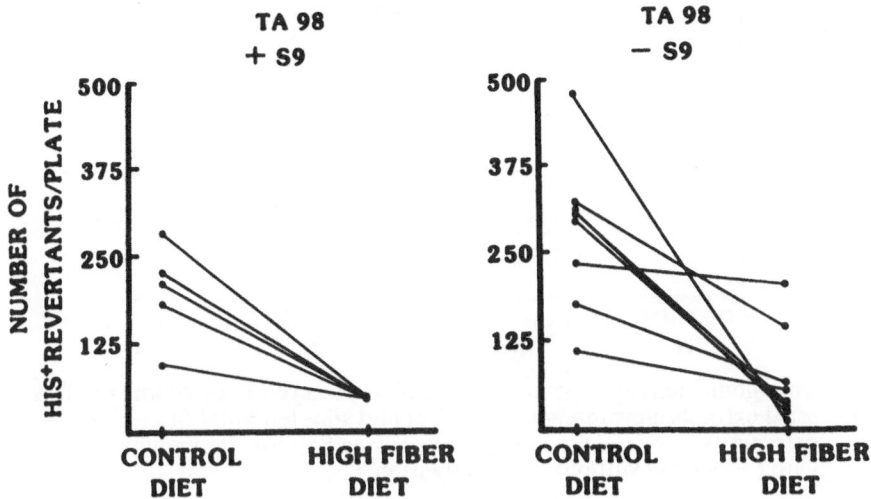

Fig. 1.   Mutagenic activity (histine-positive revertants/plate or 200mg dry feces) of fecal extracts consuming control diet and supplemental fiber diet. The extracts were tested for mutagenicity using <u>Salmonella</u> <u>typhimurium</u> TA 98 with (+ S9) and without (-S9) activation.

Table 1.   Dietary Nutrient Intake of Healthy Subjects during the Control and Supplemental Fiber Periods

	g/day			
	Control diet		Supplemental fiber diet	
Nutrients	Males (8)[a]	Females (7)	Males (8)	Females (7)
Energy, Kcals/day	2656 ± 255[b]	2405 ± 330	2489 ± 149	2274 ± 188
Total Protein	96 ± 5	94 ± 15	97 ± 7	98 ± 10
Total Lipids	120 ± 15	113 ± 29	119 ± 9	96 ± 39
Total Carbohydrates	261 ± 32	235 ± 23	255 ± 19	258 ± 18
Total Dietary Fiber	14 ± 2	17 ± 2	24 ± 2	27 ± 2

[a] Number in parentheses, number of participants.

[b] Mean ± SE.

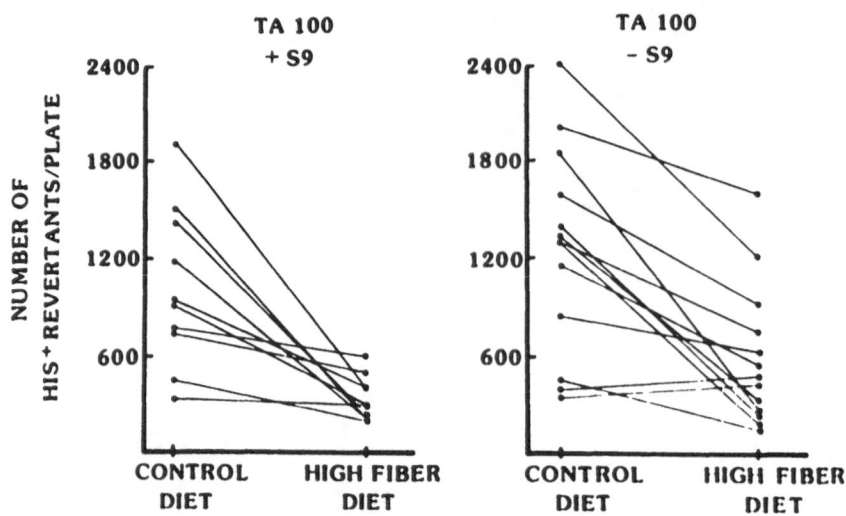

Fig. 2.   Mutagenic activity (histine-positive revertants/plate or 200mg dry feces) of fecal extracts consuming control diet and supplemental fiber diet. The extracts were tested for mutagenicity using <u>Salmonella typhimurium</u> TA 100 with (+S9) and without (-S9) activation.

EFFECTS OF TYPES OF DIETARY FIBER ON FECAL MUTAGENS

This study was designed to investigate the effect of types of dietary fiber and fiber fraction, namely wheat bran, oat fiber, and cellulose on fecal mutagens in healthy subjects from New York consuming high-fat/ moderately low-fiber diets (28). In order to identify individuals excreting significant levels of fecal mutagens, 72 healthy individuals were screened for fecal mutagenic activity using the Ames <u>Salmonella typhimurium</u>/microsomal assay system. Twenty one of them were found to excrete high levels of mutagens and recruited for the fiber intervention study. A 4-day dietary record and 2-day stool samples were collected from each volunteer before the fiber intervention study (control diet period). The subjects were randomly divided into various dietary groups and were given about 10g per day of one of the dietary fibers (micrograde roasted hard red wheat bran, Williamson's Better Basics oat fiber, or Solka-Floc BW-40FCC or BW-200FCC cellulose) in the form of bread, pasta or muffins. About 126g per day of fiber-supplemented foods were required to supply about 10g of supplemental fibers. At the end of 5 weeks on each high fiber dietary regimen, 24-hour stool samples for 2 days and 4-day dietary record were obtained. A 4-week period of control diet without supplemental fiber followed each fiber supplementation period. At the end of those 4 weeks, 24-hour stool samples for 2 days and a 4-day dietary record were obtained. Stool samples were analyzed for fecal mutagenic activity using the Ames strains TA 98 and TA 100 with or without microsomal (S9 activation).

Nutrient analysis of 4-day food records obtained before and at the end of the fiber supplemental period is summarized in Table 2. The dietary intake of total calories, fat and protein were not significantly different between the control and fiber supplemental periods. As expected, the dietary fiber intake was higher during the fiber supplemental periods than during fiber intake was higher during the fiber supplemental periods than during the control periods. The daily fresh stool output was higher during all fiber periods irrespective of types of fiber.

162

Table 2.   Daily Nutrient Intake Stool Output of Healthy Subjects During the Control and Supplemental Fiber Period

Nutrients	Control Diet	Cellulose Diet	Control Diet	Oat Fiber Diet	Control Diet	Wheat Bran Diet	Control Diet
Energy, Kcal	1654	1894	1685	1977	1653	1971	1679
Total Protein, g	78	73	66	75	86	75	64
Total Lipids, g	58	75	61	73	65	76	62
Total Carbohydrates,g	196	240	219	245	200	240	218
Total Dietary Fiber, g	18	28	17	26	17	29	16
Stool Weight, g	88	138	80	132	78	124	81

[a] Mean (n = 19).

[b] Significantly different from its control diet period, P < 0.05

[c] Significantly different from its follow-up control diet period, P < 0.05.

Table 3 shows the fecal mutagenic activity during the fiber and control diet periods. A significant decrease in the fecal mutagenic activity in Ta 98 and TA 100 with or without microsomal activation was observed during the periods of cellulose and wheat bran compared to their respective control (baseline) diet periods, whereas the oat fiber had no significant effect. Changing the diets from cellulose and wheat bran to their control follow-up periods significantly increased the fecal mutagenic activity. On the otherhand, switching from oat fiber period to its follow-up control diet period had no such effect on fecal mutagenic activity.

Table 3.   Mutagenic Activity of Fecal Extracts from Healthy Subjects Consuming Control and Supplemental Fiber Diets

	No. of His[+] Revertants/200mg mg of Feces						
Salmonella Strains	Control Diet	Cellulose Diet	Control Diet	Oat bran Diet	Control Diet	Wheat Bran Diet	Control Diet
TA98							
+S9(5)[a]	124	58	95	85	96	43	88
-S9(4)	80	19	49	29	58	21	40
TA100							
+S9(6)	534	230	358	336	411	236	340
-S9(13)	487	294	459	464	506	259	448

[a] Numbers in parenthesis, number of subjects showing positive mutagenic activity.

[b] Mean

[c] Significantly different from its control diet period diet, P < 0.05.

[d] Significantly different from its follow-up control diet period, P < 0.05.

The results generated from these studies indicate that certain dietary fibers may reduce the production and/or excretion of colonic mutagens. The effect of mutagen production may depend on the type of fiber consumed

EFFECT OF TYPES OF FIBER ON FECAL BACTERIA ENZYMES

Colonic bacteria may have a role indirectly in colon carcinogenesis through its effect in maintaining the enterohepatic circulation of bile acids, production of secondary bile acids and putative mutagenic substances, and the production of enzymes that are involved in the metabolism of procarcinogens and tumor promoters. Population studies have shown that encrease in fiber intake was associate with changes in fecal bacterial enzymes, particularly β-glucuronidase, 7α-dehydroxylase, hydroxysteroid dehydrogenase, azoreductase and nitroreductase which are involved in the production of tumor promoters and in the metabolism of procarcinogens. There is some evidence to suggest that certain dietary fibers not only stimulate the growth of certain bacteria in the gut but also modifies the activities of bacterial enzymes.

In this study, we investigated the effect of supplemental wheat bran, oat bran and corn bran on the metabolic activity of gut microflora in women using β-glucuronidase, 7α-dehydroxylase, azoreductase and nitroreductase as indices. The experimental design and protocols were similar to those described above. The supplemental fiber was given in the form of cold cereal, hot cereal or bran muffins. After 4 weeks of baseline diet data collection from each participant, they were asked to add an additional 13-15g of fiber to their regular meals so that the total fiber consumption would be about 28g per day. This dietary regimen was followed for 8 weeks after which normal baseline diet without the supplemental fiber was resumed for 4 weeks. This pattern was followed for each of the 3 dietary fibers. The fecal samples were collected for 2 days at each baseline period and at the end of each supplemental fiber period. Fecal bacterial β-glucuronidase, 7α-dehydroxylase, azoreductase and nitroreductase activities were determined.

Nutrient analysis of records obtained from each volunteer during the baseline and supplemental fiber periods indicate that the intake of total calories, fat, protein, cholesterol, and calcium were similar during the two periods. The total fiber intake during the baseline period averaged about 14g per day whereas the fiber intake during the supplemental fiber (wheat bran, oat bran and corn bran) periods averaged about 28g per day.

Fecal bacteria enzyme activities are summarized in Table 4. A significant decrease in the activities of β-glucuranidase, 7α-dehydroxylase and nitroreductase was observed during the period of wheat bran. Dietary wheat bran had no effect on azoreductase activity. Dietary oat bran decreased the activities of azoreductase and nitroreductase but had no effect on β-glucuronidase and 7α-dehydroxylase activities. Dietary corn bran had little or no effect on the activities of these enzymes.

The results of this study are compatible with the hypothesis that not all dietary fibers alter the metabolic activity of gut microflora and that their modifying effect depends on the type of fiber consumed.

ACKNOWLEDGEMENTS

These investigations were supported by USPHS Grants CA-17613 and CA-40839 from the National Cancer Institution

Table 4. Effect of Types of Dietary Fiber on Fecal Bacterial Enzyme Activities

Bacterial Enzymes	Control Diet	Wheat Bran Diet	Control Diet	Oat Bran Diet	Control Diet	Corn Bran Diet
β-Glucuronidase	1.9[a,b]	0.5	1.7	1.3	1.6	1.3
7α-dehydroxylase	35[c]	17	26	20	24	19
Nitroreductase	68[d]	30	67	46	68	52
Azoreductase	193[e]	189	220	150	186	164

[a] Mean (n = 12 women in each group).

[b] β-glucuronidase activity: mg of phenolphthalein released/g stool sample/hr.

[c] 7α-dehydroxylase activity: % cholic acid dehydroxylated/g stool sample.

[d] Nitroreductase activity: μg of aminobenzoic acid formed/g stool sample/hr.

[e] Azoreductase activity: μmoles of amaranth reduced/g stool sample.

REFERENCES

1. American Cancer Society, Cancer Facts and Figures 1988. New York: American Cancer Society (1989).

2. C. Muir, J. Waterhouse, T. Mack, J. Powell and S. Whelan. Cancer Incidence in Five Continents, Vol. 5, Scientific Publication No. 88, 1987, France

3. E.L. Wynder and T. Shigematsu. Environmental factors of cancer of the colon and rectum. Cancer, 20:1520 (1967).

4. D.P. Burkitt. Epidemiology of cancer of the colon and rectum. Cancer 28:3 (1971).

5. O.M. Jensen, R. MacLennan, and J. Wahrendorf. Diet, bowel function, fecal characteristics and large bowel in Denmark and Finland. Nutr. Cancer 4:5 (1982).

6. B.S. Reddy, A.T. Hedges, K. Laakso, and E.L. Wynder. Metabolic epidemiology of large bowel cancer: fecal bulk and contituents of high risk North American and low-risk Finnish population. Cancer 42:2832 (1987).

7. G.E. McKeown-Eyssen, and E. Bright-See. Dietary factors in colon cancer: International relationships. An update. Nutr. Cancer 7:251 (1985).

8. S.A. Bingham, D.R.R. Williams, and J.H. Cummings. Dietary fiber consumption in Britain: New estimates and their relationship.

9. O.M. Jensen. The epidemiology of large bowel cancer. In: "A Critical Evaluation," B.S. Reddy and L.A. Cohen, eds. Diet Nutrition and Cancer. CRC Press, Boca Raton, Fl. pp. 27 (1986).

10. B.S. Reddy, C. Sharma, L. Mathews, A. Engle, K. Laakso, K. Choi, P. Puska, and R. Korpella. Metabolic epidemiology of colon cancer: fecal mutagens in healthy subjects from rural Kuopio and urban Helsinki, Finland. Mutation Res. 152:97 (1985).

11. R. MacLennan, O.M. Jensen, J. Mosbech, and H. Vuori. Diet, transit time, stool weight, and colon cancer in two Scandinavian population. Am. J. Clin. Nutr., 31:S239 (1978).

12. Physiological Effects and Health Consequences of Dietary Fiber, pp. 118-134. Life Sciences Research Office. Federation of American Societies for Experimental Biology, Bethesda, NY (1987).

13. B.S. Reddy, H. Mori, and M. Nicolais. Effect of dietary wheat bran and dehydrated citrus fibe ron azoxymethane-induced intestinal carcinogenesis in Fisher 344 rats. J. Natl. Cancer Inst., 66:553 (1981).

14. H.J. Freeman, G.A. Spiller, and Y.S.A. Kim. Double blind study on the effect of purified cellulose dietary fiber on 1,2-dimethylhydrazine-induced rat colonic neoplasia. Cancer Res., 38:2912 (1978).

15. L.R. Jacobs, and J.R. Relationship between colonic luminal pH, cell proliferation, and colon carcinogenesis in 1,2-dimethylhydrazine-treated rats fed high-fiber diets. Cancer Res., 46:1727 (1986).

16. B.S. Reddy, Y. Maeura, and M. Wayman. Effect of dietary corn bran and autohydrolyzed lignin on 3,2'-dimethyl-4-aminobiphenyl-induced intestinal carcinogenesis in male F344 rats. J. Natl Cancer Inst., 71:419 (1983).

17. B.S. Reddy. Tumor promotion in colon carcinogenesis. In: Mechanisms of Tumor Promotion, Vol. 1, Tumor promotion in Internal Organs (T.J. Slaga, ed.), CRC Press, Boca Raton, FL., pp. 107 (1983).

18. M.J. Hill, B.S. Drasar, V.C. Aries, J.S. Crowther, G. Hawksworth, and R.E.O. Williams. Bacteria and etiology of cancer of the large bowel. Lancet 1:95 (1971).

19. M. Ehrich, J.A. Ashjell, R.L. Van Tassell, and T.D. Wilkins. Mutagens in the feces of 3 South Africa populations at different leels of risk for colon cancer. Mutation Res. 64:231 (1979).

20. U. Kuhnlein, D. Bergstrom, and H. Kuhnlein. Mutagens in feces from vegetarians and non-vegetarians. Mutation Res. 85:1 (1981).

21. B.S. Reddy, C. Sharma, L. Darby, K. Laakso, and E.L. Wynder. Metabolic epidemiology of large bowel cancer. Fecal mutagens in high- and low-risk population for colon cancer. A preliminary report. Mutation Res. 72:511 (1980).

22. W.R. Bruce, and P.W. Dion. Studies relating to fecal mutagens. Am J. Clin. Nutr. 33:2511 (1980).

23. T. Sugimura, S. Sato, H. Ohgaki, S. Takayama, M. Nagao, and K. Wakabayashi. Mutagens and carcinogens in cooked foods. Prog. Clin. Biol. Res., 206:85 (1986).

24. J.H. Weisburger, and W.S. Barnes. Influencce of composition of diet on the formation of mutagens. Prog. Clin. Biol. Res. 206:227 (1986).

25. H. Hayatsu, T. Hayatsu, Y. Wataya, and H. Mower. Fecal mutagenicity arising from ingestion of fried ground beef in the human. Mutat. Res., 143:207 (1985).

26. S. Takayama, Y. Nakatsuru, M. Masuda, H. Ohgaki, S. Sato, and T. Sugimura. Demonstration of carcinogenicity in F344 rats of 2-amino-3-methylimidazol [4, 5-f] quinoline from broiled sardine, fried beef, and beef extract. Gann, 75:467 (1984).

27. B.S. Reddy, C. Sharma, B. Simi, A. Engle, K. Lakso, P. Puska, and R. Korpella. Metabolic epidemiology of colon cancer: effect of dietary fiber on fecal mutagens and bile acids in healthy subjects. Cancer Res., 47:644 (1987).

28. Reddy, B.S., A. Engle, S. Katsifis, B. Simi, H. Bartram, P. Perino, and C. Mahan. Biochemical epidemiology of colon cancer: effect of types of dietary fiber on fecal mutagens, acid and neutral sterols in health subjects. Cancer Res., 49:4629 (1989).

UTILIZATION OF PURIFIED CELLULOSE IN FIBER STUDIES

Michael H. Penner and Ean-Tun Liaw

Department of Food Science and Technology
Oregon State University
Corvallis, Oregon 97331-6602

INTRODUCTION

Purified cellulose preparations are used extensively in nutrition research. This is at least partially due to the acceptance of cellulose-type fiber as an established and recommended fiber to be used in purified diets (1). The common use of cellulose in studies analyzing the effect of dietary fiber on specific physiological parameters is illustrated by noting that the majority of such studies presented at this symposium utilized cellulose as one of the experimental fibers. Although purified cellulose products are used extensively in nutritional research they are seldom characterized in terms of composition or structural properties. The lack of attention directed towards the properties of the celluloses used in various studies could promote the false assumption that all commercially available cellulose fiber products are equivalent. In this regard, the main objectives of this paper are to point out some of the potential differences between commercially available cellulose preparations, to demonstrate approaches which may be used to characterize the properties of a purified cellulose preparation, and to discuss the relevance of using purified cellulose preparations in dietary fiber studies.

RELEVANCE OF UTILIZING PURIFIED CELLULOSE PRODUCTS

The relevance of using purified cellulose as a fiber source in a given study is dependent on the question being addressed in

---

[1] Supported in part by National Institute of Environmental Health Sciences grant ES00210 and the Medical Research Foundation of Oregon.
[2] Oregon State University Agricultural Experiment Station Technical Publication Number 8997.

that particular research. Studies concerned with the consequences of refined cellulose consumption per se may have relevance with respect to understanding potential mechanisms by which other plant fibers may exert their physiological response. This is not to say that purified cellulose is equivalent to other, more complex, dietary fibers. In this type of study it is more appropriate to consider purified cellulose as an experimental or model fiber. For example, studies with rats have suggested that consumption of purified cellulose is protective against experimental colonic neoplasia relative to a fiber-free diet (2,3). If the protection observed in those studies is indeed due to purified cellulose, then an understanding of the mechanism through which the cellulose exerts its effect should prove helpful in determining how more complex food fibers may mediate colon carcinogenesis. In this respect, studies which utilize a well characterized fiber such as purified cellulose may prove enlightening.

Purified cellulose should not be considered only as an experimental fiber. The trend toward reduced calorie foods in western societies has increased the use of cellulose in the food industry, especially in bakery products. Reduced calorie breads incorporate up to 10% by weight of purified cellulose (not all commercial breads utilize cellulose as the non-caloric fiber). The amount of cellulose utilized in these products suggests that certain groups of the population, especially those utilizing low calorie products, may consume several grams of refined cellulose daily. Therefore, the physiological response to the incorporation of purified cellulose into the human diet is becoming more relevant. This trend of increased cellulose consumption will probably continue, at least into the near future, as the functional properties of refined cellulose products are fully exploited in reduced calorie foods.

The results from experimental studies with purified cellulose should not be interpreted in terms of the role of cellulose in other complex food fibers. In this respect we know very little regarding the physiological response to the cellulose component of any complex food fiber. In dietary studies with complex fibers we are limited in determining which component(s) of the fiber is influencing the physiological response we measure. Similarly, it is difficult to characterize the physical properties of a component, such as cellulose, while it remains part of the intact fiber. Digestibility is one parameter that may be used to judge the degree of similarity between cellulose in a purified state and cellulose which remains an integral part of a complex food fiber. Results from human digestibility studies (Table 1) indicate that, in general, the cellulose component of a complex food fiber is digested to a greater extent than the cellulose of a purified cellulose product. The higher digestion coefficient for the cellulose which is an integral component of a food fiber versus the cellulose commonly found in commercial purified cellulose products is strong evidence that there are distinct physical and/or chemical differences in the celluloses. This suggests that results obtained from studies with purified cellulose may not be applicable to and, in general, should not be extrapolated to the cellulose component of complex food fibers.

Purified cellulose is often used in experimental studies as a control fiber, against which other fibers are compared. The common use of cellulose in diets is largely due to the

Table 1. Apparent Digestibility of Cellulose by
Humans

Source	% Digested	Reference
Mixed diet	70	(19)
Mixed diet	26	(20)
Carrot	58	(21)
Cabbage	87	(21)
Wheat bran	23	(22)
Purified cellulose	8	(19)

recommendation of the American Institute of Nutrition to utilize a "cellulose-type" fiber in purified diets (1). Use of cellulose in this respect has generated a large amount of data on the effects of cellulose in the diet. It is likely that experimental studies with laboratory animals have generated as much or more data relative to the consumption of purified cellulose than to any other fiber type. The abundance of data available from nutritional studies with cellulose along with the perceived simplicity of the product suggest that it will continue to be used by many investigators as a "control" fiber for comparative purposes.

COMMERCIAL PREPARATION OF PURIFIED CELLULOSE PRODUCTS

The majority of products sold as cellulose fiber preparations are derived from woody plants through a pulping and bleaching process. The details of the manufacturing process likely differ between manufacturers. It is also likely that producers will modify their manufacturing processes depending on the raw materials available. The result is that one should expect to find some degree of variability in the "cellulose-type" fiber products available for nutritional studies.

A simplistic flow chart of a generalized manufacturing process is shown in figure 1. The raw material is initially debarked and chipped. The resulting chips are then subjected to a chemical pulping process designed to separate the cellulose from the other components of the wood, including lignin, other carbohydrates and trace organics. Either the sulfite or the kraft chemical pulping processes is generally used. The sulfite process is an acid cooking treatment utilizing sulfur dioxide and calcium bisulfite while the kraft process is an alkaline cooking treatment utilizing sodium hydroxide, sodium sulfide and sodium carbonate. The two processes are fundamentally different and, as expected, will result in pulps with somewhat different properties. The cellulose in the resulting pulp, either kraft or sulfite, may then be further purified through a bleaching

process. The bleached cellulose pulp is a white product with
increased brightness relative to the initial pulp. The bleaching
process generally involves a sequence of treatments which utilize
standard bleaching compounds such as chlorine, chlorine dioxide
and/or hypochlorite. The bleached cellulose pulp may then be
dried and mechanically sized.

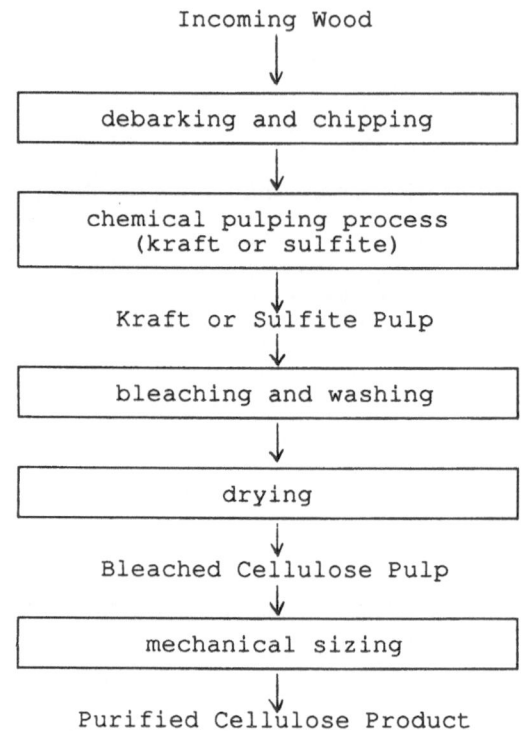

Fig. 1.   Schematic of generalized manufacturing process for
           purified cellulose products.

Independent producers of cellulose fiber products
undoubtedly differ in the processing schemes they utilize to
produce their products. The pulping, bleaching, drying and
sizing phases of the process may all be modified, resulting in
products with distinct chemical and/or physical properties. It
is important for experimental nutritionists, especially those
working on mechanisms of action of dietary fiber, to realize that
the cellulose products available for experimental diets may vary
in composition and physical properties. Variability in products
may be due to differences in the starting material used to
produce the pulp as well as to variations in and additions to the
general processing scheme depicted in Figure 1.

The fact that commercial cellulose products may differ in specific physico-chemical properties suggests that, for some types of experiments, the characterization of a cellulose preparation may be necessary for the critical evaluation of that study. The appropriate methods for the characterization of a fiber are largely dependent on the nature of the study. A logical place to start in the characterization process is to determine the percentage of the sample which is actually cellulose. The composition of cellulose preparations differ mainly in the content of residual hemicellulose and lignin. Olson et al (4) recently reported the apparent hemicellulose content of several cellulose preparations based on the quantity of monosaccharides liberated during trifluoroacetic acid hydrolysis of the samples. The hemicellulose content of the 8 samples ranged from as low as approximately 0.1% to a high of approximately 10%. Other methods involving chromatographic analysis of the hydrolyzed cellulose preparation (5) as well as gravimetric procedures (6) are available for determining the cellulose content of these products. The hemicellulose and cellulose fractions of these products are likely to differ in their extent of fermentation within the colon (7).

The cellulose molecule itself is a chemically homogeneous linear polymer composed of glucosyl residues linked ß,1-4. The molecular chains of cellulose associate through noncovalent interactions to form a physically heterogeneous supramolecular structure (8). The terms amorphous and crystalline are commonly used to classify the heterogeneous physical nature of this supramolecular complex. The crystalline regions of cellulose being highly ordered, tightly packed and of low reactivity relative to the amorphous regions. In vitro (9) and in vivo studies (10) indicate that the physical structure of cellulose influences both its reactivity in vitro and its digestibility within the large intestine. These results suggest that two cellulose preparations, each with distinct structural properties, could potentially elicit different physiological responses.

There are a variety of methods with which to assess the structural properties of cellulose (8). The choice of which method to use for a particular study depends on the equipment available to the researcher and the question being addressed in the study. Common structural parameters which may influence the metabolism of cellulose within the large intestine and which can be measured by existing techniques include the degree of polymerization, relative crystallinity index, in vitro susceptibility to cellulase degradation and small molecule accessibility. Commercially available cellulose preparations currently used for nutritional studies will undoubtedly differ in some of these properties. An example of how commercially available cellulose products may differ is provided in the following paragraphs which discuss our laboratory's analyses of selected structural and chemical properties of Avicel PH101 (FMC Corporation, Philadelphia, PA) and Solka-Floc BW-200 NF (James River Corp., Hackensack, NJ).

A fundamental property which is useful for characterizing cellulose-type preparations is the size of the component cellulose molecules. The size of a cellulose molecule is often expressed as its degree of polymerization (DP), referring to the number of glucose units in the linear molecule. A relatively rapid and inexpensive method for estimating the DP is to determine the intrinsic viscosity of the purified cellulose and to convert it to a DP value (11). The intrinsic viscosity is determined by measuring the specific viscosity of a series of dilute solutions (approximately .05 to .2 %, W/V) of cellulose in cupriethylenediamine hydroxide utilizing a Cannon-Fenske viscometer. Specific viscosities are then converted to reduced viscosities and extrapolated to zero cellulose concentration to give the intrinsic viscosity. A DP value may then be estimated by multiplying the intrinsic viscosity by 190 (11). DP values for the two cellulose preparations characterized in this study are given in Table 2. The DP values clearly indicate the average cellulose polymer in the Avicel preparation has a lower molecular weight than the corresponding cellulose polymer in the Solka-Floc preparation. The lower DP for Avicel is expected since the pulp is further hydrolyzed in mineral acid during the Avicel manufacturing process (12). These DP values, measured as described above, reflect viscosity-average molecular weights and are best interpreted as relative rather than absolute values.

The regularity of structure in cellulose molecules allows portions of the cellulose to develop crystalline regions. The extent of these crystalline regions is expected to have a significant impact on the reactivity of that preparation. It has been assumed that the extent of crystallization of cellulose in feeds affects the degree to which the cellulose component of that feed is digested (13). There are several methods with which to measure the relative extent of crystallization of cellulose. The method used in the current study is based on the diffraction of x-rays by the ordered regions of cellulose. The x-ray data are obtained by measuring the diffraction intensity of nickel-filtered CuKα radiation between Bragg angles (2θ) of 10° and 30°. The crystallinity index is then calculated using the following empirical relationship:

$$CrI(\%) = [1 - I(am)/I(002)] \times 100$$

where I(002) denotes the maximum intensity of the (002) lattice diffraction at $2\theta = 22.5°$ and I(am) is the intensity of diffraction at $2\theta = 18°$ (14). The two cellulose preparations examined in this study have similar crystallinity indexes (Table 2). Avicel appears to be somewhat more crystalline than Solka-Floc, however, the diffraction intensity from both of the preparations indicate considerable crystalline character.

The extent of digestion of a fiber within the large intestine may have a significant effect on the physiological response it elicits. It is therefore appropriate when characterizing cellulose samples to provide an estimate of the relative digestibility of the different cellulose preparations. Although in vivo digestibility studies are the optimum measure of the extent of digestion of a fiber they are not practical for routine analyses due to the complexity of these studies. In vitro methods have therefore been utilized for these purposes.

Table 2. Selected Properties of Two Cellulose Products[a,b]

Cellulose Fiber Product	Degree of Polymerization[c]	Crystallinity Index[d]	Susceptibility to Enzymatic Hydrolysis[e]	Small Molecule Accessibility[f] %
Avicel pH (10)	219 (3)	80	30.4 (5.0)	17.5 (0.2)
Solka-Floc BW 200	703 (20)	73	39.1 (0.3)	28.7 (0.1)

[a] Values reported are means with SEM in parenthesis.
[b] See text for discussion of methods.
[c] Values derived from viscosity average molecular weight determinations.
[d] Values derived from X-ray diffraction measurements.
[e] Values reflect total number of reducing sugars solubilized after 6 hr incubation with a fungal cellulase preparation.
[f] Values reflect the percentage of potentially reactive hydroxyls esterified in 90% formic acid.

Relative susceptibilities to digestion have been estimated from
in vitro studies employing mixed microbial batch culture systems
inoculated with an appropriate fecal extract (15,16). This
approach is particularly appealing for preparations containing
significant non-cellulosic carbohydrates because these
carbohydrates may also be degraded in this system, allowing the
microbial cellulase enzymes access to the cellulosic substrate.
A second approach, which is best utilized only when
characterizing cellulose preparations of high purity, is to
determine the susceptibility of the cellulose to degradation by
an appropriate cellulase enzyme system. A previous study
utilizing laboratory rats, a fungal cellulase enzyme preparation,
and three cellulose preparations derived from Avicel found a
correlation between the relative extent of enzymatic hydrolysis
of the celluloses in vitro and the relative extent to which they
were digested in vivo. (10). In the present study the relative
susceptibility of the cellulose preparations to enzymatic
degradation was estimated by incubating the products at 10% (W/V)
with 16 ug/ml cellulase (<u>Trichoderma</u> <u>viride</u> cellulase;
Calbiochem, La Jolla) in 50mM Na acetate, pH 5.0, at 40C° and
measuring the quantity of reducing sugar equivalents solubilized
after 6h. The results suggest the two cellulose products are
different with regard to their enzymatic susceptibility, Solka-
Floc being slightly more susceptible to enzymatic degradation
than Avicel under these assay conditions (Table 2). Currently,
this technique is best used as an initial step in evaluating the
similarity of cellulose samples since further studies are
necessary before one can realistically extrapolate these in vitro
results to actual in vivo digestibilities.

The ability of a fiber to interact with small molecules such
as minerals and small organics is directly related to the small
molecule accessibility of the fiber. Within cellulose fibers
there are regions which are essentially unavailable for
interaction with small molecules. The percentage of a cellulose
fiber preparation which is unavailable for interaction with small
molecules may be of relevance due to the potential for the
interaction of cellulose with physiologically important molecules
such as metabolic regulators, carcinogens and toxins. There are
several methods which may be utilized to estimate the relative
amount of inaccessible cellulose in a given cellulose preparation
(8). The formylation method of Nickerson was employed in the
present study (17). The fraction of cellulose accessible to low
molecular weight components is estimated by measuring the
percentage of the cellulose's potentially reactive hydroxyl
groups esterified after 1h in 90% formic acid at 40°C. The
number of hydroxyl groups esterified during the 90% formic acid
treatment is measured by the Eberstadt method after
saponification (18). The total number of potentially reactive
hydroxyl groups is taken as the number of hydroxyl groups
esterified under the same conditions for an equivalent amount of
soluble starch. Our results (Table 2) indicate that both
celluloses are largely inaccessible to small molecules and that
the accessibility of Avicel is roughly 1.6 fold less than that of
Solka-Floc.

SUMMARY

Purified cellulose-type fiber products are widely used in
experimental nutrition. Their use in a broad spectrum of studies
may potentially lead to the acceptance of the misconception that

the various commercially available cellulose products are equivalent. In this paper we have attempted to show that this is not the case. The comparative structural data of Table 2 and the compositional data of Olsen et al (4) provide examples which indicate that purified cellulose preparations should not necessarily be considered equivalent. Unfortunately, our current lack of understanding of how fibers are metabolized and how they may affect specific physiological parameters makes it difficult to determine which, if any, of the measurable structural and chemical properties will be of relevance for a given in vivo study. At present, it appears that researchers utilizing/ evaluating the consequences of consuming a purified cellulose-type fiber would be prudent to provide at least a limited amount of data on the properties of the cellulose preparation used in their studies. The characterization of the cellulose product may be done by a variety of methods depending on the expertise of the laboratory (8). The methods and results discussed in this paper provide an example of the type of information which may be obtained from an in vitro characterization of cellulose products.

ACKNOWLEDGMENTS

We thank Dr. Juinn-Chin Hsu for his helpful discussions and technical assistance.

REFERENCES

1. American Institute of Nutrition (1977) J. Nutr. 107, 1340-1348.
2. Freeman, H.J., Spiller, G.A., and Kim, Y.S. (1980) Cancer Res. 40, 2661-2665.
3. Freeman, H.J., Spiller, G.A., and Kim, Y.S. (1978) Cancer Res. 38, 2912-2917.
4. Olson, A.C., Gray, G.M., Chiu, M., Betschart, A.A., and Turnlund, J.R. (1988) J. Agric. Food Chem. 36, 300-304.
5. Crowell, E.P. and Burnett, B.B. (1967) Anal. Chem. 39, 121-124.
6. Robertson, J.B. and Van Soest, P.J. (1981) in "The Analysis of Dietary Fiber in Food". (W.P.T. James and O. Theander, eds.), Marcel Dekker, 123-158.
7. Ehle, F.R., Jeraci, J.L., Robertson, J.B., and Van Soest, P.J. (1982) J. An. Sci. 55, 1071-1081.
8. Rowland, S.P. and Bertoniere, N.R. (1985) in "Cellulose Chemistry and Its Applications" (T.P. Nevell and S.H. Zeronian, eds.), Ellis Horwood, 112-137.
9. Baker, T.I., Quicke, G.V., Bentley, O.G., Johnson, R.R., and Moxon, A.L. (1959) J. An. Sci. 18, 655-662.
10. Hsu, J.C. and Penner, M.H. (1989) J. Nutr. 119, 872-878.
11. American Society for Testing Materials (1986) Annual Book of ASTM Standards 15.04, 360-366.
12. Anonymous (1986) The Avicel Advantage, FMC Corp., Philadelphia, PA.
13. Van Soest, P.J. (1973) Federation Proc. 32, 1804-1808.
14. Segal, L., Creely, J.J., Martin, A.E., Jr., and Conrad, C.M. (1959) Textile Res. J. 29, 786-794.
15. Bayless, C.E. and Houston, A.P. (1984) Appl. Envir. Micro. 48, 626-632.
16. Ehle, F.R., Robertson, J.B., and Van Soest, P.J. (1982) J. Nutr. 112, 158-166.

17. Nickerson, R.F. (1951) Textile Res. J. 21, 195-202.
18. Genung, L.B. and Mallatt, R.C. (1941) Indust. Eng. Chem. 13, 369-374.
19. Slavin, J.L., Brauer, P.M., and Marlett, J.A. (1981) J. Nutr. 111, 287-297.
20. Southgate, D.A.T. and Durnin, J.V.G.A. (1970) Br. J. Nutr. 24, 517-535.
21. Nyman, M., Asp, N., Cummings, J., and Wiggins, H. (1986) Br. J. Nutr. 55, 487-496.
22. Heller, S.N., Hackler, L.R., Rivers, J.M., Van Soest, P.J., Roe, D.A., Lewis, B.A., and Robertson, J. (1980) Am. J. Clin. Nutr. 33, 1734-1744.

COMPLEXITY IN THE INTERPRETATION OF DATA DERIVED

FROM STUDIES OF DIETARY FIBER

G. A. Spiller

Health Research and Studies Center
Los Altos, California 94023

The study of the effects of dietary fiber on health needs to consider differences between the effects of natural high-fiber whole foods and the effect of isolated or purified fibers.

Two major differences exist between a diet high in high-fiber plant foods and a diet to which an isolated or concentrated fiber has been added: 1) There are many bioactive compounds normally associated with high-fiber foods that may be lost in the isolation process; and 2) a diet high in high-fiber foods supplies more protein, lipids, carbohydrates and, in general, more energy from plant foods and much less from animal products.

BIOACTIVE COMPOUNDS

The bioactive nonfibrous compounds often present in natural high-fiber foods include not only the well recognized ones, such as essential vitamins and minerals often very high in whole grains, vegetables, fruits, nuts and other unrefined plant foods, but also a large number of compounds not recognized as essential in human nutrition but that may have important physiological effects.

The potentially bioactive but nonessential compounds include saponins (common in many plant foods), omega-3 fatty acids (green leaves), alpha-tocotrienol (grains such as barley), sitosterols (common in unrefined lipid fraction of oil seeds), leaf and fruit waxes (outer coating of many leaves and fruits) and beta-carotene (green and yellow vegetables and fruits). Whole grains, nuts, beans, green leaves, other vegetables and fruits are often rich in one or more of these bioactive (but not yet recognized as essential) compounds that tend to be overlooked in favor of attributing the effect exclusively to fiber.

These associated bioactive compounds are lost during the process of isolating fiber polymers for clinical studies, with the result that the effect of fiber in whole foods is often different from that of isolated fibers of similar polymeric composition.

Table 1. Examples of Bioactive Compounds Associated with High-Fiber Plant Foods

saponins (glycosides)	serum lipids bile acid excretion
omega-3 fatty acids (alpha-linolenic)	blood coagulation serum lipids cell membrane composition
alpha-tocotrienol	serum lipids
sitosterols	serum lipids colon carcinogenesis (protection)
leaf waxes (e.g. cutins)	serum lipids? gut function? (unknown, but theoretically possible)
beta-carotene	carcinogenesis (protection)

ALTERATION IN NUTRIENT BALANCE

A high intake of whole fiber foods alters intake of other foods in a more drastic way than purified fibers do, by altering the sources, amounts and types of fats, proteins and digestible carbohydrate. A natural, high-fiber diet supplies more calories from carbohydrate and less from animal fat and protein.

This means that a diet in which grains, beans, nuts and potatoes are the main source of energy and which contains 30 g of dietary fiber, differs from a diet in which, for example, meat is the main source of energy and to which 30 g of the same kind of dietary fiber has been added.

When a concentrated or purified fiber product is *added* to the diet, no other change needs to take place. The fiber in this case works as a supplement which may have major physiological effects (e.g. guar gum on serum cholesterol or wheat bran on fecal output) but the diet does not need to be altered, i.e. it can still be a diet in which most of the calories are derived from animal products rather than plant products with all the differences we have just seen.

The clinical investigator has to make a choice, taking into account the facts we have just outlined. A study can be set up to study a concentrated fiber or a study can be designed to study high-fiber foods. The results could indeed be quite different. In the case of high-fiber foods the investigator has to be willing to allow the fiber foods to *misplace* other foods and the results should be attributed to the complex diet changes that have taken place.

CONCLUSION

We must not be tempted to attribute effects of whole high-fiber foods to the fiber fraction alone or to generalize the effect or lack of effect of an isolated polymer to the whole food when studying dietary fiber or contradictory and misleading results will be the outcome.

A modification of the 1985 classification that we had suggested at the XIII International Congress of Nutrition in Brighton, UK, should be very useful to investigators: 1) high-fiber foods, 2) fiber-enriched foods, 3) isolated or concentrated fiber (or fiber supplements).

There are many ways in which fiber can be beneficial, from the isolated form to the supplemented food to the whole, natural food. What is important in the study of fiber and in presenting results to the scientific community and to the general public is to clearly state the study design and its limitations, and to avoid improper generalizations of results obtained with purified or concentrated fiber products.

REFERENCES

Spiller, G.A., "Handbook of Dietary Fiber in Human Nutrition," CRC Press, Boca Raton, FL (1986).

Spiller, G.A. and Jenkins, D. J. A., Dietary fibre supplements, physiological and pharmacological aspects: a workshop report, in "Proceedings of the XIII International Congress of Nutrition - 1985," T. G. Taylor and N. K. Jenkins, ed., John Libbey & Company Ltd., London.

ISSUES IN DIETARY FIBER ANALYSIS

Judith A. Marlett

Department of Nutritional Sciences
University of Wisconsin-Madison
Madison, Wisconsin

INTRODUCTION

There are many reasons for measuring dietary fiber. We have three overall aims in our laboratory. First, analysis of the major sources of dietary fiber in the USA food supply would provide a data base of values that could be used clinically and epidemiologically to assess the kinds and amounts of dietary fiber consumed by individuals and populations. Second, the content and composition of dietary fiber sources can be related to their physiological effects. For example, evidence has accumulated rapidly in recent years to suggest many soluble dietary fiber sources have physiological effects primarily in the upper part of the gastrointestinal tract, whereas insoluble sources influence the physiology of the lower gut (1). Ultimately, researchers want to understand the mechanisms by which dietary fiber has a particular physiological effect; the composition of dietary fiber sources, as well as their physiochemical properties, will provide insight into understanding these mechanisms of action for dietary fiber.

A variety of methods have been promulgated for dietary fiber analysis (1,2). It has been particularly challenging for researchers in the field to follow the evolution of these methods and to identify distinguishing features of the methods that may be responsible for differences in the results obtained by different approaches. Several speakers in this symposium have addressed issues in dietary fiber analysis. One objective of this chapter is to identify and discuss differences among the analytical methods for dietary fiber. Distinguishing the differences among methods provides insight into the utility of different data bases and into the relationships between fiber composition and physiological function. There are examples of dietary fiber sources whose analysis as soluble and insoluble fractions are not consistent with what we know about their physiological effects. The second objective of this chapter is to illustrate two differences between in vitro data, i.e. data obtained by fiber analysis methods, and in vivo measurements; these differences also begin to explain some of the apparent inconsistencies between dietary fiber composition and physiological effects. In this chapter dietary fiber is defined as lignin plus those polysaccharides in plants that cannot be digested by the endogenous enzymes in the mammalian monogastric gastrointestinal tract (1).

New Developments in Dietary Fiber
Edited by I. Furda and C. J. Brine
Plenum Press, New York, 1990

DIFFERENCES AMONG METHODS OF DIETARY FIBER ANALYSIS

Most dietary fiber methods take either a gravimetric or chemical approach to analysis (1). Steps found in one or both approaches to analysis are illustrated in Figure 1. All methods recommend a non-quantitative lipid extraction if it is >5-10% of the dry sample. In a gravimetric procedure a dry food sample is subjected to chemical and enzymatic steps to extract nonfiber components (1,2). These steps usually include one or more enzymatic steps to hydrolyze starch and protein. The sample is then usually made up to 76-80% ethanol to precipitate the polysaccharides and dried. The residue weight, after correction for incompletely removed materials such as crude protein and ash, is taken as the total dietary fiber. Most fiber analysis procedures are identified by the investigator who first published the procedure. The four major gravimetric approaches to dietary fiber analysis are the neutral detergent fiber (NDF) method developed by Peter Van Soest (3,4), the total dietary fiber procedure developed under the leadership of Leon Prosky of FDA and based on the experience of Asp, Furda, and Schweizer and Würsch (1) that has been given approval by the Association of Official Analytical Chemists (AOAC) (5,6), the soluble and insoluble dietary fiber analysis method by Neils Asp of Sweden (7), and the soluble and insoluble dietary fiber analysis method developed by Roger Mongeau, Canada (8). Two modifications of the AOAC method, as discussed elsewhere in this symposium, have been proposed. One is designed to distinguish the soluble and insoluble fractions and has been (6) and is being further evaluated in interlaboratory collaborative studies. The other, proposed by Dr. Lee, uses a different buffer system, the components of which are less likely to precipitate upon exposure to 76-80% ethanol. Recently, as Dr. Van Soest presented in this symposium, he and his colleague Dr. Jerraci have developed a method that combines the use of urea with dialysis to measure total dietary fiber; this method also is being evaluated in a collaborative study.

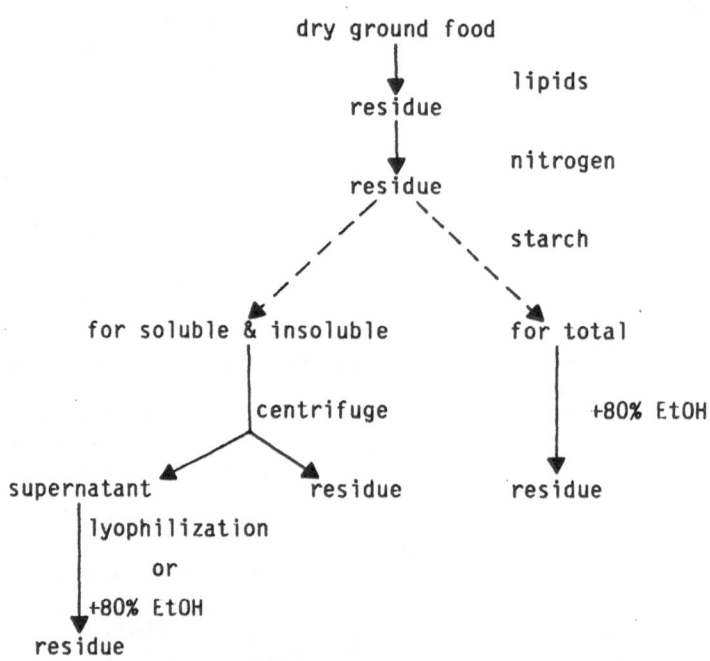

Fig. 1. Various Steps in Dietary Fiber Methods

Enzymatic and chemical steps are also used to extract nonfiber components in those methods that approach dietary fiber analysis chemically (1) (Fig. 1). Steps for protein extraction are not usually included. A residue containing the total fiber is obtained by precipitation of the fiber polysaccharides with 80% ethanol. Alternately, soluble and insoluble fiber fractions are obtained by aqueous extraction of this precipitate or by centrifugation of the sample without prior treatment with ethanol. Aliquots of the total fiber residue or of the soluble and insoluble fractions are treated with sulfuric acid to hydrolyze the fiber polysaccharides to the constituent neutral sugars (1,2). Neutral sugars can be measured indirectly with colorimetric techniques or directly by high performance liquid chromatography or by gas chromatography following derivatization. Nearly all of the glucose in insoluble fractions represents cellulose. The exception is a portion of the glucose in oats and barley fractions which represents mixed linkage ß-glucans. These are distinguished from cellulose with a specific enzymatic assay (9, 10). Other neutral sugars in soluble and insoluble fractions represent hemicelluloses. Uronic acids, which are either measured colorimetrically or as carbon dioxide following decarboxylation, represent pectic materials. The major chemical methods of fiber analysis are those of Theander (11,12) of Sweden, a method also used in our laboratory (10,13,14) and the methods of Englyst (15-17) from England which are modifications of the pioneering method of Southgate (18,19).

The evolution of several of these methods has been recently reviewed (2). Every method offers advantages and disadvantages and the same dietary fiber value will not always be obtained with different methods of analysis. The source and method of analysis of fiber data should be carefully reviewed before a fiber value is used.

Some of the differences among methods of analysis that may be responsible for the different dietary fiber values are outlined in Table 1. Only Southgate's original method (18,19) and the one developed by Professor Theander (11,12) begin fiber analysis with an alcohol step to extract simple sugars. Several gravimetric and chemical methods use the general insolubility of polysaccharides in ethanol as a means of recovering them from solutions (Fig 1). If large amounts of simple sugars remain in a sample, a portion may co-precipitate with the fiber polysaccharides during this ethanol step. These trapped simple sugars would be measured as dietary fiber by either a gravimetric or a chemical analysis and were likely the reason for the higher total dietary fiber values obtained by the AOAC procedure compared to those obtained by the Theander method for 2 of 5 foods, a one-day food composite and apples, in a recently published study (20).

Table 1. Differences Among Fiber Analysis Methods

1. Extraction of endogenous simple sugars
2. Starch hydrolysis
3. Precipitation of fiber polysaccharides with ethanol
4. Corrections of residue weight
5. Single vs. multiple sample analysis
6. Acid hydrolysis
7. Measurement of neutral sugars or uronic acids
8. Measurement of Klason lignin

Incompletely hydrolyzed starch also will be measured as dietary fiber in both gravimetric and chemical methods of analysis (Table 1). Most methods use a combination of a heat stable endo-α-amylase and an exo-1,4,α-gluco-sidase to hydrolyze starch, although there is considerable variation in hydrolysis conditions (6,12). Englyst's procedure employs a combination of porcine pancreatic α-amylase or pancreatin and pullulunase. Efficiency of this starch extraction is checked only in the detailed method 'A' developed by Theander. The use of ethanol to precipitate fiber polysaccharides could be responsible for some differences among methods. It has been reported that some polysaccharides are incompletely precipitated by ethanol (21). Given the diversity of fiber components in the human food supply it is possible that some of these components would not be recovered with ethanol precipitation, resulting in an underestimation of dietary fiber.

None of the approaches to nitrogen extraction, except the one used in the NDF method, effectively extract protein. The effect of both the protease step in the AOAC procedure and the pepsin treatment in Asp's method is to move protein from the insoluble to the soluble fiber fraction (14). Correction of the fiber residue weight either for crude protein (Nx6.25) or for ash, particularly with the current buffer system in the AOAC method, can be sources of error. Two methods utilize analyses of separate aliquots of the sample to derive dietary fiber data, the method by Mongeau (8) to measure soluble and insoluble fiber and one of the methods by Englyst (17). This type of approach assumes there is neither an overlap nor a loss of dietary fiber components among the different fractions.

In the chemical approaches to analysis, incorrect measurement of either the neutral or acidic sugars could lead to an over-or an under-estimation of the dietary fiber. Acid hydrolysis conditions are particularly important for accurate measurement of neutral sugars. We have recently found, in contrast to some earlier observations (22), that a single step dilute acid hydrolysis is not adequate to completely hydrolyze the soluble fiber fraction. The ability to measure cellobiose by HPLC is an indicator of the completeness of the hydrolysis that we use, along with the determination of recoveries of the fiber fractions. Cellobiose is not usually measured by the GC methods that are used to determine fiber-derived neutral sugars. GC analysis, however, does measure fucose and rhamnose, two sugars that co-elute with other sugars in HPLC analysis (13,23).

Any one of the above differences could lead to the different dietary fiber results that have been described (2,20). In addition, different approaches to Klason lignin could account for different results. Anderson

Table 2. Effect of Klason Lignin on Total Dietary Fiber

	Total Dietary Fiber	
	with Lignin	without Lignin
	--------% dry wt. of food--------	
Apple, peeled	9.3	8.8
Apple, unpeeled	12.3	9.2
Carrots, raw	29.7	28.8
Cucumber, peeled	15.8	15.4
Cucumber, unpeeled	21.2	18.2
Kidney Beans, canned	21.7	20.1
Oat Bran	17.4	13.5
Onions	14.0	14.0
Orange, fresh	10.0	9.5
Rice, cooked	1.7	1.1
White Wheat Bread	3.8	3.4
Wheat Bran	39.4	33.3

and Bridges (24) determined lignin as the insoluble residue after acid hydrolysis corrected for ash content. Englyst (17) does not include lignin as part of the dietary fiber complex. Most other laboratories measure lignin as the insoluble material remaining after acid hydrolysis of the insoluble fiber fraction, i.e. Klason lignin. We have calculated some of our total dietary fiber results two ways, with and without lignin (Table 2). Most of the foods whose fiber values were lowered significantly by removing lignin were those that do have lignin, i.e. wheat and oat bran, kidney beans, and unpeeled cucumber and apple. The relatively high lignin content of rice is probably an error and reflects the collection of precipitated protein with the lignin; this problem can be minimized by careful washing of the lignin residue to neutralize it and solubilize the protein. Incorporation of a protein hydrolysis step usually had a limited effect on the Klason lignin value (14).

Some methods (17,24) generate higher soluble dietary fiber values than the Theander method 'A' we and Theander use (11-14,25). Failure to include lignin in the fiber complex leads to a higher proportion of the total fiber being measured as the soluble fraction but only when lignin is a

Table 3. Effect of Klason Lignin on Distribution of Fiber Between Soluble and Insoluble Fractions

	Soluble Dietary Fiber	
	with Lignin	without Lignin
	---------% of total fiber---------	
Apple, peeled	22	23
Apple, unpeeled	16	22
Carrots, raw	5	5
Cucumber, peeled	9	9
Cucumber, unpeeled	8	9
Kidney Beans, canned	24	26
Oat Bran	42	54
Onions	3	3
Orange, fresh	27	28
Rice, cooked	11	22
Wheat Bran	4	5
White Wheat Bread	26	29

significant amount of the total fiber (Table 3). Different methods of analysis also give different proportions of soluble and insoluble fiber even when the total dietary fiber measured by the methods may be the same (14). The addition of a protease step to two starch hydrolysis steps, as is done in the AOAC method, significantly increased the proportion of the total fiber extracted as soluble fraction for kidney beans, oat bran and peas; the effect on the recovery of soluble fiber from peas and rice was more modest (14). Incorporation of a pepsin digestion, as is done in Asp's procedure, tended to increase the soluble fraction further, especially for kidney beans and macaroni. Without any protein digestion the mean soluble fiber fraction among the 5 foods was 19% of the total fiber; with the additional protease step it was 26% and with the pepsin digestion, 30% (14).

RELATIONSHIP OF IN VITRO ANALYSES TO IN VIVO MEASUREMENTS

In an effort to understand the basis for some of fiber's physiological effects many in vitro studies have been conducted because the systems can be more tightly controlled. Previous work with fiber's cation exchange and water holding capacities suggest that in vitro data are not good predictors of in vivo effects. Those fibers that have the greatest ability to absorb water generally have the least effect on stool weight (1). In fact, the work of Stephen and Cummings suggest the relationship between in vitro water holding and a fiber's effect on stool weight is an inverse one (26). Purified cellulose (27) but not pectin (28) has been shown to interfere with calcium balance in man; data from in vitro studies suggest pectin but not cellulose would interfere with mineral bioavailability (29).

Comparison of two other kinds of data, related to dietary fiber analysis, extends these observations of the limitations of using in vitro findings to predict fiber's effects in vivo. No fiber analysis method minnicks the highly integrated complex and continuous process of digestion. Masticated food enters the stomach in much larger particles than the 20-40 mesh that dry foods are ground to for use in fiber analysis. No procedure incorporates the vigorous mixing with HCl and pepsin that occurs in the stomach, in part because prolonged incubation under such conditions may hydrolyze some hemicelluloses (7,30). However, a shorter in vitro incubation with pepsin, 3h, pH 1.5, 40°C, does significantly increase the proportion of total dietary fiber extracted as the soluble fraction (Table 4) (14).

Table 4. Effect of HCL and Pepsin on Analysis of Soluble Dietary Fiber[a]

| | Soluble Dietary Fiber | |
	without treatment	with treatment
	---------------% dry wt. of food--------------	
Peas	2.3	3.4
Kidney Beans	5.2	8.9
Oat Bran	6.5	7.5
Rice	0.1	0.4
Macaroni	1.0	1.4

[a]Data derived from ref. 14.

The proportion of the total fiber from the different foods solubilized with a pepsin digestion was variable, further complicating any prediction of in vivo response from in vitro data. The more vigorous in vivo gastric process, combined with simultaneous digestion of other components of the plant matrix, e.g. lipids and starch in the small intestine, would probably generate still a greater soluble fiber fraction. The proportion of total dietary fiber that behaves functionally like soluble fiber in the stomach and small intestine is unknown at this time and is likely to be greater than what is measured as soluble fiber by any analytical method that measures the components of total dietary fiber.

Secondly, the relationship between resistant starch, that is starch unavailable to enzymatic hydrolysis but solubilized by alkali, or residual starch, starch not hydrolyzed during fiber analysis, and starch bioavailability also is unclear. We have utilized a colonectomized rat to study starch bioavailability in vivo. The amount of residual starch measured in our fiber analysis method was the same as the amount of starch recovered in the ileal effluent after preliminary test doses of the rice were fed to the rat model (Table 5). However, the bioavailability of kidney bean starch was much less than the residual starch measured during fiber analysis. Use of chromic oxide as a meal marker assured complete recoveries of the ileal effluent and the failure to detect muramic acid by GC analysis of the ileal effluent indicated that no substantial microfloral population resided in the gut of the colonectomized animals.

Table 5. In Vitro vs. In Vivo Measurement of Starch Bioavailability

	Starch	
	in ileal effluent	residual in fiber
	(% of fed)	(% dry wt. of food)
Rice	0.2[a]	0.2
Kidney Beans	11.1	1.4

[a]Mean, n=4

CONCLUSIONS

Chemical composition and in vitro behavior are but two characteristics of dietary fibers that may provide information about mechanisms of action. For example, beans are being promoted for their beneficial physiological effects, but the compositional differences between kidney beans and peas are relatively small (Table 6). Fruits have been promoted as primary sources of pectins; the data for apple and orange indicate that fruits do have more pectins than most of the other foods (Table 6). However, pectic materials

Table 6. Fiber Components in Selected Foods

	β-glucan	Cellulose	Hemicelluloses	Pectins	Total[a]
	------% of total fiber-polysaccharides-------				
Peas	–	64	25	11	22
Kidney Beans	–	46	42	13	20
Apple	–	37	26	36	10
Orange	–	23	33	43	12
Cabbage	–	36	30	32	23
Carrots	–	33	32	31	20
White Bread	–	29	69	3	4
Wheat Bran	–	28	66	6	34
Oat Bran	48	13	34	4	13

[a]% dry weight of food

189

are only ≈35-45% of the total dietary fiber in these two fruits and carrots and cabbage, two vegetables, have nearly the same amounts of pectin. There are major compositional differences between oat bran and kidney beans, given the similarities in their physiological effects, i.e. hypocholesterolemia. Except for the two brans, the amount of Klason lignin in these foods is low, 2-5% of the total dietary fiber. It is evident that much more information is needed about dietary fiber sources before mechanisms of action will be understood. Many other physiochemical properties, e.g. particle size, viscosity, structure, etc. (1,2), as well as fermentability and starch bioavailability, need to be determined.

ACKNOWLEDGMENTS

This research program is funded by National Institutes of Health grants DK21712 and CA46339, NIH Contract CN-85069, and College of Agricultural and Life Sciences, University of Wisconsin-Madison.

REFERENCES

1. Pilch S. (ed.) Physiological Effects and Health Consequences of Dietary Fiber. FASEB, Bethesda, Maryland, 1987.

2. Marlett JA. Analysis of Dietary Fiber in Human Foods, in: Third Vahouny Conference on Dietary Fiber in Health and Disease, (D. Kritchevsky, ed.) April, 1988, Washington, D.C. (In press)

3. Van Soest PJ, Wine RH. Use of detergents in the analysis of fibrous foods. IV. Determination of plant cell-wall constituents. J. Assoc. Off. Anal. Chem. 50:50-55, 1967.

4. Robertson JB, Van Soest PJ. The detergent system of analysis and its application to human foods, in: The Analysis of Dietary Fiber in Food James WPl, Theander O. (eds.) Marcel Dekker, Inc., New York, pp 123-158, 1981.

5. Prosky L, Asp NG, Furda I, et al. Determination of total dietary fiber in foods and food products: collaborative study. J. Assoc. Off. Anal. Chem. 68:677-679, 1985.

6. Prosky L, Asp, NG, Schweizer TF, et al. Determination of insoluble, soluble and total dietary fiber in foods and food products: interlaboratory study. J. Assoc. Off. Anal. Chem. 71:1017-1023, 1988.

7. Asp NG, Johansson CG, Hallmer H, Siljestrom M. Rapid enzymatic assay of insoluble and soluble dietary fiber. J. Agric. Food Chem. 31:476-482, 1983.

8. Mongeau R, Brassard R. A rapid method for the determination of soluble and insoluble dietary fiber: comparison with AOAC total dietary fiber procedure and Englyst's method. J. Food Sci. 51:1333-1336, 1986.

9. Aman P, Graham H. Analysis of total and insoluble mixed linked (1,3), (1,4)-β-D-glucans in barley and oats. J. Agric. Food Chem. 35:704-709, 1987.

10. Shinnick FL, Longacre MJ, Ink SL, Marlett JA. Oat fiber: composition vs. physiological function. J. Nutr. 118:144-151, 1988.

11. Theander O, Aman P. Studies on dietary fibers. 1. Analysis and chemical characterization of water-soluble and water-insoluble dietary fibers. Swed. J. Agric. Res. 9:97-106, 1979.

12. Theander O, Westerlund E. Studies on dietary fiber. 3. Improved procedures for analysis of dietary fiber. J. Agric. Food Chem. 34:330-336, 1986.

13. Neilson MJ, Marlett JA. A comparison between detergent and nondetergent analyses of dietary fiber in human foodstuffs, using high-performance liquid chromatography to measure neutral sugar composition. J. Agric. Food Chem. 31:1342-1347, 1983.

14. Marlett, J.A., Chester, J.G., Longacre, M.J., and Bogdanske, J.J. Recovery of Soluble Dietary Fiber is Dependent on the Method of Analysis. Am. J. Clin. Nutr. 50:479-485, 1989.

15. Englyst HN, Wiggins HS, Cummings JH. Determination of non-starch polysaccharides in plant foods by gas-liquid chromatography of constituent sugars as alditol acetates. Analyst 107:307-318, 1982.

16. Englyst HN, Hudson GJ. Colorimetric method for routine measurement of dietary fibre as non-starch polysaccharides. A comparison with gas-liquid chromatography. Food Chem. 24:63-76, 1987.

17. Cummings JH, Englyst HN. The development of methods for the measurement of dietary fibre in food, in: Morton ID. (ed.) Cereals in a European Context: First European Conference on Food Science and Technology. New York: VCH Publishers, Inc., pp 188-220, 1987.

18. Southgate DAT. Determination of carbohydrates in foods. II. Unavailable carbohydrates. J. Sci. Fd. Agric. 20:332-335, 1969.

19. Southgate DAT. Use of the Southgate method for unavailable carbohydrates in the measurement of dietary fiber, in: The Analysis of Dietary Fiber in Food, James WPT, Theander O. (ed.) Marcel Dekker, Inc., New York, pp 1-19, 1981.

20. Marlett JA, Navis D. Comparison of gravimetric and chemical analyses of total dietary fiber in human food. J. Agric. Food Chem. 36:311-315, 1988.

21. Larm D, Theander O, Aman P. Structural studies on a water-soluble arabinan isolated from rapeseed (Brassica Napus), Acta Chem. Scand., B29:1011-1014, 1975.

22. Marlett JA. Analysis of dietary fiber. Animal Feed Sci. and Tech. 23:1-13, 1989.

23. Shinnick FL, Hess RL, Fischer MH, Marlett JA. Apparent Nutrient Absorption and Upper Gastrointestinal Transit with Fiber-Containing Enteral Feedings Am. J. Clin. Nutr. 49:471-475, 1989.

24. Anderson JW, Bridges SP. Dietary fiber content of selected food. Am. J. Clin. Nutr. 47:440-447, 1988.

25. Theander O, Aman P. Analysis of dietary fibers and their main constituents, in: The Analysis of Dietary Fiber in Food, James WPT, Theander O. (eds.), Marcel Dekker, Inc., New York, pp 51-70, 1981.

26. Stephen AM, Cummings JH. Water-holding by dietary fibre _in vitro_ and its relationship to faecal output in man. _Gut_ 20:722-729, 1979.

27. Slavin JL, Marlett JA. Influence of refined cellulose on human bowel function and calcium and magnesium balance. _Am. J. Clin. Nutr._ 33:1932-1939, 1980.

28. Cummings JH, Southgate DAT, Branch WJ et al. The digestion of pectin in the human gut and its effect on calcium absorption and large bowel function. _Br. J. Nutr._ 41:477-485, 1979.

29. Marlett JA. Dietary fiber and mineral bioavailability. _IM-Int. Med. Specialist_ 5:99-114, 1984.

30. Johansson C, Aman P, Asp N et al. Enzymatic and neutral detergent fibre methods for dietary fibre analysis. _Swed. J. Agric. Res._ 12:157-161, 1982.

COLLABORATIVE STUDY OF A METHOD FOR SOLUBLE AND INSOLUBLE DIETARY FIBER

Leon Prosky

Division of Nutrition
Food and Drug Administration
Washington, D.C. 20204

INTRODUCTION

The Food and Drug Administration issued a final rule in 1987 concerning the nutrition labeling of foods with respect to calories content[1]. The Agency amended the existing food labeling regulations to provide for the exclusion of nondigestible dietary fiber when the calorie content of a food for nutrition labeling purposes is determined. In essence, the amendment allows a manufacturer to subtract the carbohydrate attributable to nondigestible fiber from the total carbohydrate content of a food, when the appropriate declaration of calorie content for that food is calculated. The Federal Register further stated that "The nondigestible dietary fiber will be determined by the method, Total Dietary Fiber in Foods, Enzymatic Gravimetric Method, First Action, in the Journal of the Association of Official Analytical Chemists (JAOAC), 68:399, 1985, as amended in JAOAC 69:370, 1986." These methods were previously published as research papers previously in the JAOAC[2,3]. Considering that soluble dietary fiber (SDF) and insoluble dietary fiber (IDF) often exhibit distinctly different physiological effects[4], the basic method was extended to give not only total dietary fiber (TDF) values, but also separate values for SDF and IDF. A previously completed interlaboratory study of a method for SDF and IDF[5] revealed that the same enzymatic-gravimetric approach accepted by the AOAC for TDF could be used for SDF and IDF. The collaborative study reported on in this chapter is not yet completed, but sufficient data to assess its value as a method for determining SDF and IDF in a variety of foods and food products have been obtained.

DEVELOPMENT AND PRINCIPLE OF THE METHOD

Discussions of the development of the enzymatic-gravimetric method for determining IDF are found in several publications[2,6,7]. The methods for IDF and SDF are outlined below.

New Developments in Dietary Fiber
Edited by I. Furda and C. J. Brine
Plenum Press, New York, 1990

Duplicate test samples of dried foods, fat-extracted if containing
>10% fat, are gelatinized with Termamyl (heat-stable-alpha-amylase),
and then enzymatically digested with protease and amyloglucosidase to
remove protein and starch.

For IDF

The residue is filtered and washed with distilled H_2O. The
filtrate and wash are saved. The residue (IDF) is washed with 95%
ethanol and acetone. After drying, the residue is weighed. One of the
duplicates is analyzed for protein, and the other is incinerated at
525°C and the ash determined. IDF is the weight of the residue less
the weight of the protein and ash present.

For SDF

Four volumes of 95% ethanol are added to the combined filtrate and
water washing to precipitate soluble dietary fiber. The precipitate is
filtered and washed with 78% ethanol, 95% ethanol and acetone. After
drying, the residue is weighed. One of the duplicates is analyzed for
protein, and the other is incinerated at 525°C and the ash determined.
SDF is the weight of the residue less the weight of the protein and ash
present.

Apparatus and Reagents

For apparatus see references 1 and 5, sec. 48.A15 and for reagents
substitute the following for Reagents (d), (h), and (i) in reference 1
and 5, sec. 43.A16.

(d) Phosphate Buffer. - 0.08M, pH 6.0. Dissolve 1.400 g Na
phosphate dibasic, anhydrous (Na_2HPO_4) (or 1.753 g dihydrate) and
9.68 g Na phosphate monobasic ($NaH_2PO_4 \cdot H_2O$) (or 10.94 g dihydrate) in
approximately 700 mL H_2O. Dilute to 1 L with H_2O. Check pH with pH
meter and adjust if necessary.

(h) Sodium Hydroxide Solution. - 0.275N. Dissolve 11.00 g NaOH ACS
in approximately 700 mL H_2O in 1 L volumetric flask. Dilute to volume
with H_2O.

(i) Hydrocholoric Acid Solution. - 0.350N. Dilute a stock solution
with known titer, e.g. 350 mL 1M HCl to one liter with distilled H_2O.

Sample Preparation

Determine insoluble and soluble dietary fiber on dry samples,
without pretreatment whenever possible. Mill dry foods with particles
>0.5 mm to 0.3-0.5 mm mesh. Homogenize and freeze-dry wet foods
before milling. If high fat content (>10%) prevents proper milling,
defat with petroleum ether (3 times with 25 mL portions/g sample)
before milling. Determine residual moisture in milled samples by
drying overnight in a 70° vacuum oven or for 5 hours in 105° air oven.
Record losses of weight due to fat and/or H_2O removal and report final
% dietary fiber on a dry matter basis.

Determination

(a) Run blank through entire procedure along with tests to measure
any contribution from reagents to residue.

(b) Weigh duplicate 1 g test portions, accurate to 0.1 mg, into 400 mL tall-form beakers. Weights should not differ by more than 20 mg. Add 50 mL pH 6.0 phosphate buffer to each beaker. Check pH with pH meter. Adjust to pH 6.0 ± 0.2 with 0.275N NaOH if necessary. Add 0.1 mL Termamyl solution.

(c) Cover beaker with Al foil and place in boiling H_2O bath 15 minutes. Shake gently at 5 minute intervals. Increase incubation time when number of beakers in boiling H_2O bath makes it difficult for beaker contents to reach temperature of 95°-100°C. Use thermometer to ensure 15 minutes at 95°-100°C. Thirty minutes should be sufficient. Cool solutions to room temperature. Adjust to pH 7.5 ± 0.2 by adding 10 mL 0.275N NaOH solution.

(d) Add 5 mg protease. Protease sticks to spatula, so it may be preferable to prepare enzyme solution (50 mg in one mL phosphate buffer) just before use and pipet 0.1 mL to each sample. Cover beaker with Al foil. Incubate 30 minutes at 60°C with continuous agitation.

(e) Cool to room temperature. Add 10 mL of 0.350M hydrochloric acid solution. Measure pH and add acid dropwise if necessary. Final pH should be 4.0-4.6. Add 0.3 mL of amyloglucosidase, cover with Al foil and incubate at 60°C for 30 minutes with continuous agitation.

Insoluble Dietary Fiber

(f) Weigh crucible containing Celite to the nearest 0.1 mg. Then wet and redistribute the bed of Celite in the crucible using a stream of distilled water from a wash bottle. Apply suction to the crucible to draw the Celite onto the fritted glass as an even mat. Filter the enzyme mixture from step (e) through the crucible into a pre-weighed suction flask.

(g) Wash residue two times with 10 mL of distilled water. Save the filtrate and water washing for determination of soluble dietary fiber.

(h) Wash residue two times with 10 mL each of 95% ethanol and two times with 10 mL each acetone. Discard washings. With some products, a gum is formed, trapping the liquid. If the surface film that develops after the addition of the wash ethanol and acetone to the Celite is broken with a spatula, filtration is improved. Long filtration times should be avoided by careful intermittent suction throughout the filtration. Normal suction can be applied at washing. Back-bubbling with air, if available, is another way of speeding filtration.

(i) Dry crucible containing residue overnight in a 70°C vacuum oven or a 105°C air oven. Cool in desiccator and weigh crucible, Celite, and residue to nearest 0.1 mg. Subtract crucible and Celite weight to determine weight of residue.

(j) Analyze the residue from one test portion of the set of duplicates for protein. Protein is probably most easily analyzed by carefully scraping the Celite and the fiber mat onto a suitable piece of filter paper which can be folded shut and inserted into the Kjeldahl flask. A piece of filter paper should be analyzed to ensure that it will not contribute to the protein value obtained. To determine the nitrogen content use the Kjeldahl analysis as specified in, "Official Methods of Analysis" of the AOAC 47.021-47.023. Use nitrogen x 6.25 for the protein factor in all cases in the collaborative study. Otherwise use the correct factor for conversion of nitrogen to protein when it is known.

(k) Incinerate second residue of the set duplicates for 5 hours at 525°C. Cool in desiccator and weigh crucible to nearest 0.1 mg. Subtract crucible and Celite weight to determine ash. Negative ash weight might occur as a consequence of insufficient drying of Celite or losses, especially at IDF filtration. Replace crucible if ash loss is greater than 3-5 mg. Always use actual ash weights, positive or negative, in the calculation.

Soluble Dietary Fiber

(1) Adjust the weight of the combined filtrate from step (g) insoluble fiber procedure to 100 g with distilled water. (The density of the filtrate is close to one.) Transfer the solution to beaker or erlenmyer. Add 4 volumes (400 mL) of 95% ethanol preheated to 60°C. Rinse the suction flask with part of the ethanol. Allow precipitate to form at room temperature for 60 minutes.

(m) Weigh crucible containing Celite to nearest 0.1 mg then wet and redistribute the bed of Celite in the crucible using a stream of 78% ethanol from a wash bottle. Suction is then applied to the crucible to draw the Celite onto the fritted glass as an even mat. When fiber is filtered, i.e., step (n), the Celite effectively separates the fiber from the fritted glass of the crucible allowing for easy removal of the crucible contents.

(n) Filter the diluted enzyme digest from step (1) through crucible. Discard filtrate.

(o) Wash residue successively with three 20 mL portions of 78% ethanol, two 10 mL portions of 95% ethanol, and two 10 mL portions of acetone. With some products, a gum is formed, trapping the liquid. If the surface film that develops after the addition of the sample to the Celite is broken with a spatula, filtration is improved. Long filtration times should be avoided by careful intermittent suction throughout the filtration. Normal suction can be applied at washing. Back-bubbling with air, if available, can be used to speed filtration.

(p) Proceed with step (i) through (k) of insoluble dietary fiber method.

Calculations

Calculations have been simplified by the use of new data sheets and equations. Use of separate data sheets for blanks and tests and the new equations for calculation of dietary fiber have rectified all of the problems associated with the calculations. (Figure 1).

Test Samples Analyzed

Portions of fifteen foods (seven blind duplicates and one standard containing 4.3 - 5.4% insoluble dietary fiber and 1.5 - 2.7% soluble dietary fiber) were sent to thirty-nine laboratories for analysis of soluble and insoluble dietary fiber. The foods, unknown to the analysts, were:

1.	Cabbage	9.	Chick peas	16.	Raisins
2.	Carrots	10.	Brussels sprouts	17.	Mission figs*
3.	French beans	11.	Barley	18.	Calimyrna figs*
4.	Kidney beans	12.	Rye flour	19.	Prune*
5.	Butter beans	13.	Turnips	20.	Apple*
6.	Okra	14.	Soybran[+]	21.	Peach*

Items 1-13 were purchased either fresh, canned or as a dried product at the local supermarket. Item 14 was graciously supplied by Solnut, Inc., Hudson, Iowa; item 15 by Vitamins, Inc., Chicago, Illinois; item 16 by the California Raisin Advisory Board, San Francisco, California; items 17 and 18 by the California Fig Advisory Board, Fresno, California; and items 19-22 by Vacu-Dry, Santa Rosa, California. All products were homogenized in a Cuisinart, lyophilized and ground in a Microjet 10 Centrifugal Mill (Quartz Technology, Inc., Westbury, NY) to a uniform size of 350 um. Items 2 and 17-22 were extracted three times each with ten volumes of 85% methanol, to remove sugars, so the final material could be lyophilized to a dry material. The final material was ground in a Microjet 10 Centrifugal Mill as described above.

All test samples were placed ·in 25 ml scintillation vials with screw caps and a number was taped to each vial.

RESULTS AND DISCUSSION

In three previous collaborative studies[2,3,5] it was determined that the enzymatic-gravimetric method was apparently the most practical and simplest way to carry out the total dietary fiber analysis, i.e., the weight of the residue minus the weight of protein and ash. Table 1 shows the individual analysis obtained on nine test samples (blind duplicates) by nine collaborators. All TDF values were used in the final statistics for rye bread, rice, wheat bran, corn bran, and whole wheat flour; for white wheat flour, the results of collaborator 7 were not used because the value was a Cochran and Grubbs outlier[3]. The results for potatoes (collaborator 1) and for oats (collaborator 7) were also not used; both were Cochran outliers. A Cochran outlier indicates that the laboratory did not check itself; a Grubbs outlier indicates that the laboratory did not check the values of the other laboratories, both at a probability level of 0.01. Most problems occurred with the soy isolate; two laboratory determinations were discarded because of filtration problems.

The measures of precision for determining TDF are shown in Table 2. Seven test samples that had TDF values of greater than 3% had CV values of less than 10%. When foods contained less than 1.5% TDF, the CV was about 66%. The two products with low fiber content, namely 1.42 and 1.04% for soy isolate and rice, which have large CV values, may indeed be unimportant because of their low fiber content. The method seems to give good results at the 3% TDF level or higher. On the basis of these findings, it was recommended to the AOAC that the enzymatic-gravimetric method be adopted as an official first action. This recommendation was accepted and the method was adopted as an official final action in 1986[8]. In 1988 modifications were introduced in the official method for TDF and were adopted by the AOAC. The changes included changes in concentration of buffer and base and the use of hydrochloric acid instead of phosphoric acid. Further, an interlaboratory study of a method for soluble and insoluble fiber was conducted. The results showed excellent agreement between the TDF determined independently and

Table 1. Collaborative Results (Blind Duplicates) of Determination of TDF by the Enzymatic-Gravimetric Method

Coll.	Soy Isolate	White Wheat Flour	Rye Bread	Potatoes	Rice	Wheat Bran	Oats	Corn Bran	Whole Wheat Flour
1	3.18[a]	2.40	6.22	5.41[b]	0.91	42.32	10.30	85.14	10.54
	0.00	2.41	6.51[a]	5.75	0.35	41.86	11.66	86.32[c]	11.96
2	0.73[d]	2.63	5.61[a]	7.30	0.42[a]	40.64	9.99	83.81[c]	12.02
		3.08	8.07	8.08	2.18	41.41	10.93	84.04	11.49
3	1.34	2.40	6.75	6.02	0.54	42.79	11.30	86.90	13.48
	1.23	2.40	7.05	6.70	0.55	42.32	11.79	87.96	13.15
4	1.19	2.79	6.30	7.85	1.28	44.01	10.98	86.70	12.70
	1.18	2.78	6.40	6.94	1.14	41.24	10.65	86.76	12.28
5	1.17	2.90	7.02	7.19	0.74	44.38	10.87	87.21	12.62
	0.00	2.91	6.35	7.38	0.82	41.75	11.25	86.80	12.43
6	0.75	2.67	6.11	7.06	0.90	42.34	10.98	87.59	13.03
	0.78	3.02[b,e]	6.44	6.81	1.42	42.19	10.29[b]	87.84	13.62
7	2.52[d]	3.09[b,e]	6.60	6.99	0.27	43.43	12.99[b]	88.25	12.58
		8.42	6.34	7.15	0.77	44.31	12.48	87.89	13.15
8	2.25	3.14	6.88	7.22	1.91	43.60	11.80	87.31	13.50
	1.99	3.00	7.34	7.90	1.74	44.20	10.75	88.42	12.68
9	2.70	3.08	6.35	7.97	1.49	41.96	11.70	87.47	13.11
	1.76	2.89	6.58	7.51	1.26	43.00	11.05	87.14	12.91

[a]Cochran outlier, results used.
[b]Cochran outlier, results eliminated.
[c]Grubbs outlier, results used.
[d]Erroneous results due to filtration problems.
[e]Grubbs outlier, results eliminated.

Reprinted from the "J.Assoc.Off.Anal.Chem. 68:678 (1985).

AOAC DIETARY FIBER METHOD   BLANK DATA SHEET
BLANKS

	a		b		c		d	
Crucible + Celite + tare Weight (mg)								
Crucible + Celite + Residue Weight (mg)								
Residue Weight (mg)	R_1	R_2	R_1	R_2	R_1	R_2	R_1	R_2
Protein (mg) P								
Crucible + Celite + Ash weight (mg)								
Ash weight (mg) A								
Blanks (mg)								
Mean Blank (a+b+c+d)÷4 (mg) B								

$$\text{Blanks (mg)} = \frac{R_1 + R_2}{2} - P - A$$

AOAC DIETARY FIBER METHOD   SAMPLE DATA SHEET

	Sample				Sample			
	1		2		1		2	
1. Sample Weight (mg)	m_1	m_2	m_1	m_2	m_1	m_2	m_1	m_2
2. Crucible + Celite tare weight (mg)								
3. Crucible + Celite Residue weight (mg)								
4. Residue Weight (mg)	R_1	R_2	R_1	R_2	R_1	R_2	R_1	R_2
5. Protein (mg) P								
6. Crucible + Celite + Ash weight (mg)								
7. Ash Weight (mg) A								
8. Mean Blank (mg) B								
9. Dietary Fiber (%)								

$$\text{Dietary Fiber (\%)} = \frac{\dfrac{R_1 + R_2}{2} - P - A - B}{\dfrac{m_1 + m_2}{2}} \times 100$$

Fig. 1.   Sample and blank data sheets and equations for calculation of blank and percent dietary fiber.

the TDF determined by summing the soluble dietary fiber (SDF) and insoluble dietary fiber (IDF)[5]. This would eliminate the necessity of determining TDF independently except when TDF alone is desired. A new method for recording data and calculating the % TDF was also accepted by the AOAC. (Figure 1). This data sheet has been adapted for col- lecting data and calculating the % IDF and % SDF. For SDF, the mean blank equals (a+c)/2 and for IDF the mean blank equals (b+d)/2. The shortcoming of the study was that too few products were used and those that were used contained less than 1 % SDF. Therefore, more products with a variety of matrices were used in the present study of a method for SDF and IDF.

Table 2.   Measures of Precision for Determining TDF

Sample	Dietary Fiber	Repeatability CV, %	Reproducibility CV, %
Soy Isolate	1.42	66.25	66.25
White Wheat Flour	2.78	5.55	9.80
Rye Bread	6.58	3.94	5.29
Potatoes	7.25	5.66	7.49
Rice	1.04	45.62	53.71
Wheat Bran	42.65	2.33	2.66
Oats	11.03	5.30	5.30
Corn Bran	86.86	0.56	1.56
Whole Wheat Flour	12.57	3.67	5.92

Reprinted from the J. Assoc. Off. Anal. Chem. 68:679 (1988)

The summary of results of the present study, although incomplete, are shown in Tables 3 and 4. With approximately half the collaborators reporting their results the measures of precision for determining IDF and SDF are shown in these two tables. The statistics were done with all values reported by the collaborators. No values were rejected for statistical considerations. The average IDF ranged from 3% for barley to 59% for soybran (Table 3). The repeatability (RSD_r) of the determination for the twenty-two foods analyzed ranged from 1.1% for apples to 10.5% for barley, which is considered excellent for this analysis. The variation among the laboratories (RSD_R) ranged from 2.0% for apples to 20.9% for calimyrna fig, which was also very good when one considers that half the foods had RSD_R of less than 10% and an additional 30% had RSD_R of less than 15%. The measures of precision for the determination of SDF are presented in Table 4. The products analyzed had an average SDF which ranged from 1.4% for chick peas to 33.4% for prunes. The repeatability (RSD_r) values varied from 2.3% for apples to 39.7% for chick peas with approximately 75% of the laboratories reporting less than 15% variation in their results. The reproducibility (RSD_R) values show that approximately 65% of the laboratories had an RSD_R of ≥20%. The preliminary results indicate that the differences in the SDF results between the laboratories is probably due to something that happens when the IDF fraction is filtered away from the TDF fraction. Since preliminary results[5] from an earlier study indicate that TDF analysis done independently yields the same results when compared with TDF values obtained by summing IDF and SDF, it may be more practical to obtain SDF values, after deter- mining TDF and IDF, by difference.

Table 3. Measures of Precision for Determining IDF

Product	Average IDF, %	No. of Coll-aborators	Repeat-ibility RSD$_r$, %	Reproduc-ibility RSD$_R$ %
Cabbage	21.71	8	4.19	8.12
Carrots	32.63	10	5.59	12.06
French Beans	26.05	8	3.52	5.57
Kidney Beans	13.95	12	5.69	7.17
Butter Beans	17.19	8	1.93	9.09
Okra	22.38	13	4.88	14.23
Onion	13.60	9	5.18	11.80
Parsley	34.42	10	3.56	14.16
Chick Peas	17.06	11	10.15	15.85
Brussels Sprouts	30.29	13	2.34	7.79
Barley	4.27	11	10.49	14.82
Rye Flour	11.66	12	5.10	8.60
Turnips	21.34	9	6.90	17.83
Soybran	59.00	10	1.36	2.78
Wheat Germ	15.72	8	4.75	6.43
Seedless Raisins	49.85	7	5.44	19.72
Mission Figs	33.19	4	3.11	12.18
Calimyrna Figs	43.50	4	6.19	20.86
Prunes	46.18	6	6.11	19.44
Apples	54.38	3	1.08	1.96
Peaches	39.51	4	2.62	6.92
Apricots	44.64	4	4.00	9.60

Table 4. Measures of Precision for Determining SDF

Product	Average SDF, %	No. of Coll-aborators	Repeat-ibility RSD$_r$, %	Reproduc-ibility RSD$_R$ %
Cabbage	5.43	8	12.15	30.66
Carrots	11.02	9	12.25	17.41
French Beans	11.03	8	3.79	7.28
Kidney Beans	3.39	12	19.71	21.79
Butter Beans	3.10	11	10.99	23.17
Okra	11.83	11	9.41	15.35
Onions	3.60	9	30.96	39.84
Parsley	5.13	9	21.83	56.86
Chick Peas	1.35	11	39.07	45.55
Brussels Sprouts	6.13	14	10.31	21.85
Barley	3.77	9	14.42	37.67
Rye Flour	3.31	13	14.76	20.42
Turnips	9.32	8	10.38	27.14
Soybran	7.05	9	15.44	15.44
Wheat Germ	1.95	8	21.25	43.55
Seedless Raisins	14.17	5	9.42	45.46
Mission Figs	10.84	4	11.32	13.58
Calimyrna Figs	18.37	4	5.15	17.17
Prunes	33.42	6	6.56	28.51
Apples	18.56	3	2.34	13.17
Peaches	27.30	4	3.09	11.61
Apricots	26.53	4	4.24	18.74

Acknowledgements

The author expresses appreciation to the following collaborators who participated in this study:

G. Conti, General Foods Corporation, White Plains, NY
D. Sullivan and J. A. Stassi, Hazleton Laboratories, Madison, WI
J. F. Ashton, Australasian Food Research Laboratories, Cooranbong, Australia
F. Alstin and E. Bogren, Tecator Laboratories, Hoganas, Sweden
N-G. Asp, University of Lund, Lund, Sweden
A. Sorensen, National Food Agency, Soborg, Denmark
P. A. Burdaspal, Centro Nacional De Alimentacion y Nutricion, Madrid, Spain
D. W. McBride, Woodson-Tenent Laboratories, Des Moines, IA
J. G. Faugere, Labortoire Municipale Ville De Bordeaux, Bordeaux, France
F. Fidanza, Institute of Nutrition, Perugia, Italy
D. G. Oakenfull, CSIRO Food Resarch Laboratory, New South Wales, Australia
W. Frolich and A. Gulliksen, Matforsk, Oslo, Norway
D. C. Mugford, Bread Research Institute of Austrlia, North Ryde, Australia
B. A. Lewis, Cornell University, Ithaca, NY
R. Sapp, Con Agra Consumer Frozen Food Co., Batesville, AR
P. Dysseler, D. Hoffem and N. Vanlaethem, C.E.R.I.A.-I.I.F.-I.M.C., Brussels, Belgium
G. Rice, General Foods Corporation, Battle Creek, MI
B. Hsieh, The Quaker Oats Company, Barrington, IL
G. S. Ranhotra and J. Gelroth, American Institute of Baking, Manhattan, KS
M. Katan, Agricultural University, Wageningen, The Netherlands
J. Prodolliet, Nestec Ltd., Vevey, Switzerland
T. Howes, Shaklee Corp., Hayward, CA
C. Lintas, Instituto Nazionale della Nutrizione, Italy
R. Mongeau, Health and Welfare Canada, Ottawa, Ontario, Canada
R. Fein and T. Mulder, Amway Corporation, Ada, MI
D. Gamblin, Ralston Purina Co., St. Louis, MO
T. F. Schweizer, Nestec Ltd., Vevey, Switzerland
G. Testolin, Universita di Milano, Italy
W. Seibel and E. Rabe, Federal Research Center for Grain and Potato Processing, Detmold, FGR
M. Kliauga, Good Housekeeping Institute, New York, NY
B.O. Schneeman and J. Tietyen, University of California, Davis, CA
   Appreciation is also extended to Richard Albert and Anne Wilson, Food and Drug Administration, who conducted the statistical evaluation of the data.

References

1.   Nutrition labeling of foods; Calorie content, Federal Register, 52:28690 (1987).

2.   L. Prosky, N-G. Asp, I. Furda, J. W. DeVries, T. F. Schweizer, and B. F. Harland, Determination of total dietary fiber in foods, food products, and total diets: interlaboratory study, J. Assoc. Off. Anal. Chem. 67:1044 (1984).

3. L. Prosky, N-G. Asp, I. Furda, J. W. DeVries, T. F. Schweizer, and B. F. Harland, Determination of total dietary fiber in foods and food products: collaborative study, J. Assoc. Off. Anal. Chem. 68:677 (1985).

4. D. A. T. Southgate, The relation between composition and properties of dietary fiber and physiological effect, in: "Dietary Fiber Basic and Clinical Aspects," G. V. Vahouny and D. Kritchevsky, eds., Plenum Press, New York and London (1986).

5. L. Prosky, N-G. Asp, T. F. Schweizer, J. W. DeVries and I. Furda, Determination of insoluble, soluble, and total dietary fiber in foods and food products: interlaboratory study, J. Assoc. Off. Anal. Chem. 71:1017 (1988).

6. L. Prosky and B. F. Harland, Dietary fibre methodology, in: "Dietary Fibre, Fibre-Depleted Foods and Disease," H. Trowell, D. Burkitt and K. Heaton, eds., Academic Press, Orlando and London (1985).

7. N-G. Asp and C-G. Johansson, Dietary fibre analysis, Nutr. Abs. Rev. Clin. Nutr., 54:735 (1984).

8. Vitamins and other nutrients, J. Assoc. Off. Anal. Chem. 69:370 (1986).

NON-STARCH POLYSACCHARIDES (DIETARY FIBER) AND RESISTANT STARCH

Hans N. Englyst and John H. Cummings

MRC Dunn Clinical Nutrition Centre
100 Tennis Court Road
Cambridge CB2 1QL, U.K.

INTRODUCTION

Progress in dietary fiber research has been slow because of the lack of an agreed definition of fiber, and the development of different techniques for its measurement. Two main methods have emerged which, by adopting contrasting approaches, include different components of the diet as dietary fiber.

Enzymatic gravimetric methods measure dietary fiber as the residue remaining on a filter after treatment of the food with starch and protein degrading enzymes, and correction for ash and nitrogen content. The procedures of Asp et al. (1983), Schweizer and Wursch (1979) and Furda (1981) and their adaptations to an AOAC procedure (Prosky et al. (1984) are examples of this technique.

Enzymatic chemical methods aim to identify and measure dietary fiber as its chemical components. The procedures of Southgate (1969), and Englyst et al. (1982) are examples of this technique. In the Englyst procedure starch is completely removed and dietary fiber measured as the non-starch (non -α-glucan) polysaccharides (NSP) by gas liquid chromatography of constituent sugars after acid hydrolysis. The residue measured as dietary fiber by the AOAC procedure consists of NSP, some retrograded starch, lignin, maillard reaction products and a number of unidentified components of plant and animal origin. The unpredictable nature of the residue of the AOAC method, and the inclusion of retrograded starch as dietary fiber has led to confusion as to the amount and physiological importance of fiber and starch in food.

DEFINITION OF DIETARY FIBER

The original concept of dietary fiber was based on an association between the intake of unrefined plant foods and the absence of some Western diseases in African peoples (Hipsley, 1953; Burkitt, 1969; Trowell, 1972). The refining of plant foods leads to the removal of cell-wall material and the term dietary fiber was introduced to describe this material. Southgate has pointed out (1980) that much confusion could have been avoided if the term cell wall material had not been replaced by the less descriptive and less precise term dietary fiber.

New Developments in Dietary Fiber
Edited by I. Furda and C. J. Brine
Plenum Press, New York, 1990

205

Hipsley's concept of dietary fiber was of material derived from the plant cell-wall which remained in feces (1953). In 1972 Trowell defined dietary fiber as the skeletal remains of plant cells that are resistant to digestion by the enzymes of man. However such a definition, whilst focussing on the plant cell wall as the major component of fiber, does not form the basis for an analytical method because it includes the physiological concept of resistance to digestive enzymes in man. In order to give the analyst a proper objective therefore in 1978 we suggested that dietary fiber should be measured as the non-starch polysaccharides in plant foods (Cummings, 1980; Englyst, 1981).

It has been argued that dietary fiber should be defined as the food components resisting digestion in the small intestine. Such a definition would include NSP, lactose (in many ethnic groups), raffinose, stachyose and other oligosaccharides, a small amount of lignin, organic anions such as oxalate and tartrate, maillard reaction products, some fat, hair, bone, grit and other insoluble materials and, depending on food processing, a variable amount of protein. Moreover, in the process of developing the Englyst method for measurement of dietary fiber as NSP a type of starch resistant to digestion in the small intestine of man was identified (Englyst et al., 1982). This led to the suggestion that some starch should be included as dietary fiber (Asp and Johansson, 1984). Subsequently it was shown that starch was the major carbohydrate to escape digestion for many foods (Englyst and Cummings, 1985,1986,1987). In addition, lactulose, polydextrose, neosugar and similar synthetic products resistant to pancreatic amylase would be included as dietary fiber. Such a broad definition is in clear contrast to the original concept of dietary fiber.

Resistance to digestion of foodstuffs in the small intestine depends not only on food processing and chemical structure and modification but on a number of highly variable factors not limited to the food itself. These include chewing, transit through the stomach and small intestine and the amount and type of other food components in the meal as a whole. The extent of digestion in the small intestine is therefore highly variable for an individual and even bigger differences in digestibility are seen between individuals or between population groups. Large differences in small intestinal digestibility are also seen between man and animals. Defining dietary fiber as material resisting digestion in the small intestine is not analytically realistic or a scientifically meaningful alternative to defining dietary fiber as NSP.

In 1985 Trowell, when writing about the evolution of the dietary fiber concept, stated that "starch was never named at this time or subsequently as a constituent of dietary fibre" and concluded "At the present time there is considerable agreement concerning the principal constituents of dietary fibre. They are all polysaccharides, mainly cellulose, hemicellulose and pectic substances, conveniently designated non-starch polysaccharides (NSP)". By 1987 it was proposed by Trowell, Southgate and ourselves that dietary fiber should be defined for the purposes of food labelling as NSP since this gives the best index of plant cell-wall polysaccharides, is precisely measurable, and is in keeping with the original concept of dietary fiber (Englyst et al., 1987d).

METHODS FOR MEASUREMENT OF DIETARY FIBER

The AOAC Procedure

The method currently known as the AOAC procedure (Prosky et al., 1984, 1985,1988) is summarised below.

Duplicate samples of dried foods, fat-extracted if containing > 5% fat, are gelatinized in buffer with Termamyl (heat-stable alpha-amylase), and then enzymatically digested with protease and amyloglucosidase. Four volumes of 95% ethanol are then added to precipitate soluble polysaccharides. The total residue is obtained by filtration and then washed with 74% then 95% ethanol, and acetone. The residue is dried and then weighed. One of the duplicates is analysed for N and the other is incinerated at 525°C and ash determined. Total dietary fiber is the weight of the residue less the weight of the protein (N x 6.25) and ash. A modification to the method to allow soluble and insoluble dietary fiber to be measured has been suggested (Prosky et al., 1988).

The Englyst Method

This method aims to determine dietary fiber as NSP. It gives values for soluble and insoluble dietary fiber which, if required, are further separated into cellulose and non-cellulosic polysaccharides (NCP) with values for constituent sugars. The method has been described in full (Englyst et al., 1982; Englyst and Cummings, 1984; Englyst and Hudson, 1987; Englyst and Cummings, 1988; Englyst et al., 1990) and is summarized below.

Measurement of total NSP by gas-liquid chromatography (Fig. 1). Starch (including resistant starch) is dispersed with dimethyl sulfoxide and then hydrolysed by incubation with pancreatin and pullulanase. The starch-free residue is precipitated with ethanol, hydrolysed with sulfuric acid and neutral sugars released on hydrolysis are measured as alditol acetates by gas-liquid chromatography (GLC). Uronic acids are measured by colorimetry. Total NSP is then calculated as the sum of the individual neutral sugars and uronic acids.

Rapid colorimetric measurement of total NSP (Fig. 2). Up to and including the hydrolysis with sulfuric acid the procedure is identical to that for measurement of total NSP by GLC. After the sulfuric acid stage the hydrolysate is neutralized with sodium hydroxide, dinitrosalicylate reagent added and the mixture heated for 15 minutes at 100°C and absorbance for test samples and standards read at 530 nm. A value for total NSP is obtained directly as the reducing sugars measured in the hydrolysate. If desirable, separate values for glucose and uronic acids may be obtained.

Measurement of soluble and insoluble NSP. In the procedure for measurement of total NSP precipitation of soluble NSP with ethanol is replaced by a 1 hour extraction with phosphate buffer, pH 7, at 100°C. Otherwise the procedure for measurement of soluble and insoluble NSP is identical to that for total NSP and it is equally suitable with GLC or colorimetric measurement of constituent sugars. The values obtained represent insoluble NSP. Soluble NSP is obtained as the difference between total NSP and the insoluble NSP.

Separation of NSP into cellulose and non-cellulosic polysaccharides. If dispersion of cellulose with 12M sulfuric acid is omitted from the procedure for total NSP, and is replaced with direct hydrolysis with 2M sulfuric acid only the non-cellulosic polysaccharides (NCP) will be hydrolysed. A value for cellulose is obtained as the difference between the glucose content of total NSP and that of NCP.

The GLC procedure requires a gas chromatograph and some analytical skills. It can be carried out within one and a half days. The rapid colorimetric procedure is virtually a single tube method; it does not require any special skills or equipment other than a spectrophotometer, and can be carried out within an 8 hour working day, which is approximately half the time required by the AOAC procedure.

207

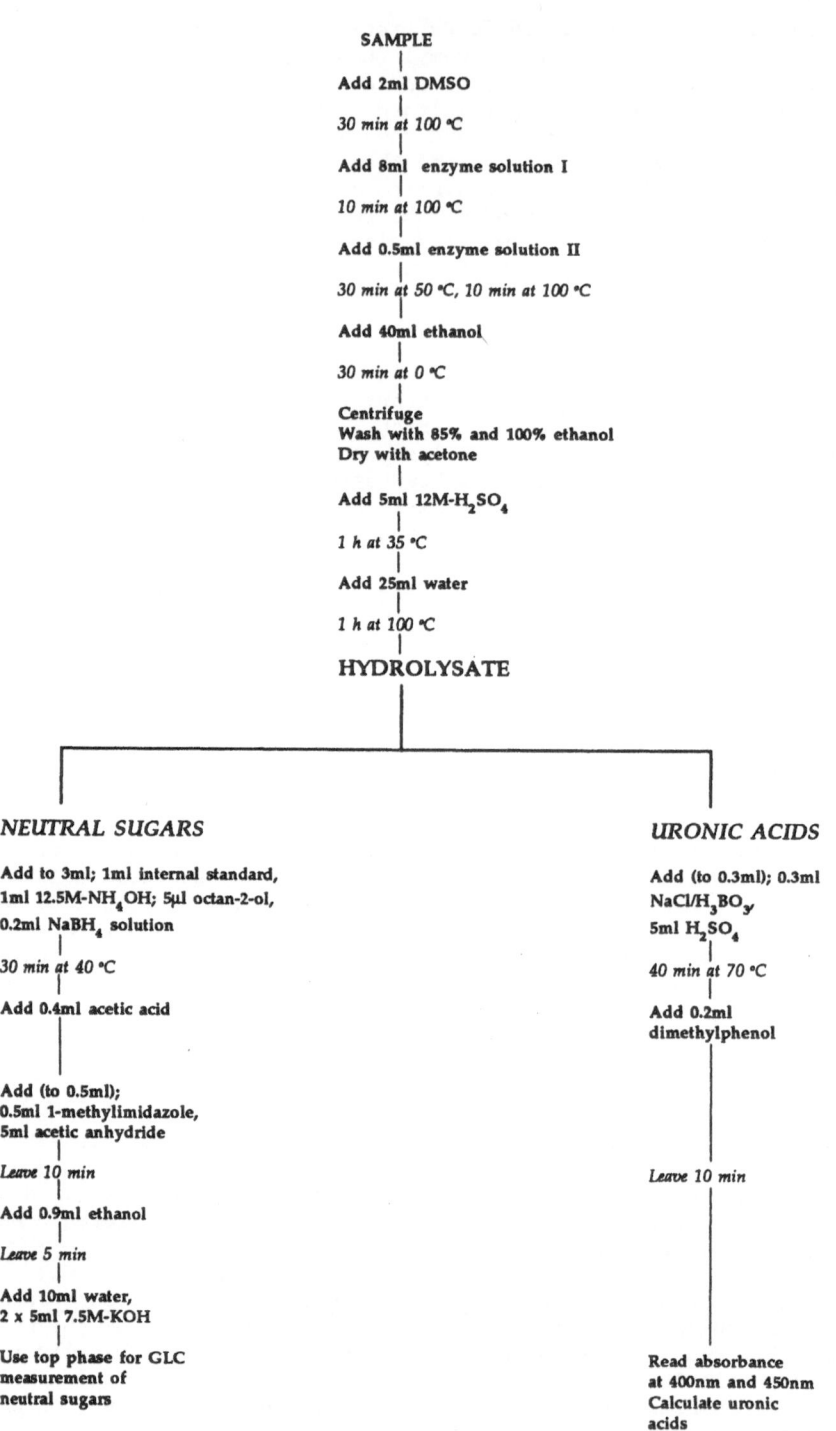

SAMPLE

Add 2ml DMSO

30 min at 100 °C

Add 8ml enzyme solution I

10 min at 100 °C

Add 0.5ml enzyme solution II

30 min at 50 °C, 10 min at 100 °C

Add 40ml ethanol

30 min at 0 °C

Centrifuge
Wash with 85% and 100% ethanol
Dry with acetone

Add 5ml 12M-H_2SO_4

1 h at 35 °C

Add 25ml water

1 h at 100 °C

HYDROLYSATE

NEUTRAL SUGARS

Add to 3ml; 1ml internal standard,
1ml 12.5M-NH_4OH; 5µl octan-2-ol,
0.2ml $NaBH_4$ solution

30 min at 40 °C

Add 0.4ml acetic acid

Add (to 0.5ml);
0.5ml 1-methylimidazole,
5ml acetic anhydride

Leave 10 min

Add 0.9ml ethanol

Leave 5 min

Add 10ml water,
2 x 5ml 7.5M-KOH

Use top phase for GLC
measurement of
neutral sugars

URONIC ACIDS

Add (to 0.3ml); 0.3ml
NaCl/H_3BO_3
5ml H_2SO_4

40 min at 70 °C

Add 0.2ml
dimethylphenol

Leave 10 min

Read absorbance
at 400nm and 450nm
Calculate uronic
acids

Total DF = Neutral sugars + Uronic acids.   Soluble DF = Total DF - Insoluble DF.

For measurement of insoluble DF, replace the 40ml of ethanol with 40ml of pH 7 buffer and extract for 30 min at 100°C.

Fig. 1   DETERMINATION OF TOTAL, SOLUBLE AND INSOLUBLE DIETARY FIBRE
BY THE ENGLYST GLC PROCEDURE

SAMPLE
|
Add 2ml DMSO
|
30 min at 100 °C
|
Add 8ml enzyme solution I
|
10 min at 100 °C
|
Add 0.5ml enzyme solution II
|
30 min at 50 °C, 10 min at 100 °C
|
Add 40ml ethanol
|
30 min at 0 °C
|
Centrifuge
Wash with 85% and 100% ethanol
Dry with acetone
|
Add 5ml 12M-H_2SO_4
|
1 h at 35 °C
|
Add 25ml water
|
1 h at 100 °C
|
Add to 1ml:
 0.1ml glucose solution
 1ml 4M-NaOH
 2ml Kit colour reagent
|
15 min at 100 °C
|
Add 25ml water
|
Read the absorbance at 530nm
Calculate total dietary fibre

Soluble DF = Total DF - Insoluble DF.

For measurement of insoluble dietary fibre, replace the 40ml of
ethanol with 40ml of pH 7 buffer and extract for 30 min at 100°C

Fig. 2   **ENGLYST PROCEDURE FOR RAPID DETERMINATION OF
TOTAL, SOLUBLE AND INSOLUBLE DIETARY FIBRE
BY COLORIMETRY**

DIETARY FIBER VALUES OBTAINED BY THE ENGLYST PROCEDURE

Detailed results for the dietary fiber in 178 fruit and vegetables and in 114 cereal and cereal products using the Englyst procedure have been published recently (Englyst et al., 1988, 1989). Table 1 gives dietary fiber values in a selection of plant foods. Values for total, soluble and insoluble dietary fiber can be obtained by both the GLC and the rapid colorimetric procedure. However by using GLC, fiber can be further separated into cellulose and non-cellulosic polysaccharides (NCP) and values for the individual constituent sugars obtained.

Table 1 shows that wheat bran has a high proportion of the total NSP in the form of cellulose and an insoluble arabino-xylan. Only a small amount of uronic acid is found. In white bread, cornflakes and potato, the NCP-glucose is low, corresponding to the low content of β-glucans in these products. This is a clear indication that starch has been completely removed in the analytical procedure and therefore does not interfere in this method even with high starch foods. Oats and rye have a higher proportion of total NSP in the form of soluble material, measured mainly as NCP-glucose and originating from β-glucan. It should be noted that rye also has a high content of soluble NCP-pentose.

Cereal products contain more NCP xylose than arabinose. This is in contrast to fruit and vegetables where arabinose, especially soluble NCP-arabinose, is the predominant pentose. Fruit and vegetables generally have a high content of uronic acids, originating from pectin. This is in contrast to cereals where in general only traces of uronic acids are found. NSP from both fruit and vegetables and cereal products may have a high content of cellulose but the amount is generally highest in fruit and vegetables.

The detailed information about dietary fiber obtained by GLC analysis is important in the interpretation of physiological studies and for epidemiological work, but is not always necessary for routine work and food labelling. Accurate values for total, soluble and insoluble dietary fiber can be obtained by the more rapid colorimetric method. It has been shown in a large number of samples that there is good agreement between dietary fiber measured by GLC and colorimetry (Englyst and Hudson, 1987; Englyst et al., 1987b). The GLC procedure is also valid for analysis of gut contents and fecal samples (Englyst, 1985). The rapid colorimetric assay has been developed and validated for foodstuffs only, especially plant products.

Comparison of Results obtained by the Englyst and the AOAC Procedures

In Table 2 a comparison is shown between dietary fiber values obtained by the Englyst and the AOAC procedures. The AOAC procedure gives results which are on average about 62% greater than those of the Englyst procedure. For the low fiber products, Cornflakes and Rice Krispies, the values obtained by the AOAC procedure are three times more than those obtained by the Englyst procedure. The amount by which individual results differ is highly variable and unpredictable. The ratio between the AOAC and the Englyst results ranges from 1.3 to 3.6.

There are two main reasons for the higher values by the AOAC procedure. Firstly the AOAC technique fails to remove completely all starch, especially that retrograded during food processing and sample preparation. In 1984 Asp and colleagues reported an increase in apparent dietary fiber content, as measured by the enzymatic-gravimetric technique, in wheat and rye breads when compared to the corresponding flours. The increase was mainly in the crumb and when GLC analysis was applied to the 'fiber' residue, it could largely be accounted for by the presence of glucose, originating from starch (Table 3).

210

Table 1. Detailed Results by GLC

NON-STARCH POLYSACCHARIDES (NSP) IN PLANT PRODUCTS

| Sample | Total g/100g Fresh Weight | Total g/100g Dry Weight | Cellulose | Non-Cellulosic Polysaccharides | | | | | | | |
				Rha	Fuc	Ara	Xyl	Man	Gal	Glu	U.Ac.
Wheat Bran, Arjuna											
Soluble NSP	3.8	4.2	-	t	t	1.3	1.8	0.1	0.2	0.6	0.2
Insoluble NSP	32.2	36.9	8.2	t	t	8.6	15.9	0.2	0.6	2.4	1.0
DM=0.88 Total NSP	36.0	41.1	8.2	t	t	9.9	17.7	0.3	0.8	3.0	1.2
White Bread, Mothers Pride											
Soluble NSP	0.9	1.6	-	t	t	0.5	0.8	t	0.1	0.2	t
Insoluble NSP	0.7	1.1	0.2	t	t	0.3	0.4	0.1	t	0.1	t
DM=0.60 Total NSP	1.6	2.7	0.2	t	t	0.8	1.2	0.1	0.1	0.3	t
Rye Bread, Imported											
Soluble NSP	3.7	6.7	-	t	t	1.7	3.2	t	0.2	1.6	t
Insoluble NSP	3.6	6.6	1.2	t	t	1.8	2.6	0.1	0.1	0.7	0.1
DM=0.55 Total NSP	7.3	13.3	1.2	t	t	3.5	5.8	0.1	0.3	2.3	0.1
Oatmeal, medium ground											
Soluble NSP	3.7	4.1	-	t	t	0.2	0.2	t	0.1	3.4	0.2
Insoluble NSP	2.9	3.3	0.4	t	t	0.7	1.1	0.1	0.1	0.8	0.1
DM=0.90 Total NSP	6.6	7.4	0.4	t	t	0.9	1.3	0.1	0.2	4.2	0.3
Cornflakes, Kelloggs											
Soluble NSP	0.4	0.4	-	t	t	t	0.2	t	t	0.1	0.1
Insoluble NSP	0.5	0.5	0.3	t	t	0.1	0.1	t	t	t	t
DM=0.96 Total NSP	0.9	0.9	0.3	t	t	0.1	0.3	t	t	0.1	0.1
Potato, Old, cooked Flesh only, boiled 20 mins. EM=0.82 DM=0.20											
Soluble NSP	0.7	3.8	-	0.1	t	0.3	t	t	1.9	0.6	0.9
Insoluble NSP	0.5	2.6	2.0	t	t	0.1	0.1	t	0.3	t	0.1
Total NSP	1.2	6.4	2.0	0.1	t	0.4	0.1	t	2.2	0.6	1.0
Beans, French, cooked Whole beans, no stalks, boiled 10 mins. EM=0.98 DM=0.10											
Soluble NSP	1.3	12.7	-	0.2	t	1.1	0.2	0.2	2.6	0.1	8.3
Insoluble NSP	1.8	17.7	11.1	0.1	t	1.2	1.5	1.2	1.5	0.5	0.6
Total NSP	3.1	30.4	11.1	0.3	t	2.3	1.7	1.4	4.1	0.6	8.9
Carrots, raw Flesh only, EM=0.76 DM=0.12											
Soluble NSP	1.4	11.4	-	0.7	t	1.7	t	0.1	3.0	t	5.9
Insoluble NSP	1.0	8.1	6.4	t	t	0.3	0.3	0.3	0.4	0.1	0.3
Total NSP	2.4	19.5	6.4	0.7	t	2.0	0.3	0.4	3.4	0.1	6.2
Cabbage, Winter January King, raw Inner leaves. EM=0.58 DM=0.12											
Soluble NSP	1.4	11.8	-	0.7	t	3.1	0.1	0.1	1.5	0.1	6.2
Insoluble NSP	1.5	12.6	8.0	t	t	1.5	0.9	0.4	1.2	t	0.6
Total NSP	2.9	24.4	8.0	0.7	t	4.6	1.0	0.5	2.7	0.1	6.8
Tomato, fresh, raw Flesh, skin and seeds EM=1.00 DM=0.06											
Soluble NSP	0.4	7.4	-	0.2	t	0.5	0.1	t	1.0	0.2	5.4
Insoluble NSP	0.7	11.4	7.5	0.1	t	0.4	0.9	1.3	0.7	0.2	0.3
Total NSP	1.1	18.8	7.5	0.3	t	0.9	1.0	1.3	1.7	0.4	5.7
Apples, Golden Delicious Raw with skin, no core EM=0.86 DM=0.14											
Soluble NSP	0.7	5.4	-	0.2	0.1	1.0	0.1	0.1	0.4	0.1	3.4
Insoluble NSP	1.0	7.1	4.2	0.1	0.1	0.7	0.7	0.2	0.6	0.2	0.3
Total NSP	1.7	12.5	4.2	0.3	0.2	1.7	0.8	0.3	1.0	0.3	3.7
Oranges, raw Flesh only, no pith, no pips EM=0.74 DM=0.14											
Soluble NSP	1.4	9.8	-	0.3	t	1.9	0.1	0.1	1.4	0.1	5.9
Insoluble NSP	0.7	5.2	3.4	t	t	0.3	0.5	0.3	0.4	t	0.3
Total NSP	2.1	15.0	3.4	0.3	t	2.2	0.6	0.4	1.8	0.1	6.2

Rha = Rhamnose;   Fuc = Fucose;   Ara = Arabinose;   Xyl = Xylose;
Man = Mannose;   Gal = Galactose;   Glu = Glucose;   U.Ac = Uronic Acid
DM = Dry matter;   EM = Edible matter

Table 2. Comparison of Dietary Fiber Values for Cereal Foods
          Obtained by the Englyst and AOAC Procedures

| | g/100g dry matter | |
Food	AOAC[a]	Englyst
White bread	6.6	3.4[b]
Graham bread	9.3	4.7[b]
Wholemeal rye bread	15.0	11.7[b]
Cornflakes	2.9	0.9[c]
Porridge oats	11.5	7.8[c]
Rice Krispies	1.8	0.5[c]

[a] From Jacobsen and Leth, 1987
[b] From Hansen and Englyst, 1988
[c] From Englyst et al., 1989

No starch was detectable in the dietary fiber fraction for any of the flours,
but all bread samples contained starch in this fraction, equivalent to
between 9 and 31% of analysed 'dietary fiber'. An interlaboratory study
between the Dunn in Cambridge and the Food Chemistry Department in Lund,
Sweden showed that cornflakes contained significant starch in the gravimetric
residue. 2.8g/100g cornflakes measured as resistant starch by the Englyst
technique, and 1.4g as resistant starch by GLC analysis of the gravimetric
residue (Englyst et al., 1987a). Other workers have also shown increases in
dietary fiber content of foods on cooking as measured by the AOAC method.
Ranhotra and Gelroth (1988a,b) recorded increases in total dietary fiber
(AOAC) of 26 and 34% during bread and cracker making respectively, and of
soluble fiber of 57 and 56%. They attributed these changes to the inclusion
of starch in the residue. Much of the difficulty reported by laboratories
using the AOAC method (Prosky et al., 1984,1988) relates to the handling of
starchy foods. Coefficients of variation of over 50% have been observed for
rice. The failure of the AOAC method with a commonly consumed food such as
rice and its complete inability to cope with a purified source such as
fiberform (Prosky et al., 1988) must give rise to anxiety about its use on a
wider range of foodstuffs, and especially processed and mixed diets.

Secondly the AOAC procedure includes a variety of unspecified materials
that are referred to as 'lignin'. When accurately measured, lignin is
quantitatively insignificant in most human foods. The inclusion of material
measuring as lignin, such as tanins and maillard and other degradation

Table 3. GLC Analysis of the Dietary Fiber from a Bread Crumb
          and the Corresponding Flour

| | Polysaccharides - mg/100mg original dry matter | | | | | | |
	Arabinose	Xylose	Mannose	Galactose	Glucose	Klason lignin	Total
Flour	1.3	1.0	tr.	0.2	1.7	0.4	4.5
Bread	1.1	0.8	0.3	0.2	3.5	1.0	6.8

From Johansson et al., 1984

Table 4. Dietary Fiber in g/100g dry matter

| | AOAC | NSP (Englyst) | Composition of NSP | | | | | |
			Ara	Xyl	Man	Gal	Glu	U.Acid
Cheeseburger								
Vegetable fraction	5.0	2.8	0.6	0.8	0.2	0.2	0.8	0.2
Non-vegetable fraction	3.8	0.1	-	-	t	t	0.1	-
Hamburger								
Vegetable fraction	5.4	2.8	0.6	0.9	0.2	0.2	0.7	0.2
Non-vegetable fraction	2.5	0.2	-	-	t	t	0.2	-
Big								
Vegetable fraction	6.1	4.8	0.9	1.2	0.3	0.4	1.5	0.5
Non-vegetable fraction	4.3	0.1	-	-	t	t	0.1	-
Fishburger								
Vegetable fraction	4.6	1.9	0.5	0.6	0.1	0.2	0.4	0.1
Non-vegetable fraction	5.1	1.9	0.4	0.5	0.1	0.2	0.7	t
Giant								
Vegetable fraction	5.2	2.2	0.5	0.6	0.1	0.3	0.5	0.2
Non-vegetable fraction	5.1	0.1	-	-	-	t	0.1	-

products formed during food processing, may however result in significant overestimation of dietary fiber. Johansson and colleagues (1984) have reported the Klason lignin content of white flour to be 0.4%, but 1% in bread made from this flour (Table 3). It seems unlikely that white flour contains 0.4% of lignin and clearly lignin cannot be produced by baking. The material measured as lignin is more likely to be maillard reaction products and such material is not a legitimate part of dietary fiber. The vulnerability of the AOAC gravimetric procedure to the effect of food processing has been well documented by Dysseler (1988). They have shown that when using the AOAC procedure a substantial amount of dietary fiber measuring substances was produced simply by heating a solution of milk powder to $100^{\circ}C$ or a glucose-glycine solution to $80^{\circ}C$. The amount formed was related to the duration of heating.

In another study in which the AOAC and Englyst methods were compared in five Belgian fast foods, much higher values for dietary fiber were again seen with the AOAC procedure. Average dietary fiber was 3.5g per serving AOAC and 1.3g by the Englyst method (Table 4). On separating the foods into plant and non-plant fractions 37% of the fiber measured by the AOAC procedure was found in the non-plant fraction. With the exception of fishburgers only traces of dietary fiber were measured in the non-vegetable portion when using the Englyst procedure. The detailed composition of the constituent sugars of NSP show that some cereal had been used in the manufacture of the fishfingers (Deelstra et al., 1989). The measurement of substantial amounts of 'fiber' in non plant material raises considerable doubts about the value of the AOAC method.

Slightly better agreement is found for total dietary fiber values for

Table 5. Total and Soluble Dietary Fiber (DF); a Comparison of the AOAC and Englyst Procedures

| | Total DF[*] | | % Soluble DF | |
	AOAC[**]	Englyst	AOAC[**]	Englyst
Carrot	33.2	19.5	19.8	41.5
Swede	50.5	35.8	31.2	46.0
Peas	26.3	24.0	17.1	30.0
Beans, green	32.8	29.9	28.0	42.8
Brussel sprouts	34.0	31.3	30.0	52.1

[*]
[**] g/100g dry matter
by Nyman et al., 1987

vegetables obtained by the AOAC and the Englyst procedures (Table 5), although AOAC results are again considerably higher. Furthermore a much higher proportion of total dietary fiber is measured as soluble material by the Englyst procedure.

A most important, but often overlooked factor, when determining the solubility of plant material is the pH during extraction. Table 6 shows values for NSP remaining insoluble after extraction of a carrot preparation, at various pHs for 1 hour and 100°C. Virtually all the uronic acids are extracted at pH 7 and for rhamnose, arabinose and galactose only one third of the material insoluble at pH 5 remains so at pH 7. The NSP-glucose in carrot consists mainly of cellulose which is highly insoluble and therefore not affected by pH. Overall therefore, following extraction at pH 7, 63% of carrot NSP is measured as soluble dietary fiber. However, if extracted at pH 5 only 29% is soluble. In the AOAC procedure soluble dietary fiber is extracted at pH 4.7 whilst in the Englyst procedure it is extracted at 7 which explains why the Englyst procedure gives higher values for soluble dietary fiber.

Table 6. Extraction of Non-Starch Polysaccharides at Different pH

| | Non-Starch Polysaccharides | | | | | | | |
	Rham-nose	Arabi-nose	Xylose	Mannose	Galac-tose	Glucose	Uronic Acids	Total
Carrot:								
Unextracted	2.63	6.19	1.13	1.34	9.22	21.27	31.98	73.96
Extracted at:								
pH 4	2.12	5.01	1.12	1.22	8.00	21.26	14.00	52.83
pH 5	2.16	5.36	1.08	1.16	7.67	20.43	14.56	52.42
pH 6	1.68	4.48	1.06	1.21	6.85	21.15	6.40	42.83
pH 7	0.40	1.70	1.04	1.24	2.49	20.64	0.10	27.61
pH 8	0.46	1.47	1.06	1.32	2.04	20.48	0.00	26.83

NSP sugars released from carrot by Seaman hydrolysis of unextracted materials and materials extracted at pH 4, 5, 6, 7 and 8. All values are given as mg/100mg alcohol insoluble materials.

214

During development of the Englyst procedure, various pH's were used for the extraction of soluble dietary fiber because of the pH effect on solubility (Table 6). A pH of 7 was selected for extraction of soluble dietary fiber because it included the major fraction of soluble material and is also the most physiological since small bowel pH is around 6-7 after a meal (Fordtran and Locklear, 1966; Englyst et al., 1982; Cummings et al., 1987). The pH used for extraction of soluble dietary fiber in the AOAC procedure has been optimized for the enzymatic hydrolysis of starch with amyloglucosidase. The marked effect of pH on the solubility of fiber is not addressed in papers describing this method. The result is a significant underestimation of soluble dietary fiber. It is important that any interpretation of soluble fiber data is done in the light of the method used. Ideally conditions of pH, time and temperature should be standardized.

COLLABORATIVE TRIALS OF METHODS FOR MEASUREMENT OF DIETARY FIBER

In view of the public demand for nutritional labelling for dietary fiber the UK Ministry of Agriculture, Fisheries and Food (MAFF) in 1982 decided to survey available methods and to organize collaborative trials to assess their utility.

Twenty-two analysts took part in the initial trial in which six methods including the Neutral Detergent Fiber (Robertson and Van Soest, 1977), the AOAC-Asp (Prosky et al., 1984) and the Englyst (Englyst et al., 1982) procedures were compared. The results from the initial trial were considered by an advisory panel consisting of representatives from the food industry, research establishments and MAFF. The panel agreed that the Englyst method appeared to be the most accurate and informative and that this method should be studied further (Cummings et al., 1985). Following the first trial the Englyst procedure was tested in a further two trials. In preparation for these trials, the method was simplified by eliminating the need for special equipment and reducing the time of analysis by more than 50%. Furthermore in the third MAFF trial a colorimetric method (Englyst and Hudson, 1987) for the measurement of constituent sugars was introduced as an alternative to GLC.

With the modifications incorporated the Englyst GLC method proved to be much more robust in the hands of the analysts than any previous method in the three trials. Good repeatability (for the 'blind duplicates') was observed (r 0.65 to 3.03) and much better reproducibility between laboratories was seen (R 1.16 to 3.83). The colorimetric Englyst procedure stood up well in the trial (Englyst et al., 1987b). The conclusion of the three MAFF studies has been that dietary fiber should be measured, for the purposes of food labelling, as non-starch polysaccharides by the Englyst procedure.

A fourth trial, organized by MAFF, testing the Englyst procedure (GLC and colorimetric) for the measurement of soluble and insoluble dietary fiber in cereal, fruit and vegetables is in progress.

At the time of the first MAFF collaborative trial an AOAC (Association of Official Analytical Chemists) method was being subjected to a parallel trial in the USA (Prosky et al., 1984). At this time it was a new method for measuring dietary fiber and is a combination of enzymatic and gravimetric procedures mainly based on the work of Asp (Asp and Johansson, 1981). The trial was designed to assess the reproducibility of the method amongst participating laboratories and not to select an accurate method since only the one technique was chosen for study. The absolute values obtained by the modified Asp procedure were not questioned by the organizers of the trial.

Following a further two trials the modified Asp method for total dietary fiber has been approved official final action by AOAC but the unacceptably high reproducibility coefficients for soluble dietary fiber require further study. Currently an EC trial is underway to compare the latest versions of the AOAC procedure and the Englyst colorimetric method. Further trials are planned in the USA in 1990, this time to compare the AOAC and Englyst procedures.

DIGESTIBILITY OF NSP AND STARCH IN THE SMALL INTESTINE OF MAN

Having developed a method for measurement of NSP and identified a fraction of starch resistant to pancreatic amylase in vitro (Englyst et al., 1982) the digestibility of these fractions has been addressed in a series of studies in man (Englyst and Cummings, 1985,1986,1987). The human gut is relatively inaccessible for study, but the ileostomy subject provides a unique opportunity to observe digestion in the small intestine.

In these studies NSP from white bread, cornflakes, oats, banana and potato was almost completely recovered in the ileostomy effluent (Table 7). The amount of NSP available for fermentation in the large intestine can therefore be measured simply by determining the amount of NSP in the food. The lack of digestibility of NSP was expected. It was however surprising to discover that a substantial amount of starch also escaped digestion in the small intestine.

Table 8 shows the result of studies of the digestibility of starch in ileostomy subjects. For cereal products more starch than that present as retrograded amylose in the food is recovered in the ileostomy effluent. Banana does not contain any retrograded amylose. However when feeding banana to ileostomists up to 89% of the starch escapes digestion. The starch in freshly cooked potato is well digested. Cooled cooked potato contains 3% retrograded amylose but 12% of the starch resists digestion and is recovered in ileostomy effluent. Reheated potato is digested better than the cooled potatoes but not as well as the freshly cooked.

It is clear from these and other studies (Cummings and Englyst, 1990) that a substantial amount of starch escapes digestion in the small intestine and that only a small proportion of this starch is in the form of retrograded amylose, originally defined as a type of resistant starch (Englyst et al., 1982). It is the retrograded amylose which the AOAC procedure aims to include as dietary fiber (Sievert and Pomeranz, 1989). Measurement of retrograded amylose alone will however not represent the amount of starch actually resisting digestion in the small intestine.

All starch resisting digestion in the small intestine, including that originally defined as resistant starch, is subject to bacterial fermentation

Table 7. Recovery of NSP from Ileostomy Effluent

	NSP (g)	
	Fed	Recovered
White Bread	2.3	2.3
Cornflakes	0.6	0.7
Oats	6.6	6.3
Bananas	2.1	2.0
Potato	3.3	3.4

From Englyst and Cummings, 1985,1986,1987

in the large intestine with the production of volatile fatty acids. Based on this physiological property all starch escaping digestion in the small intestine should be grouped together as resistant starch (RS). Reserving the term RS for the type of resistant starch fraction first identified (i.e. retrograded amylose) is misleading since this fraction is only a small proportion of the starch which actually escapes digestion in the small intestine.

FACTORS INFLUENCING STARCH DIGESTION

Most of the starch which reaches the colon is not totally resistant to pancreatic amylase, but for one reason or another its hydrolysis is retarded so that it is not completely digested during its passage through the small intestine. The reasons for this incomplete digestion may be separated into intrinsic factors (i.e. properties of the starchy food itself) and extrinsic factors.

Intrinsic Factors

Physical inaccessibility. When starch is contained within undisrupted plant structures such as whole or partly milled grains and seeds it will resist digestion. Here, cell walls entrap starch and prevent its complete swelling and dispersion (Wursch et al., 1986) thus delaying or preventing its hydrolysis with pancreatic amylase in the small intestine. In humans, whole or coarsely milled grains of wheat, maize and oats have been shown to elicit smaller plasma insulin responses than finely ground flours (Heaton et al., 1988). It has also been observed (Englyst, 1985) that after a meal of sweetcorn, peas and beans, up to 20% of fecal solids may be starch contained in recognizable, undigested food. Other examples of foods where the gross physical structure may retard hydrolysis, and result in incomplete digestion of starch in the small intestine, are parboiled white rice (Wolever et al., 1986) and spaghetti (Hermansen et al., 1986).

Resistant starch granules. In the plant, starch is present in granules where it exists in a partially crystalline form. The crystal structure of starch within the granule can be of X-ray diffraction pattern A, B or C as distinguished by Katz (1934). The X-ray pattern may be explained by envisaging a double-helical conformation which packs into a hexagonal array.

Table 8. Digestion of Starch in the Small Intestine of Man

	g Fed	(%RS)[*]	Starch % Recovered in Ileostomy Effluent
White Bread	62	(2)	3
Oats	58	(0)	2
Cornflakes	74	(4)	5
Banana	19	(0)	89
Freshly cooked Potato	45	(2)	3
Cooled, cooked Potato	47	(3)	12
Reheated Potato	47	(3)	8

[*] in form of retrograded amylose

From Englyst and Cummings, 1985, 1986, 1987

In the A-type starch, the helices are loosely packed and the centre of the array is occupied by a further helix whereas in the B-type the array is closely packed and the centre is occupied by water (Wu and Sarko, 1978a,b; Gidley, 1987).

The actual crystalline structure of the starch granule is suggested to depend on the chain length of amylopectin. The A, B and C type starches are reported to contain amylopectin molecules with average chain lengths of 23-29, 30-44 and 26-29 respectively (Hizukuri, 1985). A is the normal pattern for cereal starch granules, B is typical of potato, amylomaize and retrograded starch, and C (a combination of A and B patterns) is characteristic of certain pea and bean starches. In general starch granules showing X-ray diffraction patterns B or C tend to be the most resistant to pancreatic amylase, although the degree of resistance is dependent on the plant source (Fuwa et al., 1980). Cooking disrupts the granules and facilitates the hydrolysis of the starch contained within them. Thus the resistant crystal structure of potato starch granules is of no nutritional importance when potatoes are cooked before consumption. The highly resistant nature of banana starch granules is of more nutritional significance since bananas are often eaten raw.

Retrograded starch. Controlled drying of a heated starch gel can produce any of the A, B or C X-ray diffraction patterns depending on the temperature (Katz, 1937). Crystallization at higher temperature and lower water content will favor the A and lower temperature and high water content the B pattern. Therefore on cooling, gelatinized starchy foods will retrograde and develop X-ray diffraction pattern B (Katz, 1934,1937; Wu and Sarko, 1978a). During retrogradation, solubility of the starch molecule decreases and so does its susceptibility to hydrolysis by acid and enzymes (Sievert and Pomeranz, 1989).

Chain length and linearity are important factors affecting retrogradation. The longer the starch chains, the greater the number of interchain hydrogen bonds formed. The extent of crystalline bonding in amylopectin is limited by the branch length. Therefore amylopectin retrogrades to a lesser extent than amylose, and retrograded amylopectin is not so firmly bound as retrograded amylose (Sterling, 1978).

Pure amylose retrograded with the B pattern can be solubilized only by autoclaving (Wu and Sarko, 1978a). The resistance to dissolution is due to the extensive network of intra- and interhelical hydrogen bonds that stabilize the double helical structure of crystalline amylose. In mixtures, amylopectin has an inhibitory effect and the properties described for pure α-amylose are enhanced in proportion to the content of amylose (Wu and Sarko, 1978a). Dissolution of amylose and amylopectin gels require temperatures of 153 and 59°C respectively (Ring et al., 1987).

Retrograded starch may be separated into that redispersed at 100°C (mainly retrograded amylopectin) and that resistant to dispersion in boiling water (mainly retrograded amylose). Studies in man suggest that the mainly retrograded amylose fraction virtually completely resists digestion in the small intestine of man (Englyst and Cummings, 1985,1987).

Other factors. Other factors which are intrinsic to starchy foods have been shown to affect α-amylase activity in vitro. These include specific amylose-lipid complexes (Holm et al., 1983), native α-amylase inhibitors (Shainkin and Birk, 1970), and various non-starch polysaccharides which may have a direct effect on enzyme activity (Dunaif and Schneeman, 1981). However, it is not clear to what extent these factors affect the digestibility of starch in vivo.

<u>Extrinsic Factors</u>

The digestibility of starch within the small intestine is also influenced by a number of highly variable factors which are independent of crystallinity and physical form of the starchy food itself. These include the extent of chewing, transit time of the food along the small intestine, the concentration of amylase available for breakdown of the starch, the amount of starch and the presence of other food components which might retard enzymatic hydrolysis. It is not possible to predict with any certainty the extent to which such extrinsic factors may influence the digestion of a starchy food. Thus, as is the case for dietary fiber, a physiological definition is not a realistic basis for an in vitro analytical technique.

Despite limitations in terms of describing the in vivo digestibility of starch, an in vitro analytical technique may be used to obtain a reproducible measure of the potential of various starchy foods to be slowly or incompletely digested in the small intestine.

IN VITRO NUTRITIONAL CLASSIFICATION OF STARCH

Total starch (TS) is measured as the glucose released from a milled or homogenized sample, gelatinized at 100°C, dispersed with KOH and then incubated with pancreatin and amyloglucosidase.

The fractionation of starch into the subfractions shown in Table 9 is based on the incubation of starchy foods with pancreatin and amyloglucosidase under specified conditions.

<u>Rapidly digestible starch (RDS)</u>. This is measured as the glucose released from a sample after 20 minutes. It consists mainly of amorphous and dispersed starch.

<u>Slowly digestible starch (SDS)</u>. Slowly digestible starch is measured as the glucose released from a sample after 120 minutes minus RDS. It consists mainly of starch with X-ray diffraction patterns A and C, but it also includes any B type starch (granular or retrograded) hydrolysed within 120 minutes.

Table 9.   In Vitro Nutritional Classification of Starch

Type of Starch	Example of Occurrence	Probable Digestion in Small Intestine
RAPIDLY DIGESTIBLE STARCH	Freshly cooked starchy food	Rapid
SLOWLY DIGESTIBLE STARCH	Most raw cereals	Slow but complete
RESISTANT STARCH		
1. Physically inaccessible starch	Partly milled grain and seeds	Resistant
2. Resistant starch granules	Raw potato and banana	Resistant
3. Retrograded starch	Cooled, cooked potato, bread and cornflakes	Resistant

Resistant starch (RS). A value for the total amount of RS is obtained as TS - (RDS + SDS). RS1 may be measured as the increase in glucose released within 120 minutes obtained by milling or homogenizing the food sample. RS2 is measured as TS - RS3 and the glucose released after 120 minutes for a milled or blended sample. RS3 is measured as the starch in a milled sample that resists dispersion in boiling water and hydrolysis with pancreatic amylase and pullulanase as described previously (Englyst et al., 1982). It mainly consists of retrograded amylose.

The starch fractions described here are determined solely by properties inherent in the food sample and are not the result of variable physiological factors. The fractions therefore represent reproducible measurements by which starchy foods can be compared and validated within a nutritional framework.

DISCUSSION

The development of two techniques, which include different components of the diet in their measurement of dietary fiber has led to considerable confusion in research and food labelling of dietary fiber. The AOAC procedure includes part of the starch retrograded during food processing and also "lignin" measuring substances as dietary fiber. Higher values are therefore obtained by the AOAC procedure but the physiological effect of the material measured by this technique is unknown and may be different from that of dietary fiber measured solely as NSP. Different recommended dietary amounts (RDAs) are therefore required for dietary fiber measured by the two methods and no single conversion factor to NSP can be applied to values obtained by the AOAC procedure.

Starch included in the AOAC procedure consists mainly of retrograded amylose. One argument advanced is that this type of starch escapes digestion in the human small intestine as does dietary fiber. When however the digestibility of various types of starch is tested in man (Englyst and Cummings, 1985;1986;1987) it becomes clear that retrograded amylose, i.e. RS3, represents only a small proportion of the total amount of starch escaping digestion in the small intestine. There is no reason to single out RS3 as dietary fiber.

In addition to NSP and various types of starch, other substances escape digestion in the human small intestine; for example, some protein, lactose (in many ethnic groups), raffinose, stachyose and free sugars. This characteristic is therefore not unique to NSP and RS3 and is not a rational basis for the inclusion of any type of starch with dietary fiber.

It has been suggested that RS3 may have a fecal bulking effect and therefore should be included as dietary fiber. However, virtually all the food components that escape digestion in the small intestine, including the three types of RS, a substantial amount of protein and free sugars, are expected to have bulking effects, either directly or through fermentation and bacterial growth (Cummings and Englyst, 1987; Englyst et al., 1987c; Macfarlane and Englyst, 1986). Bulking effect is therefore not unique to NSP and RS3 and thus not a reason to include this type of starch as dietary fiber.

Current interest in RS lies in the fact that it is technically possible to increase the amount to 20% or more. We have produced white bread with more than 20% of the starch in the form of RS3. The amount of RS3 formed during processing of starchy foods is controlled by a number of factors including the source of starch, water content, pH, heating temperature and time, number of heating and cooling cycles, freezing and drying. The amount of RS3 in starchy foods is therefore in the hands of those who prepare foods, either at home or in the food industry.

The presently small amount of RS3 in foods is nutritionally insignificant. However if RS3 were to be included as dietary fiber the RS content of some types of food could be raised substantially, with the aim of boosting sales through a claim of high dietary fiber content. Inclusion of RS3 would also cause considerable problems for food labelling. For example, different apparent "dietary fiber" values will be obtained for the same food if it is hot, cold, frozen or dried. The apparent "dietary fiber" content will also change if test samples are heated, cooled, frozen or dried before analysis. One or more of these treatments are always used prior to analysis by the AOAC method. Dietary fiber values obtained by the AOAC method thus do not reflect the RS content in food as eaten, but include RS produced by storage and pretreatment of the sample. For example the RS measured in a cooled or dried potato sample will not be present in the potato when eaten hot.

The dependence of RS on food processing makes dietary fiber values obtained by the AOAC and other methods which include part of the RS as dietary fiber, highly difficult or impossible to defend. In contrast dietary fiber values measured as NSP are independent of food processing and storage and therefore easy to defend. The AOAC method does not measure a specified component of the diet defined either chemically or nutritionally.

There are other potential problems in having a dietary fiber method which includes RS. A single food may well yield a range of results for dietary fiber if RS is included in the value, depending on the method of cooking, cooling, freezing and other treatments in food preparation and storage. This makes the construction of food tables for dietary fiber difficult and the calculation of the fiber content of individual dishes prepared from the same food or mixed diets impossible. Dietary fiber measured as NSP is not affected by food processing so the amount of fiber in processed foods and mixed diets can be calculated from the recipe and food table values.

Starch slowly or incompletely digested in the small intestine may very well be an important food component in its own right. We have therefore proposed that starch for nutritional purposes be classified into rapidly digestible starch (RDS), slowly digestible starch (SDS) and resistant starch (RS) as shown in Table 9. Resistant starch represents all the starch that escapes digestion in the small intestine and not only the small proportion included as dietary fiber by the AOAC procedure. There is currently a complete reappraisal of the site and rate of starch digestion in man being undertaken in research centres around the world. This is a new and exciting development in digestive physiology and we believe it may have implications for health that are at least equal to the fiber story. No sound reason can be put forward for confusing these distinct, although overlapping areas of science by including some starch as dietary fiber. To do so would confuse the public as to the amount and importance of fiber and starch in food, and would not be in the best interests of the food industry.

In addition to the inclusion of retrograded starch the main reason for the high values for dietary fiber by the AOAC procedure is the inclusion of lignin measuring substances. Accurately measured lignin would be quantitatively insignificant, but the material measured as lignin may result in large overestimations of dietary fiber especially in processed foods.

Lignin is not a carbohydrate and its physiological significance (in animal studies) is very different from that of NSP. Lignin should therefore not be measured with NSP and it is not included in the analytical definition of dietary fiber (Englyst et al., 1987d). Lignin is quantitatively a minor component in the human diet and it is difficult to determine (Cummings et al., 1985). None of the present dietary fiber methods, such as the AOAC, that include lignin can in fact justify the values on strict chemical grounds. These methods simply isolate a collection of materials including Maillard reaction products, which are better referred to as substances measuring as lignin. Lignification of plant material may influence the properties of NSP but information about this aspect is only obtained when lignin is measured accurately and separately from NSP.

If it is shown in the future that lignin is an important food component a case must be made for its separate measurement. Values for NSP and lignin should never be grouped together since this will invalidate both measurements.

Another argument advanced in favor of the AOAC procedure is that it is simple and quicker than the Englyst procedure. In fact it is not so. The times required to complete the AOAC and the Englyst GLC procedure are very similar. The Englyst colorimetric procedure is however much quicker allowing total soluble and insoluble dietary fiber to be measured within an 8 hour working day or in approximately half the time required to measure dietary fiber by the AOAC procedure.

The Englyst procedure is virtually a single tube technique and no special skills or equipment, other than a spectrophotometer, are required for the rapid colorimetric version. A kit form is now available from Novo Biolabs.

The present recommendations by the UK Ministry of Agriculture, Fisheries and Food to the EEC and guidelines on dietary fiber labelling have been drafted following a series of three large multicentre collaborative trials of methods for measurement of dietary fiber carried out by the Ministry (Cummings et al., 1985; Englyst et al., 1987a,b) and as a result of recent research in the field. The conclusion of these trials has been that dietary fiber should be defined for the purposes of food labelling as non-starch polysaccharide (NSP).

REFERENCES

Asp, N-G., and Johansson, C-G., 1981, Techniques for measuring dietary fiber; principal aims of methods and a comparison of results obtained by different techniques. In: The Analysis of Dietary Fiber in Food (James, W.P.T., and Theander, O., Eds.). Marcel Dekker Inc., New York, pp.173-189.

Asp, N-G., and Johansson, C-G., 1984, Dietary fibre analysis. Nutr. Abstr. Rev. Clin. Nutr., 54: 735-752.

Asp, N-G., Johansson, C-G., Hallmer, H., and Siljestrom, M., 1983, Rapid enzymatic assay of insoluble and soluble dietary fiber. J. Agric. Food Chem., 31: 476-482.

Burkitt, D.P., 1969, Related disease - related cause. Lancet, 2: 1229-1231.

Cummings, J.H., 1980, Some aspects of dietary fibre metabolism in the human gut. In: Food and Health - Science and Technology (ed. Birch, G.A.). Applied Science Publishers, London, pp.441-458.

Cummings, J.H., and Englyst, H.N., 1987, Fermentation in the human large intestine and the available substrates. Am. J. Clin. Nutr., 45: 1243-1255.

Cummings, J.H., and Englyst, H.N., 1990, Starch fermentation in the human large intestine. Can. J. Physiol. Pharmacol. (in press).

Cummings, J.H., Englyst, H.N., and Wood, R., 1985, Determination of dietary fibre in cereals and cereal products - collaborative trials. Part I: Initial trial. J. Assoc. Off. Anal. Chem., 23: 1-35.

Cummings, J.H., Pomare, E.W., Branch, W.J., Naylor, C.P.E., and Macfarlane, G.T., 1987, Short chain fatty acids in human large intestine, portal, hepatic and venous blood. Gut, 28: 1221-1227.

Deelstra, H., Van Dael, P., Van Cauwenbergh, R., Englyst, H.N., and Cummings, J.H., 1989, Determination of dietary fiber in total diets. A comparison of the AOAC and the Englyst method. Proceedings of the 5th European Conference on Food Chemistry, Versailles, France, pp. 137-141.

Dunaif, G., and Schneeman, B.O., 1981, The effect of dietary fiber on human pancreatic activity in vitro. Am. J. Clin. Nutr., 34: 1034-1035.

Dysseler, P., 1988, Impact of heat treatment on dietary fibre values. Paper presented to the EC Meeting on Dietary Fibre: Definition and Methods of Analysis, Cambridge, UK, September 1988.

Englyst, H., 1981, Determination of carbohydrate and its composition in plant materials. In: The Analysis of Dietary Fiber in Food (James, W.P.T., and Theander, O., Eds.). Marcel Dekker Inc., New York, pp. 71-93.

Englyst, H.J.N., 1985, "Dietary polysaccharide breakdown in the gut of man," Ph.D. Thesis, University of Cambridge, U.K.

Englyst, H.N., and Cummings, J.H., 1984, Simplified method for the measurement of total non-starch polysaccharides by gas-liquid chromatography of constituent sugars as alditol acetates. Analyst, 109: 937-942.

Englyst, H.N., and Cummings, J.H., 1985, Digestion of the polysaccharides of some cereal foods in the human small intestine. Am. J. Clin. Nutr., 42: 778-787.

Englyst, H.N., and Cummings, J.H., 1986, Digestion of the carbohydrates of banana (Musa paradisiaca sapientum) in the human small intestine. Am. J. Clin. Nutr., 44: 42-50.

Englyst, H.N., and Cummings, J.H., 1987, Digestion of polysaccharides of potato in the small intestine of man. Am. J. Clin. Nutr., 45: 423-431.

Englyst, H.N., and Cummings, J.H., 1988, An improved method for the measurement of dietary fibre as the non-starch polysaccharides in plant foods. J. Assoc. Off. Anal. Chem., 71: 808-814.

Englyst, H.N., and Hudson, G.J., 1987, Colorimetric method for routine measurement of dietary fibre as non-starch polysaccharides. A comparison with gas-liquid chromatography. Food Chemistry, 24: 63-76.

Englyst, H.N., Bingham, S.A., Runswick, S.A., Collinson, E., and Cummings, J.H., 1988, Dietary fibre (non-starch polysaccharides) in fruit, vegetables and nuts. J. Hum. Nutr. Dietet., 1: 247-286.

Englyst, H.N., Bingham, S.A., Runswick, S.A., Collinson, E., and Cummings, J.H., 1989, Dietary fibre (non-starch polysaccharides) in cereal products. J. Hum. Nutr. Dietet., 2: 253-271.

Englyst, H.N., Cummings, J.H., and Wood, R., 1987a, Determination of dietary fibre in cereals and cereal products - collaborative trials. Part II: Studies of a modified Englyst procedure. J. Assoc. Publ. Analysts, 25: 59-71.

Englyst, H.N., Cummings, J.H., and Wood, R., 1987b, Determination of dietary fibre in cereals and cereal products - collaborative trials. III. Study of further simplified procedures. J. Assoc. Publ. Analysts, 25: 73-110.

Englyst, H.N., Hay, S., and Macfarlane, G.T., 1987c, Polysaccharide breakdown by mixed populations of human faecal bacteria. FEMS Microbiol. Ecol., 95: 163-171.

Englyst, H.N., Quigley, M., Hudson, G.J., and Cummings, J.H., 1990, Measurement of non-starch polysaccharides (dietary fibre) by GLC, HPLC and colorimetry. (in press).

Englyst, H.N., Trowell, H.W., Southgate, D.A.T., and Cummings, J.H., 1987d, Dietary fiber and resistant starch. Am. J. Clin. Nutr., 46: 873-874.

Englyst, H., Wiggins, H.S., and Cummings, J.H., 1982, Determination of the non-starch polysaccharides in plant foods by gas-liquid chromatography of constituent sugars as alditol acetates. Analyst, 107: 307-318.

Fordtran, J.S. and Locklear, T.W., 1966, Ionic constituents and osmolality of gastric and small intestinal fluids after eating. Am. J. Dig. Dis., 11: 503-521.

Furda, I., 1981, Simultaneous analysis of soluble and insoluble dietary fiber. In: The Analysis of Dietary Fiber in Food (James, W.P.T., and Theander, O., Eds.). Marcel Dekker Inc., New York, pp. 163-172.

Fuwa, H., Takaya, T., and Sugimoto, Y., 1980, Degradation of various starch granules by amylases. In: Mechanisms of Saccharide Polymerization and Depolymerization (Marshall, J.J., ed.), Academic Press, New York, pp.

Gidley, M.J., 1987, Factors affecting the crystalline type (A-C) of native starches and model compounds: a rationalisation of observed effects in terms of polymorphic structures. Carbohydrate Res., 161: 301-304.

Hansen, I., and Englyst, H.N., 1988, Dietary fibre in some Scandinavian breads. Naringsforskning Arg. 32: 108-112.

Heaton, K.W., Marcus, S.N., Emmett, P.M., and Bilton, C.H., 1988, Particle size of wheat, maize and oat test meals: effects on plasma glucose and insulin responses and on the rate of starch digestion in vitro. Am. J. Clin. Nutr., 47: 675-682.

Hermansen, K., Rasmussen, O., Arnfred, J., Winther, E., and Schmitz, O., 1986, Differential glycaemic effects of potato, rice and spaghetti in Type 1 (insulin dependent) diabetic patients at constant insulinaemia. Diabetalogia, 29: 358-361.

Hipsley, E.H., 1953, Dietary 'fibre' and pregnancy toxaemia. Br. Med. J., 2: 420-422.

Hizukuri, S., 1985, Relationship between the distribution of the chain length of amylopectin and the crystalline structure of starch granules. Carbohydrate Res., 141: 295-306.

Holm, J., Bjorck, I., Ostrowska, S., Eliasson, A-C., Asp, N-G., Larsson, K., and Lundquist, I., 1983, Digestibility of amylose-lipid complexes in-vitro and in-vivo. Starch/Starke 35: 294-297.

Jacobsen, J.S., and Leth, T., 1987, Overvagningssystem for neringsstoffer, brod og cerealier. Centrallaboratoriets afdeling A/Aalborg Landsdelslaboratorium Levnedsmiddelstyrelsen Publication No. 14.

Johansson, C-G., Siljestrom, M., and Asp, N-G., 1984, Dietary fibre in bread and corresponding flours - formation of resistant starch during baking. Z. Lebensm. Unters Forsch., 179: 24-28.

Katz, J.R., 1934, X-ray investigation of gelatinization and retrogradation of starch and its importance for bread research. Bakers Weekly, 81: 34-37.

Katz, J.R., 1937, The amorphous part of starch in fresh bread, and in fresh pastes and solutions of starch. Recl. Trav. chim. Pays-Bas Belg., 18: 55-59.

Macfarlane, G.T., and Englyst, H.N., 1986, Starch utilization by the human large intestinal microflora. J. Appl. Bacteriol., 60: 195-201.

Nyman, M., Palsson, K-E., and Asp, N-G., 1987, Effects of processing on dietary fibre in vegetables. Lebensm. Wiss. u. Technol., 20: 29-36.

Prosky, L., Asp, N-G., Furda, I., Devries, J.W., Schweizer, T.F., and Harland, B.F., 1984, Determination of total dietary fiber in foods, food products and total diets: interlaboratory study. J. Assoc. Off. Anal. Chem., 67: 1044-1052.

Prosky, L., Asp, N-G., Furda, I., Devries, J.W., Schweizer, T.F., and Harland, B.F., 1985, Determination of total dietary fiber in foods and food products: collaborative study. J. Assoc. Off. Anal. Chem., 68: 677-679.

Prosky, L., Asp, N-G., Schweizer, T.F., Devries, J.W., and Furda, I., 1988, Determination of insoluble, soluble and total dietary fiber in foods and food products: interlaboratory study. J. Assoc. Off. Anal. Chem., 71: 1017-

Ranhotra, G. and Gelroth, J., 1988a, Soluble and total dietary fiber in white bread. Cereal Chem., 65: 155-156.

Ranhotra, G. and Gelroth, J., 1988b, Soluble and insoluble fiber in soda crackers. Cereal Chem., 65: 159-160.

Ring, S.G., Colonna, P., I'Anson, K.J., Kalichevsky, M.T., Miles, M.J.,
    Morris, V.J., and Orford, P.D., 1987, The gelation and crystallisation
    of amylopectin. Carbohydrate Res., 162: 277-293.
Robertson, J.B., and Van Soest, P.J., 1977, 69th Annual Meeting of the
    American Society of Animal Science, Madison, WI.
Schweizer, T.F., and Wursch, P., 1979, Analysis of dietary fibre. J. Sci.
    Food Agric., 30: 613-619.
Shainkin, R., and Birk, Y., 1970, α-Amylase inhibitors from wheat. Isolation
    and characterization. Biochim. Biophys. Acta, 221: 502-513.
Sievert, D. and Pomeranz, Y., 1989, Enzyme-resistant starch. I.
    Characterization and evaluation by enzymatic, thermoanalytical and
    microscopic methods. Cereal Chem., 66: 342-347.
Southgate, D.A.T., 1969, Determination of carbohydrates in foods. II.
    Unavailable carbohydrates. J. Sci. Food. Agric., 20: 331-335.
Southgate, D.A.T., 1980, What is dietary fibre? Food Technol. N.Z., 15: 7-9.
Sterling, C., 1978, Textural qualities and molecular structure of starch
    products. J. Texture Studies 9: 225-255.
Trowell, H., 1972, Ischaemic heart disease and dietary fibre. Amer. J. Clin.
    Nutr., 25: 926-932.
Trowell, H., 1985, Dietary fibre; a paradigm. In: Dietary Fibre, Fibre-
    Depleted Foods and Disease (Trowell, H.W., Burkitt, D., and Heaton,
    K.W., Eds.). Academic Press, London, pp. 1-20.
Wolever, T.M.S., Jenkins, D.J.A., Kalmusky, J., Jenkins, A., Giordano, C.,
    Giudici, S., Josse, R.G., and Wong, G.S., 1986, Comparison of regular
    and parboiled rices: explanation of discrepancies between reported
    glycemic responses to rice. Nutrition Research, 6: 349-357.
Wu, H-C., and Sarko, A., 1978a, The double-helical molecular structure of
    crystalline B-amylose. Carbohydrate Res., 61: 7-25.
Wu, H-C., and Sarko, A., 1978b, The double-helical molecular structure of
    crystalline A-amylose. Carbohydrate Res., 61: 27-40.
Wursch, P., Del Vedovo, S., and Koellreutter, B., 1986, Cell structure and
    starch nature as key determinants of the digestion rate of starch in
    legume. Am. J. Clin. Nutr., 43: 25-29.

DELIMITATION PROBLEMS IN DEFINITION AND ANALYSIS OF DIETARY FIBER

Nils-Georg Asp

Applied Nutrition
Chemical Center, Lund University
Box 124, S-221 00 Lund, Sweden

INTRODUCTION

The debate on dietary fiber definition and analysis has been much focused recently on "resistant starch" and lignin (1-3). However, a number of other delimitation problems have to be considered as well, including several more or less enzyme resistant fractions of starch.

Precipitation with 78-80 percent ethanol (v/v) is used in all current methods for analysis of dietary fiber, in order to separate fiber poly-saccharides from low molecular weight sugars and starch hydrolysis products. Although polysaccharides are generally insoluble in alcohols at this strength, the delimitation is obviously an arbitrary one. The cut-off for linear polysaccharides is usually considered to be around DP (degree of polymerization) 10, but some polysaccharides may be soluble in spite of a much higher DP (4).

In gravimetric methods the dietary fiber residue is associated with non-fiber material such as undigestible protein and minerals. Current enzymic gravimetric methods employ correction for protein (Kjeldahl N x 6.25) and ash to be compatible with the most generally accepted fiber definition as polysaccharides and lignin, that are not digested in the human small intestine (5). Undigestible Maillard reaction products will also appear in the gravimetric residues, and the nitrogen content of these products determines to what an extent they will be accounted for in the protein correction.

Gravimetric residues may also contain some fiber associated organic substances such as tannins, saponins, phytates and other organic salts, that must be taken into consideration when discussing the delimitation of dietary fiber - conceptually and analytically.

STARCH

Until recently, starch was believed to be completely digestible in the small intestine. Therefore, undigestible polysaccharides were thought to be equivalent to non-starch polysaccharides, and the dietary fiber definition was often expressed as non-starch polysaccharides plus lignin. The discovery of an enzyme resistant starch fraction called resistant starch (6), that was shown to pass the small intestine undigested (7,8) made the demarcation between dietary fiber and starch less clearcut, and initiated the debate on a physiologically meaningful delimitation.

Human experiments with various techniques have shown that resistant starch as defined originally accounts for only a fraction of the total malabsorbed starch (9). There are a number of other starch fractions that need to be taken into account. These are listed in Table 1. The term "resistant starch" (RS) is now often used for all malabsorbed starch, but will be restricted here as originally defined (see below).

Starch with covalent bindings not found in native starches

In modified food starches, groups are introduced to improve for instance gel formation and stability. This can be expected to make such starches more or less enzyme resistant, both in vivo and in vitro. Heat processing may also create new covalent bindings by formation of anhydro sugars that add to starch molecules (10).

Table 2 shows enzymic starch and dietary fiber determinations in native and variously modified potato starch (11). Enzymic starch analysis using Termamyl and amyloglucosidase, and glucose oxidase assay of liberated glucose gave only 71 percent recovery in acetyl distarch phosphate (ADSP) and 58 percent in hydroxypropyl distarch phosphate (HPDSP). Pretreatment with 2 M KOH for 30 min at room temperature, however, increased the recovery of ADSP to 97 percent, but did not influence that of HPDSP. Analysis with the enzymic, gravimetric method of Asp et al. (12) did not show significant amounts of dietary fiber in any of the samples.

Dry heat treatment of wheat starch at 180°C for 2 or 4 hours diminished the analyzed starch both without and with previous KOH treatment, as shown in Table 3 (13). Also in this case no "dietary fiber" was formed as analyzed with the enzymic, gravimetric method.

Table 1.   Starch fractions presenting delimitation
            problems in fiber definition and analysis.

1. Starch with covalent bindings not found in native
   starch
   a. Modified food starches
   b. Starch reacted at heat processing

2. Resistant starch, i.e. starch available to amylases
   only after solubilization in DMSO or alkali
   (retrograded amylose)

3. Amylose-lipid complexes
   (not determined as fiber with methods using Termamyl
   or prolonged incubation with excess of other amylases)

4. Ungelatinized starch
   a. Highly digestible in vivo (e.g. cereal starches)
   b. Poorly digestible in vivo (e.g. potato, banana,
   high amylose corn starch)

5. Physically inaccessible starch
   a. Large particles of foods
   b. Intact cellular structure (leguminous seeds)

Thus, modification of starches - both intentionally and by heat treatment - can diminish starch estimates by enzymic methods without formation of dietary fiber. Obviously, the enzyme resistant fragments resisting amylase degradation are small enough to be soluble in 78 percent ethanol and therefore not recovered as dietary fiber.

On the other hand, processing of wheat by extrusion cooking, autoclaving, steaming-flaking and roller drying at realistic conditions did not change starch or dietary fiber content significantly (14).

Table 2. Analysis of starch and dietary fiber in native and modified, raw potatoe starches (g/100 g d.m.) (Östergård, Björck, Gunnarsson, 1988)

Sample	Starch determination		Dietary fiber[3]
	Enzymic[1]	D:o + NaOH[2]	
Potato starch	97	99	< 0.4
D S P	95	100	< 0.4
A D S P	71	97	< 0.4
H P D S P	58	60	< 0.4

DSP =Distarch phosphate, ADSP =Acetylated distarch phosphate, HPDSP =Hydroxypropyl distarch phosphate
1 Holm et al, 1986 (Termamyl + amyloglucosidase)
2 Pretreatment with 2 M KOH, 30 min, room temp.
3 Enzymic, gravimetric method of Asp et al, 1983

Table 3. Analysis of starch and dietary fiber in heat treated wheat starch (g/100 g d.m.) (Siljeström, Björck, Westerlund, 1989)

Duration of heating at 180°C	Starch determination		Dietary fiber[3]
	Enzymic[1]	D:o + KOH[2]	
0 h	97.4	98.2	-
2 h	95.5	97.4	-
4 h	90.6	94.3	0.3

1 Holm et al, 1986 (Termamyl + amyloglucosidase)
2 Pretreatment with 2 M KOH, 30 min, room temp.
3 Enzymic, gravimetric method of Asp et al, 1983

Resistant starch

The term "resistant starch" (RS) was originally defined by Englyst et al. (6) as a starch fraction that resists digestion with extensive amylase/ pullulanase treatment as used in the Englyst method for dietary fiber analysis. It can be hydrolyzed after solubilization with KOH or dimethyl- suphoxide (DMSO).

There is now evidence that RS is strongly retrograded amylose with intermediate chain length (DP around 60-65) (15,16). Both animal and human experiments have shown that RS defined in this way passes the small intestine. Thus, this fraction represents the minimum amount of mal-absorbed starch, and it is readily fermented in the large bowel. Thus, RS has physiological properties generally attributed to dietary fiber.

RS is recovered as dietary fiber in both gravimetric and GLC methods not using any DMSO solubilization step. It can be determined easily in gravimetric residues as an important step in dietary fibre characteri-zation.

Amylose-lipid complexes

Amylose-lipid complexes are relatively resistant to amylase degradation in vitro. Use of Termamyl at elevated temperature, however, as well as prolonged incubation with excess amylase, hydrolyse the complexes. They seem to be more or less completely absorbed in the small intestine, although at a significantly slower rate than free amylose (17).

Ungelatinized starch

Most starchy foods are heat treated before consumption, but the starch is not always fully gelatinized. Cereal starches, however, seem to be absorbed extensively also without gelatinization. On the other hand, potato starch and banana starch are incompletely absorbed in the raw state (18).

Analysis methods for starch and dietary fiber include heating or alkali solubilization steps, in order to ensure complete starch degrada-tion. Therefore, it is not possible to differentiate digestible and un-digestible fractions of raw starch granules.

Physically inaccessible starch

The importance of physically inaccesible starch, either due to large food particles or to intact cellular structures, has been demonstrated for the glycemic response (19). However, little is known regarding the effect on total digestibility. Methods need to be developed for simulating in vivo digestibility.

ALCOHOL SOLUBLE SACCHARIDES

The raffinose family of α-galactosides in leguminous seeds is a well recognized group of undigestible oligosaccharides, that are alcohol soluble and therefore not determined as dietary fiber.

Fructans have recieved considerable interest recently. The Jerusalem artichok contains 20-29 percent and onions 65 percent fructans with DP 3 to more than 100 (20). Wheat and rye contain about 1 percent fructans, that are undigestible in the rat small intestine (21). Half of these cereal fructans have DP 5 or less (22).

The low molecular weight fructans are readily soluble in alcohol and therefore not determined as dietary fiber. High molecular weight fractions that precipitate in 78 percent alcohol would be recovered as dietary fiber with gravimetric methods. GLC methods, however, do not determine fructose as a dietary fiber monomer.

Analysis of sugar beet fiber

Analysis of a sugar beet fiber preparation (Fibrex[R], Fibrex AB, Arlöv, Sweden) revealed peculiar delimitation problems related to an exceptionally high alcohol solubility of some dietary fiber poly-saccharides. When analyzing dietary fiber with the method of Asp et al (12) and summing up this fiber estimate with protein, fat, ash and sucrose analyses, a deficit of 8-11 percent was seen (Table 4). The AOAC method (23) gave similar results, indicating that the acid pepsin step in the method of Asp et al. did not cause losses. Exclusion of Termamyl in the analysis also did not change the result.

Increase of the ethanol concentration from 78 to 82 or 84 percent increased the dietary fiber estimate 1 to 1.5 percent, whereas increased precipitation time did not change the fiber value.

The filtrate after 78 percent ethanol precipitation in the gravi-metric dietary fiber analysis according to Asp et al. was evaporated and hydrolyzed according to Theander et al. (24). GLC determination showed typical dietary fiber monomers, and the dominating components were arabinose (1.7-2.1 percent of the analyzed Fibrex preparation), galactose (1.5-3.5 percent) and uronic acids (1.0-1.9 percent).

To estimate the molecular size distribution of the alcohol soluble polysaccharides, the filtrate was separated on a Sephadex G-75 column (Fig 1). A considerable fraction of the total carbohydrates eluted early, corresponding to DP more than 100. The pentoses eluted mainly in the early fraction A, whereas the hexoses were recovered only in the late low-molecular weight fraction B. The uronic acids were distributed throughout the chromato- gram. Table 5 shows the monomeric composition of the high-molecular weight fraction (A). Arabinose constituted 51 percent,

Table 4. Composition of 4 samples of sugar beet fiber (Fibrex)
(g/100 g d.m.)

Sample	TDF (Asp et al)	Protein (Nx6.25)	Fat	Ash	Sucrose	Sum
1	72.3	10.6	0.3	3.5	3.4	89.8
2	71.4	12.2	0.3	3.0	5.1	91.7
3	71.5	10.8	0.3	3.1	3.7	89.1
4	70.8	11.2	0.3	3.0	5.0 9	90.0

1, Unmilled; 2, Coarsly milled; 3, Intermediately milled;
4, Finely milled brands.

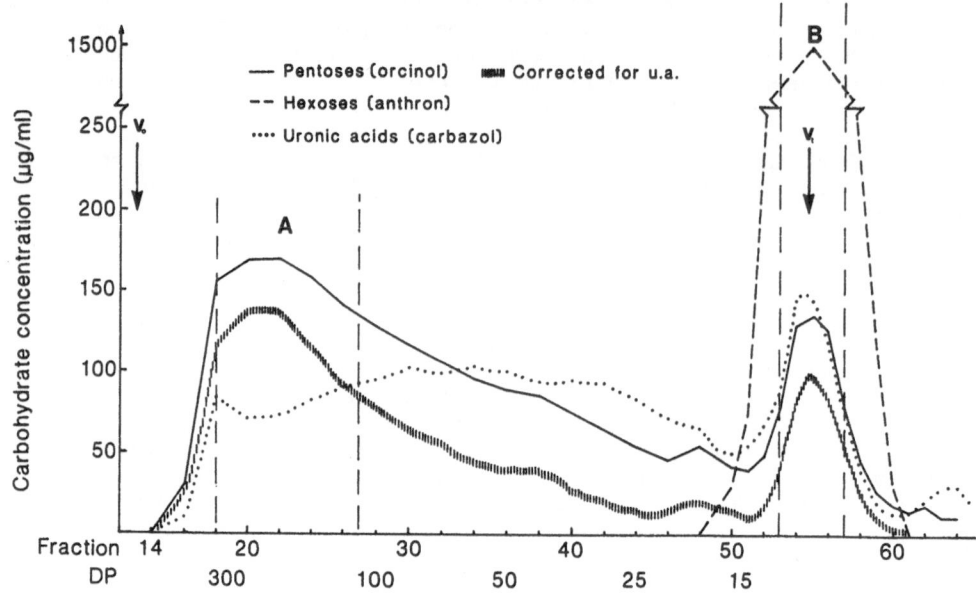

Fig. 1. Gel permeation chromatogram on Sephadex G-75 (column dimensions 1 m x 1.5 cm, elution with 0.05 M sodium phosphate pH 7.0) of the filtrate after alcohol precipitation of soluble dietary fiber according to Asp et al. (12). The estimated degree of polymerization is indicated below the graph.

representing the highly branched araban present as a main component. Uronic acids accounted for 36 percent, showing high molecular weight pectic substances in the filtrate.

The samples were also analyzed with the Englyst method, and these results are compared with the gravimetric estimates in Table 6. When taking into account that the material contains about 4 percent Klason lignin (data from producer), there was an excellent agreement between the two methods regarding insoluble fiber. The soluble fiber estimates, on the other hand, were on an average 9.4 percent higher with the Englyst method. This is the only material so far reported to give consistently higher dietary fiber estimates with the Englyst method than with an enzymic gravimetric method.

Table 5. Relative monomeric composition of high molecular weight fraction (A) of material from beet fiber soluble in 78 % ethanol

Arabinose	51 %
Xylose	5 %
Galactose	3 %
Glucose	4 %
Uronic acids	36 %

232

Table 6. Comparison of enzymic gravimetric (Asp et al) and GLC (Englyst)
assay on 4 samples of Fibrex (same as in Table V).

Sample	Soluble fiber		Insoluble fiber		Total fiber	
	Asp	Englyst	Asp	Englyst	Asp	Englyst
1	20.2	34.4	52.1	50.0	72.3	84.4
2	22.5	30.1	48.9	45.5	71.4	75.6
3	21.5	28.1	50.0	47.8	71.5	75.9
4	21.8	30.1	49.0	43.0	70.8	73.1

Since both types of methods employ alcohol precipitation to recover
soluble components, one step used in the enzymic, gravimetric assay must
have made some components more soluble in alcohol. Probably the proteolytic treatment, released polysaccharides from protein complexes. The
highly branched araban and also pectic substances in this material
obviously have quite high solubility in alcohol.

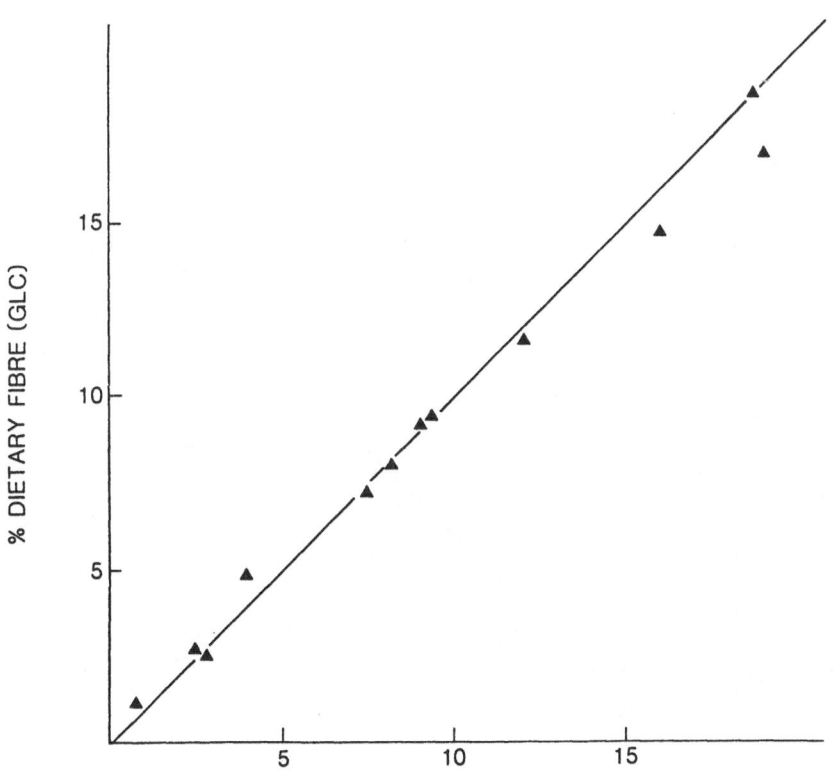

Fig. 2.   Comparison of enzymic gravimetric dietary fiber assay according
to Asp et al (12) and GLC determination according to Theander et
al. method C (24) in flours of wheat, barley, rice, maize and
sorghum with low or high extraction rate. Data form ref. (28).

OTHER DELIMITATIONS

It is often claimed that Maillard reaction products may contribute to gravimetric fiber estimates. When protein corrections are employed, however, this is generally not the case, since even advanced Maillard reaction products retain most of the nitrogen of the reacted proteins (25). Klason lignin estimates, on the other hand, increase due to Maillard reactions (26). In normally processed foods, however, such a contribution is quantitatively insignificant.

Tannins have been reported to account for a significant portion of the dietary fiber residue, when carob pods are analyzed with enzymic, gravimetric methods (27). In ordinary foods this contribution is very small. Whenever tannins are expected, the gravimetric residues can be analyzed for their contribution.

Phytate and organic salts may give some, but generally insignificant contributions to gravimetric fiber determinations corrected for protein and ash. Whenever quatitatively important, they can be analyzed in the residue, or extracted from the raw material before analysis.

AGREEMENT BETWEEN GRAVIMETRIC AND GLC METHODS

Most of the delimitation problems mentioned above are common to different types of fiber analysis methods and some are unique to the gravimetric approach. Fig. 2 and 3 compares dietary fiber estimates with the enzymic, gravimetric method of Asp et al. and component analysis

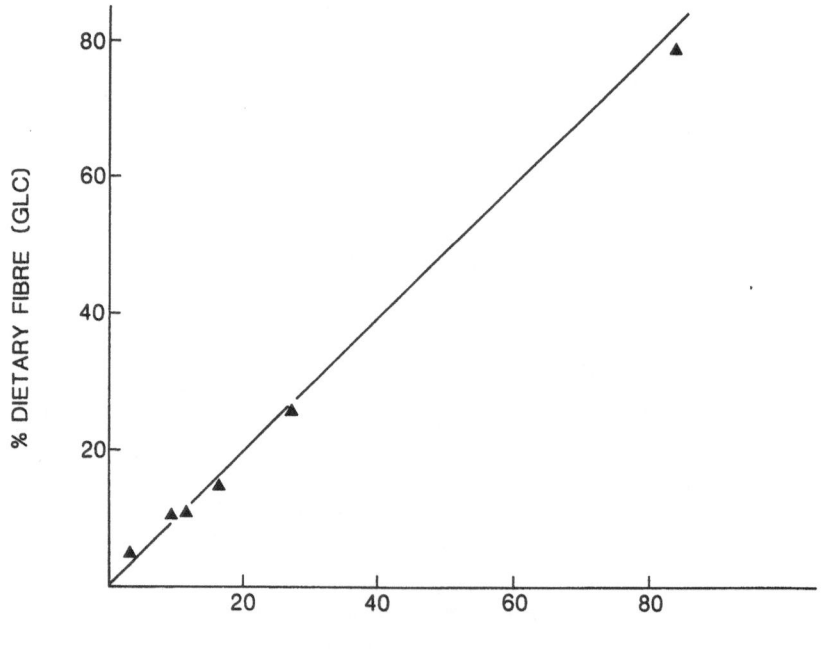

Fig. 3. Comparison of enzymic, gravimetric dietary fiber assay according to Asp et al (12) and GLC determination according to Theander et al. Method C (24) in different botanical fractions of oats. Data from ref. (29).

essentially according to Theander et al., method C (24). There was an excellent agreement when analyzing seven different cereals with quite variable dietary fiber content. We also obtained very good agreement for variously processed wheat products (14) and for raw and processed vegetables (30). This means that all the material in enzymic gravimetric residues corrected for protein and ash could be accounted for as typical dietary fiber polysaccharides and lignin.

CONCLUSIONS

There are a number of different delimitation problems in definition and analysis of dietary fiber. "Resistant starch", as originally defined, is retrograded amylose that is not absorbed in the samll intestine. Nutritionally, therefore, it shares the characteristics of non-starch polysaccharides and should be classified as dietary fiber. Other starch fractions, that may be more or less undigestible, needs further studies physiologically and are probably not included in dietary fiber analysis with current methods.

Some polysaccharides, such as the highly branched arabans in sugar beet fiber, are soluble in 78% ethanol after proteolytic treatment and therefore not analyzed as dietary fiber with enzymic, gravimetric methods.

Organic material recovered in the dietary fiber residue in enzymic, gravimetric analysis may give some, but usually quantitatively unimportant contribution to the protein and ash corrected fiber estimate. Whenever suspected, they can be analyzed in the residue.

There is generally a very good agreement between enzymic, gravimetric analysis and GLC analysis including resistant starch and lignin in both unprocessed and processed samples of cereals and vegetables.

REFERENCES

1. Englyst HN, Trowell H, Southgate DAT, Cummings JH. Dietary fiber and resistant starch. Am J Clin Nutr 1987;46:873-4.
2. Asp N-G, Furda I, DeVries JW, Schweizer TF, Prosky L. Dietary fiber definition and analysis. Am J Clin Nutr 1988;47:688-9.
3. Englyst H, Cummings JH. Reply to N-G Asp et al. Am J Clin Nutr 1988;47:690-1.
4. Southgate DAT. Determination of food carbohdyrates. Applied Science Publishers Ltd, London, 1976:25.
5. Trowell H, Southgate DAT, Wolever TMS, Leeds AR, Gassull MA, Jenkins DJA, Dietary fibre redefined. Lancet 1976;1:967.
6. Englyst H, Wiggins HS, Cummings JH. Determination of the non-starch polysaccharides in plant foods by gas-liquid chromatography of constituent sugars as alditol acetates. Analyst 1982;107:307-18.
7. Englyst HN, Cummings JH. Digestion of the polysaccharides of some cereal foods in the human small intestine Am J Clin Nutr 1985;42:778-87.
8. Björk I, Nyman M, Pedersen B, Siljeström M, Asp N-G, Eggum B. On the digestibility of starch in wheat bread - Studies in vitro and in vivo. J Cereal Sci 1986;4:1-11.
9. Englyst HN, Cummings JH. Resistant starch, a "new" food component: a classification of starch for nutritional purposes. In: Morton ID (ed). Cereals in a Europe context. Ellis Horwood Ltd, Chichester, England 1987;221-33.

10. Theander O, Westerlund E. Studies on chemical modicifactions of heat-processed starch and wheat flour. Starch/Stärke 1987; 39:88-93.
11. Östergård K, Björck I, Gunnarsson A. A study of native and chemically modified potato starch. Part I: Analysis and enzymic availability in vitro. Starch/Stärke 1988;40:58-66.
12. Asp N-G, Johansson C-G, Hallmer H, Siljeström M. Rapid enzymatic assay of insoluble and soluble dietary fiber. J Agric Food Chem 1983;31:476-82.
13. Siljeström M, Björck I, Westerlund E. Transglycosidation reactions following heat treatment of starch - Effects on enzymatic availability. Starch/Stärke 1989;41:95-100.
14. Siljeström M, Westerlund E, Björck I, Holm J, Asp N-G, Theander O. The effects of various thermal processes on dietary fibre and starch content of whole grain wheat and white flour. J Cereal Sci 1986;4:315-23.
15. Berry CS, Anson K, Miles MS, Morris VJ, Russel PL. Physical chemical characterisation of resistant starch from wheat. J Cereal Sci 1988;8:203-6.
16. Siljeström M, Björck I, Eliasson A-C. Characterization of resistant starch from autoclaved wheat starch. Starch/Stärke 1989 (in press).
17. Holm J, Björck I, Ostrowska S, Eliasson A-C, Asp N-G, Larsson K, Lundquist I. Digestibility of amylose-lipid complexes in-vitro and in-vivo. Starch/Stärke 1983;9:294-7.
18. Cummings JH, Englyst HN. Fermentation in the human large intestine and the available substrates. Am J Clin Nutr 1987;45:1243-55.
19. Würsch P, Del Vedovo S, Koellrentter B. Cell structure and starch nature as key determinants of the digestion rate of starch in legumes. Am J Clin Nutr. 1986;43:25-9.
20. Pollock CJ, Hall MA, Roberts DP. Structural analysis of fructose polymers by gas-liquid chromatography and gel filtration. J Chromatogr 1979;171:411-15.
21. Nilsson U, Björck I. Availability of cereal fructans and inulin in the rat intestinal tract. J Nutr 1988;118:1482-6.
22. Nilsson U, Dahlqvist A. Cereal fructosans; Part 2 - Characterization and structure of wheat fructosans. Food Chem 1986;22:95-106.
23. Prosky L, Asp N-G, Furda I, De Vriew JW, Schweizer TF, Harland BF. Determination of total dietary fiber in foods and food products: collaborative study. J Assoc Off Anal Chem 1985;68;677-9.
24. Theander O, Westerlund E. Studies on dietary fiber: 3 Improved procedures for analysis of dietary fiber. J Agric Food Chem 1986;34:330-6.
25. Theander O. Advances in the chemical characterization and analytical determination of dietary fibre components. In: Birch GC, Parker KJ (eds) Dietary fibre. Applied Science Publishers Ltd, 1983;77-93.
26. Theander O. Chemistry of dietary fibre components. Scand J Gastroent 1987;22, Suppl 129:21-28.
27. Saura-Calixto F. Effect of condensed tannins in the analysis of dietary fiber in carob pods. J Food Sci 1988;53:1769-1771.
28. Nyman M, Siljeström M, Pedersen B, Bach Knudsen KE, Asp N-G, Johansson C-G, Eggum O. Dietary fiber content and composition in six cereals at different extraction rates. Cereal Chem 1984;61:14-19.
29. Nyman M, Asp N-G. Fermentation of oat fiber in the rat intestinal tract: a study of different cellular areas. Am J Clin Nutr 1988;48:274-8.
30. Nyman M, Pålsson K-E, Asp N-G. Effects of processing on dietary fibre in vegetables. Zsch Lebensm Wiss Techn 1987;20:29-36.

MODIFICATIONS OF THE AOAC TOTAL DIETARY FIBER METHOD

Sungsoo C. Lee and Veronica A. Hicks

Science and Technology Center
Kellogg Company, Battle Creek, MI

Dietary fiber is generally defined as plant components that are indigestible by the human digestive enzyme system (1). Thus, total dietary fiber (TDF) includes nonstarch polysaccharides, lignin, and associated substances (1). Determination of the TDF requires an analytical method which measures all of the above components of TDF. The method should be simple, reliable, fast, and inexpensive enough to be performed on a routine basis, with large numbers of samples. The Association of Official Analytical Chemists (AOAC) employed a three-step enzymatic digestion for removal of starch and protein (2, 3). The AOAC method is simple and easy to run. Total dietary fiber is precipitated with 4 volumes of 95% ethanol, filtered and dried. Residues are then corrected for protein and ash to derive TDF values. The reported assay variability is relatively high (2,3). In this study, we modified the AOAC method in an attempt to improve the precision of the assay.

The phosphate buffer was replaced by MES-TRIS buffer (0.05M MES - 0.05M TRIS), pH 6.0. With the new buffer system, 0.205N NaOH and 0.325 N HCL were used to adjust the pH. The MES stands for 4-morpholineethanesulfonic acid. Samples in MES-TRIS buffer were preheated at 95-100°C for 20 minutes prior to enzymatic digestion steps. The rest of the analytical procedure was the same as the AOAC procedure. The analytical scheme of the modified procedures is shown in Figure 1. To compare the AOAC procedure and the modification, ten food samples were analyzed in duplicate for four days by each method, eight assays per sample per method. All the assays utilized enzymes from Sigma TDF assay kit. The TDF values were calculated by the following protocol for both methods.

$$\%TDF = \frac{\left(\dfrac{R_1 + R_2}{2}\right) - P - A - B}{\left(\dfrac{M_1 + M_2}{2}\right)} \times 100$$

where R_1 = Residue weight 1 from M_1

R_2 = Residue weight 2 from M_2

M_1 = Sample weight 1

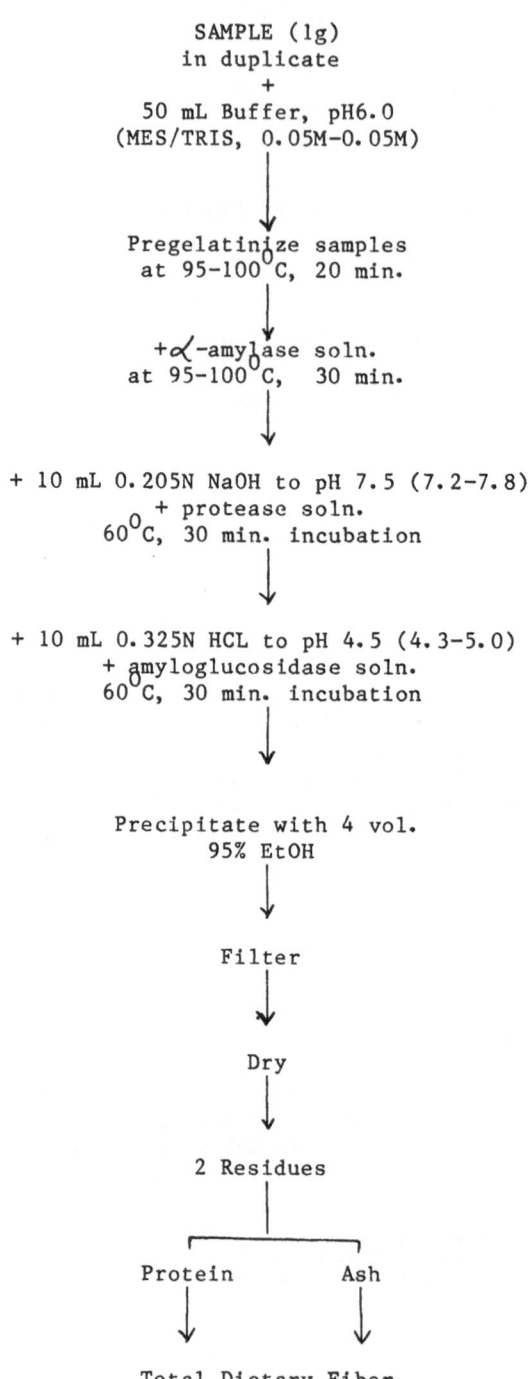

SAMPLE (1g)
in duplicate
+
50 mL Buffer, pH6.0
(MES/TRIS, 0.05M-0.05M)

Pregelatinize samples
at 95-100°C, 20 min.

+α-amylase soln.
at 95-100°C, 30 min.

+ 10 mL 0.205N NaOH to pH 7.5 (7.2-7.8)
+ protease soln.
60°C, 30 min. incubation

+ 10 mL 0.325N HCL to pH 4.5 (4.3-5.0)
+ amyloglucosidase soln.
60°C, 30 min. incubation

Precipitate with 4 vol.
95% EtOH

Filter

Dry

2 Residues

Protein   Ash

Total Dietary Fiber

Figure 1. ANALYTICAL SCHEMES FOR MODIFIED TDF PROCEDURES

M_2 = Sample weight 2
A = Ash weight from R_1
P = Protein weight from R_2

$$B = Blank = \frac{BR_1 + BR_2}{2} - P - A$$

BR = Blank residue

The modifications developed for determining TDF in food samples proved to be highly satisfactory. A review of the data indicates that mean TDF values, as measured by the AOAC and the modified procedure, are in agreement (Table 1). However, variability of the modification was less than half of the AOAC procedure (Table 1). The AOAC procedure showed satisfactory precision for carrots, wheat bran, and corn bran. However, barley and oat products showed relatively high relative standard deviation (RSD). The modification showed good precision for all the products with the exception of white wheat flour. The overall relative standard deviation for ten products was 10.0% for the AOAC and 4.3% for the modification. Statistical analyses of the variance estimates showed that variability of the two methods were significantly different for carrots, apples, barley, and oat cereal (Table 2).

The higher assay variability of the AOAC procedure was probably caused by buffer co-precipitation in ethanol. Figure 2 shows blank residue weights before protein and ash correction by the two buffer systems at various ethanol concentrations. The amount of precipitate in the phosphate buffer increased as the ethanol concentration increased. Small changes in ethanol concentration around or above 78% caused drastic changes in amounts of phosphate buffer precipitate. In phosphate buffer, calcium ions form calcium phosphates, which are sparingly soluble in water, but insoluble in ethanol and thus precipitate. However, the MES-TRIS buffer did not show much precipitation and the precipitation was not sensitive to ethanol concentration. It is noteworthy that blank values are more varied with the AOAC procedure than with the modification (8.5 $\pm$ 5.9 vs 8.2 $\pm$ 2.8 mg, respectively).

Table 3 shows that larger amounts of ash were recovered in the TDF residue prepared by the AOAC procedure than from the residues prepared by the modification. This is in agreement with the previous finding that a high amount of ash was found in the soluble fiber fractions of cereal products, which could be attributable to calcium phosphate precipitation by ethanol (4). It is suggested here that calcium phosphate precipitation could also be influenced by presence of calcium in many foods. In addition to inherent calcium in food samples, the enzyme preparation was stabilized with calcium ions. Calcium ions help in the formation of intra-molecular disulfide bridges, which help stabilize conformation of α-amylase (5). Finally, the avoidance of chelators such as phosphate and citrate is desirable as free calcium ions are required to maintain the stability of the enzyme at elevated temperatures (6).

The impact of assay precision on decision making in food industry is a relevant consideration. Table 4 summarizes minimum detectable differences for each product by both methods (7). A minimum detectable difference is defined as minimum difference required between any 2 samples to say that they are significantly different. The data indicated that the modification

Table 1.   Percent TDF in Food Samples[1],[2]

	% TDF (Mean ± S.D.)		RSD	
	AOAC	Modification	AOAC	Modification
carrot	25.9 ± 1.20	25.2 ± 0.34	4.6%	1.3%
apple	12.2 1.16	11.5 0.43	9.5	3.7
barley	13.2 1.66	13.4 0.45	12.6	3.3
oat cereal	11.7 1.35	10.8 0.38	11.4	3.5
oat bran	14.7 1.69	14.7 0.81	11.4	5.5
oat flour	9.2 0.98	9.1 0.44	10.6	4.9
corn bran	90.0 0.63	89.9 0.34	0.7	0.4
rye bread	6.9 0.56	· 7.3 0.30	8.2	4.1
whole wheat bread	6.6 0.59	6.3 0.45	8.9	7.1
white wheat flour	2.8 0.64	2.7 0.26	22.4	9.6

[1] % TDF is based on dry weight.
[2] All the numbers represent mean ± S.D. of 8 assays; 2 assays/day/sample
     were conducted for 4 days.

Table 2.   Variance Estimates by 2 TDF Methods

	Methods		
	AOAC	Modification	P-Value
carrot	1.58	0.12	0.05
apple	1.50	0.19	0.10
barley	3.02	0.20	0.01
oat cereal	2.05	0.15	0.05
oat bran	3.05	0.86	NS
oat flour	1.04	0.27	NS
corn bran	0.40	0.13	NS
rye bread	0.36	0.10	NS
whole wheat bread	0.38	0.23	NS
white wheat flour	0.45	0.07	NS

Table 3.   Ash Content (mg) Present in the TDF Residue[1]

	AOAC mean ± S.D.		Modification mean ± S.D.	
carrot	43.5 ± 11.4		10.4 ± 3.5	
apple	19.5 13.8		0.7 0.7	
barley	23.0 4.6		6.6 2.3	
oat cereal	31.9 9.5		8.7 2.6	
oat bran	35.7 7.5		17.7 1.6	
oat flour	25.1 1.2		9.6 2.1	
corn bran	21.5 4.0		3.0 1.4	
rye bread	20.1 2.3		1.0 1.1	
wheat bread	19.2 2.7		1.2 1.6	
white bread flour	9.7 3.8		0.3 0.6	

[1] Data represent mean ± S.D. of 8 ash values.

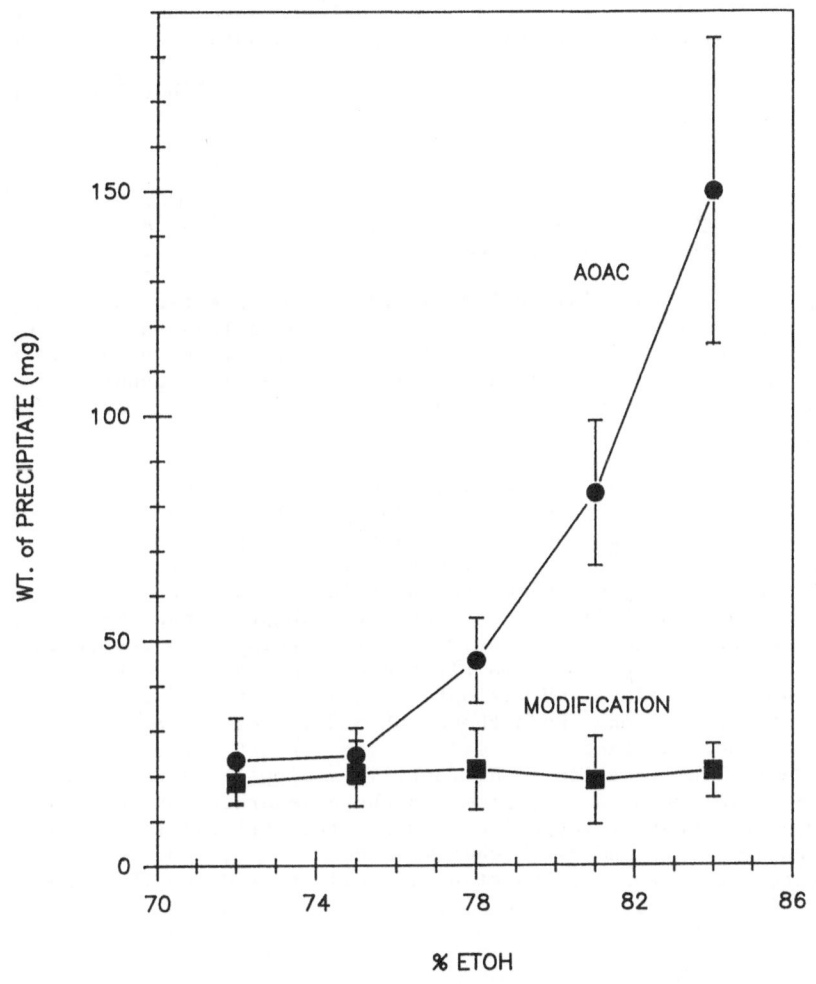

Figure 2.   COMPARISON OF ALCOHOL PRECIPITATION

FOR DIFFERENT BUFFER SOLUTIONS

The residue weights (before protein and ash correction) were measured for blanks at various ethanol concentration in two buffer systems (MES-TRIS vs. phosphate).   Each buffer system had three enzymes at normal concentrations.

was more powerful in differentiating samples. For example, if we have 2 apple samples, one of which showed 8% TDF and the other showed 10% TDF. Single analysis by the modification will be sufficient to decide that TDF levels of these 2 samples are different. But it will take more than 4 replicate assays by the AOAC method to reach a similar conclusion. The higher the assay variability, the more number of replicates are required for reliable decision making. Thus, the better the precision, the more potential savings in labor costs. The issue of assay precision, as related to potential labor savings, takes on added importance where multitudes of samples are analyzed daily. For example, food manufacturers monitor incoming raw materials in order to consistently achieve targeted levels of dietary fiber in finished foods. If minimum dietary fiber levels are specified for each ingredient, a precise assay can minimize mistakes such as described below:

1. Accepting raw materials which should not be accepted, by errorneously getting higher fiber values than actual. In this case, finished products may not reach the target level of dietary fiber.

2. Rejecting raw materials which should not be rejected, by errorneously getting lower values than actual. This could lead to shut-down of production line if no suitable raw materials are on hand. To avoid such potential disruption of production schedules, manufacturers may resort to holding unnecessarily large inventories of raw materials.

In the second phase of the study, we tried to simplify the method in order to improve the efficiency. The analytical schemes of the shortened procedure is shown in Figure 3. The samples were pregelatinized in MES buffer at pH 5.2 at 95°C for 20 minutes. The temperature was cooled down to 60°C. The samples were simultaneously incubated with α-amylase and amyloglucosidase and followed by digestion with protease. There was no pH and temperature change during the incubation, this could result in significant time savings when multitudes of samples are to be analyzed daily. In the procedure, pH was optimized at 5.2 for α-amylase and amyloglucosidase action. Even though pH 5.2 is not the optimum pH for protease action, the fact that residues are corrected for protein can compensate for the reaction condition. This simplified procedure was able to generate similar mean TDF values to those generated by the AOAC procedure (Table 5). Although it looks encouraging, the variability for barley, oat cereal, wheat flour, and rye breads were relatively high. Overall RSD was approximately 8%. Further investigation will be made on the simplified procedure in order to improve the precision.

Summary and Conclusions

1. The TDF values measured by the modification showed excellent agreement with those determined by the AOAC method.
2. The estimated assay variabilities of the modification were significantlly less than those of the AOAC TDF method.
3. The TDF method can be simplified with the simultaneous incubation of a sample with α-amylase and amyloglucosidase, followed by subsequent protease incubation without changing pH and temperature. The precision of this shortened method is comparable to that of the AOAC method.

Table 4. Minimum Detectable Difference as % TDF[*1]

METHODS	AOAC		MODIFICATION	
No. of Analyses	1	4	1	4
		(2/day x 2 days)		(2/day x 2 days)
Carrot	5.0	3.5	1.4	1.0
Apple	4.8	3.4	1.7	1.2
Barley	6.9	4.9	1.8	1.3
Oat cereal	5.7	4.0	1.5	1.1
Oat bran	6.9	4.9	3.7	2.6
Oat flour	4.0	2.9	2.1	1.5
Corn bran	2.5	1.8	1.4	1.0
Rye bread	2.4	1.7	1.2	0.8
Whole wheat bread	2.4	1.7	1.9	1.3
White wheat flour	2.7	1.9	1.1	0.8

[*1] The probability of type I error (α) and type II error (β) were 0.05
and 0.2, respectively.
Minimum detectable difference was calculated by using the following
equation:

$$ MDD = \frac{(\sigma_B^2}{r_1} + \frac{\sigma_W^2)}{r_2} (z_{\alpha/2} + z_\beta)^2 $$

Where σ_B = variance components between day
σ_W = variance components within day
r_1 = number of days assays performed
r_2 = number of assays performed within day
$z_{\alpha/2}$ = value from standard normal distribution for specified
z_β = value from standard normal distribution for specified

Sample (1 g) in duplicate
+
50 ml MES buffer, 0.05 M, pH 5.2
↓

Pregelatinized starch
95°C, 20 min.
↓

α-amylase + AMG
60°C, 40 min.
↓

Protease
60°C, 30 min.
↓

ppt w/4 vol. EtOH

ash   protein

Total Dietary Fiber

FIGURE 3. SIMPLIFIED TDF PROCEDURE

Table 5. TDF Values by the Simplified Method

	% TDF		% RSD
	Mean	S.D.	
Carrot	25.8 ±	0.51	2.0
Apple	11.7	0.31	2.6
Barley	12.6	1.29	10.3
Oat cereal	10.7	0.69	6.5
Oat flour	9.7	1.72	7.7
Corn bran	89.1	1.36	1.5
Rye bread	6.9	0.97	13.3
Wheat bread	6.5	0.70	10.8
White wheat flour	3.1	0.48	15.6

*1 % TDF is based on dry weight.
*2 All the numbers represent mean ± S.D. of 8 assays.

REFERENCES

1. H. Trowell, Dietary Fiber Definitions, Am. J. Clin. Nutr. 48:1079 (1988)

2. L. Prosky, et al., Determination of Total Dietary Fiber in Foods and Food Products: Collaborative study, J. Assoc. Off. Anal. Chem. 68:677 (1985)

3. L. Prosky, et al., Determination of Total Dietary Fiber in Foods and Food Products: Interlaborative study, J. Assoc. Off. Anal. Chem. 71:1017 (1988)

4. T. F. Schweizer & P. Wurch, (1979) J. Sci. Food Agric. 30, 613-619.

5. N. Ramasesh, K. R. Sreekantiah, & V. S. M. Murphy, Purification and characterization of a thermophilic -amylase of Aspergillus niger von Tiegham, (1982) Starke 34, 274 (1982).

6. R. M. Faulks, & S. B. Timms, A rapid method for determining the carbohydrate component of dietary fiber, Food Chem. 17, 273.

7. R. G. D. Steel, & J. H. Torrie, J. H., Principles and Procedures of Statistics. A Biomedical Approach. McGraw-Hill Book Co., New York, N.Y. (1980)

IMPROVED METHODS FOR ANALYSIS AND BIOLOGICAL CHARACTERIZATION OF

FIBER

J.L. Jeraci and P.J. Van Soest

Department of Animal Science
Cornell University
Ithaca, NY   14853

ABSTRACT

Dietary fibers are not uniform, chemically or in their nutritive and biological properties, the only common ground being their resistance to mammalian digestive enzymes. The AOAC method for total fiber is subject to inferences from ash, protein, tannins and resistant starches. These interferences can be reduced by urea enzymatic dialysis. The measurement of soluble and insoluble fiber is nutritionally relevant, since physical properties greatly modify dietary effects of fiber. Insoluble fiber is conveniently measured as neutral-detergent fiber. This procedure has been improved by reducing the starch interference and the time of analysis. Physical and biological properties of dietary fiber can be measured by using relevant procedures for hydration capacity, metal ion exchange capacity and rate of fermentation. The lignin and tannin content modify the characteristics of dietary fiber.

INTRODUCTION

The various dietary fiber procedures have been developed with the purpose of determining the content of fiber in foods or measuring the physiological and nutritional effects of these fibers. Various sources of dietary fiber have a heterogenous composition because of the different subcomponents and physical forms of the fiber. These fiber sources have well-attested differences in their physiological and nutritional effects. Thus, further characterization of dietary fiber besides that of a total value seems warranted. Further, it might be suggested that the subcomponents of dietary fiber are so diverse, that they may be regarded as valid nutritional entities to be analyzed for their own sake.

Thus the development of dietary fiber methodology has taken two directions: (1) the definition and measurement of a total dietary fiber value, and (2) the characterization of the subcomponent composition with a view towards its physical and biological properties. The objective of this paper is to review the development of the methods at Cornell University, which address the determination of various kinds of dietary fiber as well as their physiological and nutritional characterization. Improvements leading to the - various methods are also discussed.

Total Dietary Fiber

Although there is an AOAC official method for total dietary fiber (Proskey et al., 1984), a rugged, quick and precise total dietary fiber method is still needed that avoids some of the problems in the existing method. Prosky et al. (1984) described the history that led to the development of the AOAC total dietary fiber method. This method is based on the definition: "Dietary fiber consists of remnants of the plant cells resistant to hydrolysis by the alimentary enzymes of man, which includes hemicelluloses, celluloses, lignins, nondigestible oligosaccharides, pectins, gums and waxes" (Van Soest and McQueen 1973; Trowell 1976).

Compared to the earlier method (Prosky et al., 1984), the AOAC-1988 (Prosky et al., 1988) method gives some improvement in interlaboratory replication and in the coefficient of variation of the food samples. Other problems with the AOAC total dietary fiber method include resistant starch in the total dietary fiber, a relatively large crude protein correction (CP), a large ash correction, and problems with filtration. Grace and Gordon (1986) have suggested that the method can be improved by the use of other enzymes. Other modifications to the procedure have been suggested by Lee et al. (1988), Li and Andrews (1988) which are aimed at decreasing the quantity of the correction factors for crude protein and ash, or elimination of the protease step, respectively. Routine analysis of samples with the AOAC-1988 method is labor intensive and requires a substantial investment in equipment.

A new method for total dietary fiber has been developed at Cornell (Jeraci et al., in press a) which utilizes 8 M urea to hydrate and extract starch and other water-soluble polysaccharides, a heat-stable amylase and dialysis to remove starch, and a protease digestion of plant proteins. This method is simpler and relatively economical in use of chemicals, equipment, and labor. The method has been used to determine soluble dietary fiber and insoluble dietary fiber (Jeraci et al., in press a; Jeraci et al., in press b). A schematic outline of the method is shown in Figure 1. The essential principle of the method is that amylases are active in 8M urea at room temperature, allowing the gelatinization and hydrolysis of starch without extra heating. The urea amylase digestion is dialyzed to remove interfering end products as well as to remove urea.

Comparison of the AOAC and UED Methods for Total Dietary Fiber

TDF for the samples by the AOAC method was usually the same or higher than the values obtained for the UED method (Table 1). The correction values for CP and ash were higher for the AOAC-1985 and AOAC-1988 methods than for the UED method. The CP and ash contamination of samples that formed gels during the alcohol precipitation step appeared to be particularly high with the AOAC method.

The reagents in the UED method do not contain ß-glucanase activity. The ß-glucan content of the sample assayed by the Glatter method (Glatter et al., 1988), did not change as the different UED reagents were used, but as expected the residue decreased as starch and protein components were extracted. The yield of water soluble ß-glucans under the conditions of the UED method is comparable to that extracted by water at 100°C for 1 h for samples in which endogenous enzymes had been inactivated (Jeraci et al., in press a).

Removal of Starch

The UED method with either of the heat-stable amylases (No. A-0164, Sigma Chemical Co., St. Louis, MO 63178 or Termamyl 120L, Novo Laboratories, Inc. Wilton, CT 06897) removed practically all starch. Heat-stable α-amylases are active in 8 M urea at 20°C and no significant differences in the TDF yields

246

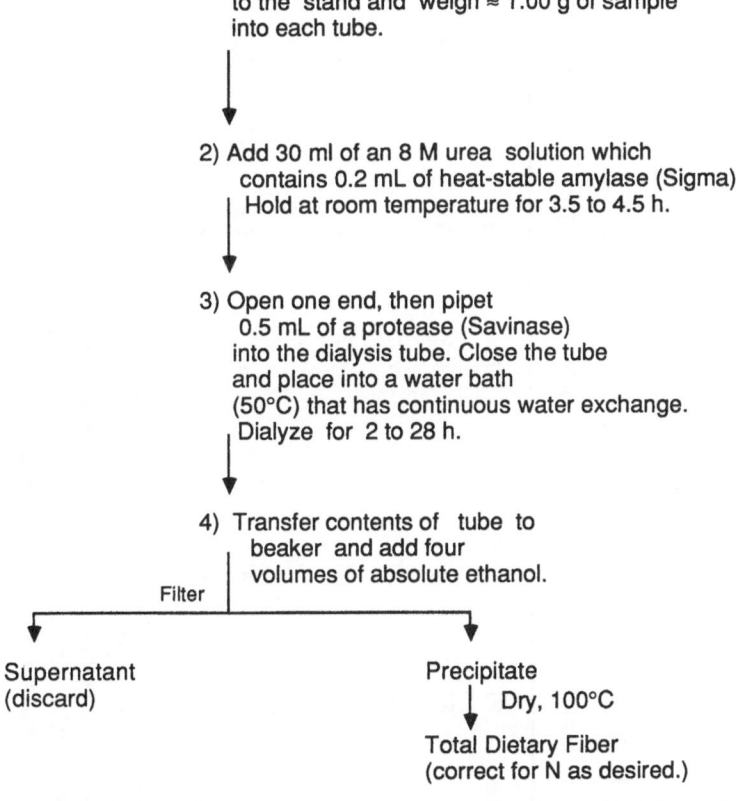

1) Attach four dialysis tubes
   to the stand and weigh ≈ 1.00 g of sample
   into each tube.

2) Add 30 ml of an 8 M urea solution which
   contains 0.2 mL of heat-stable amylase (Sigma)
   Hold at room temperature for 3.5 to 4.5 h.

3) Open one end, then pipet
   0.5 mL of a protease (Savinase)
   into the dialysis tube. Close the tube
   and place into a water bath
   (50°C) that has continuous water exchange.
   Dialyze for 2 to 28 h.

4) Transfer contents of tube to
   beaker and add four
   volumes of absolute ethanol.

Filter

Supernatant
(discard)

Precipitate
   Dry, 100°C

Total Dietary Fiber
(correct for N as desired.)

Figure 1. Flowchart of the urea enzymatic dialysis method for the
determination of total dietary fiber (Jeraci et al, in press a).

Table 1. Composition of Various Food Samples using the AOAC Total Dietary Fiber Method and the Urea Enzymatic Dialysis Method (UED)[a,b].

Food Sample	% Total Dietary Fiber		% Crude Protein[c]		%Ash[d]	
Method:	AOAC	UED	AOAC	UED	AOAC	UED
Sweet Green Peas	21.1	20.9	5.1	3.1	1.4	1.0
Broccoli	36.1	34.0	9.3	5.8	3.9	1.1
Baby Lima Beans	20.9	20.0	4.6	3.1	1.0	0.0
Green Beans	30.0	30.0	4.0	3.6	6.1	2.7
Kale	35.5	32.5	11.1	6.9	9.5	2.3
Okra	41.8	40.6	10.1	5.2	6.6	2.7
Summer Squash	21.4	20.2	4.5	3.3	3.1	0.8

[a] Adapted from Jeraci et al., in press a.
[b] Corrected for water. Crude protein determined by Kjeldahl analysis.
[c] Crude protein in the AOAC residues is significantly different than the crude protein in the UED residues (P< 0.05).
[d] Ash in the AOAC residues is significantly different than the ash in the UED residues (P< 0.01).

were observed between the two heat-stable amylases in the UED method. These conditions and incubation times of 3.5 or 4.5 h were found to be effective in removing starch from foodstuffs (Jeraci et al., in press a). The reagent blanks were the same for both amylases. The dialysis step in the UED method may minimize the effect of ash levels coming from amylases.

Removal of Protein

The minimum and maximum times needed for the Savinase 8.0 L in the protease-dialysis step (2-28 h) was studied using corn meal. No significant differences in TDF, CP, ash and reagent blanks were detected after 8 h. No significant difference in TDF and reagent blanks were observed among the 2 and 8 h times, however protease-dialysis can be adjusted to maximize the number of samples analyzed or to minimize the CP correction factor as desired (Jeraci et al 1989 a).

One objective of the UED method was to decrease the CP correction, thus minimizing the error in treating all residual N as CP (N x 6.25). In this objective, the UED method is a significant improvement compared to the AOAC-1988 method (Table 1). Currently there is no general consensus that the nitrogen in the TDF must represent CP only and that the reported TDF must be nitrogen (especially CP) free. Until these issues have been resolved, the researcher can choose the length of time for the protease step (2 h to 28 h) and apply the appropriate correction (Jeraci et al., in press a). Most of the data reported here are based on 5 to 16 h protease treatment.

It is apparent from the foodstuffs investigated thus far that the UED method affords several advantages for TDF. Thus, the UED method (1) removes essentially all starch, (2) generates smaller correction factors for CP and ash, (3) gives better quantitative recovery of purified and semi-purified dietary fiber products, (4) can handle more samples for a similar investment of labor and equipment, (5) does not expose the foodstuff to temperatures above 50°C and (6) requires only one reagent solution and two enzymes.

Soluble Fiber and Pectin

Soluble dietary fiber represents gums and other gel-like polysaccharides that are resistant to mammalian digestive enzymes. The inclusion of these soluble components into the definition of fiber has come about in a large part because of the investigations of monogastric and human nutritionists. Evidence exists to show that many soluble polysaccharides are relatively fermentable and have gel-like properties that reduce the rate of sugar absorption, and may also bind with fatty acids and cholesterol. The unique behavior of the these soluble components can justify their inclusion as a separate entity within the dietary fiber complex.

Pectin

The relationship of pectin to the soluble or insoluble portions of fiber has involved some confusion. Pectin as a part of plant cell wall may not be immediately soluble, but require such alterations as demethylation and/or debranching; viz. beta elimination reactions. Such alterations are not improbable within the range of gastrointestinal conditions. Most pectin is removed (ca 90%) by extraction with neutral detergent. This extraction is coincident with the observation that gut microbes can rapidly ferment up to 100% of pectin -this is in contrast to that of hemicellulose and cellulose. Human balance studies also indicate complete colonic fermentation of pectin (Cummings et al., 1979). The rapid fermentation probably occurs because the peptic digestion that removes protein occurs at a low pH and the hydration of the pectic substance in a low pH solution promotes demethylation. Many older procedures of cell wall preparation first used an acid pepsin digestion that resulted in the subsequently easy extraction of pectin. The current total dietary fiber procedures do not truly mimic gastric digestion, in that neutral protease systems are the exclusive treatment, whereas in the gut a peptic acid digestion precedes the alkaline sequences.

The neutral-detergent fiber (NDF) method has been criticized for not recovering pectin which has been regarded by some as part of the cell wall matrix. While a botanical argument can be made, the evidence from fermentation with gut microorganisms is that pectin is unique in being completely and rapidly fermentable. Pectin also possesses a very high cation exchange - at least in the demethylated form. Our view is that pectin deserves its recognition and might be determined as its own entity. The method (Blumencrantz and Asboe-Hansen, 1973) using meta hydroxydiphenyl is specific for galacturonic acid and is a relatively simple procedure that can be conducted along with other fiber procedures.

Insoluble Fiber

Insoluble fiber in food represents the cross-linked matrix of the plant cell wall and is most conveniently measured as neutral-detergent fiber (NDF), which includes the cellulose, hemicellulose and lignin as the major components. The original NDF method was applied to forages, and its subsequent application to starchy foods and feeds revealed interference of starch, thus presenting difficulties for the unmodified ND method. Therefore, various modifications with amylases have resulted (Table 2). Relatively pure amylases that are stable to heat and active in the detergent solution are not readily available. Many of the commercial amylases in use contain side activities including hemicellulase, ß-glucanase and protease (Mascarenhas-Ferreira et al., 1983). In some of the modified methods the sample is incubated overnight at 20° to 35°C with amylase. With this increase in incubation time, contaminating enzymes in the semi-pure amylases can degrade other components in the food leading to low values, whereas inclusion of unwanted starch leads to high values (Mascarenhas-Ferreira et al., 1983). It appears that the Bacillis subtilis amylase (Sigma type XI) contains sufficient hemicellulase to affect values of wheat bran. Termamyl 120 L (Novo Bio Laboratories) appears more satisfactory. In some modifications (Table 3), the heat-tolerant B. subtilis amylase (up to 80°C) was added during and after the refluxing step to effectively remove the starch (Robertson and Van Soest, 1981; Mertens, 1988). These modified methods increased the assay time by only 5-20 minutes, but increased the number of steps in the ND method. Currently a heat-stable α-amylase (Termamyl, T120) is used to partially degrade starch in both the brewing industry and in non-detergent chemical methods for dietary fiber (Prosky et al., 1985; Theander and Esterlund, 1986; Jeraci et al., in press a). The heat stable α-amylase has been used effectively in the ND method (Jeraci et al., in press c). The use of high temperature with a short term amylase treatment has the advantages of minimizing the effect of unwanted side activities and a more rapid procedure.

In the original ND method, starch removal was facilitated by using 2-ethoxyethanol. However, 2-ethoxyethanol (ethylene glycol monoethyl ether or Cellosolve) is now recognized as a health risk. Therefore, 2-ethoxyethanol should be replaced by a safer reagent. Triethylene glycol is suggested (Jeraci et al., in press c). Even with the use of efficient amylases, an addition of either 2 ethoxyethanol or triethylene glycol seems necessary.

Comparison of Modified Neutral-Detergent Methods

We have compared three neutral-detergent procedures using either B. Subtilis, amylase (Robertson and Van Soest 1981; Mertens 1988), or Termamyl

Table 2. Modifications of NDF

Source	Conditions	Mascarenhas-Ferreira reference
VanSoest and Wine 1968	ND boil 1 hr minus sulfite and decalin	A
Fonnesbeck 1976	Pepsin pH 2.2 40h at 55°C Boil Detergent pH 3.5	-
Taverner & King 1975	ND boil 2 hr	-
Schaller 1976	pretreatment with[a] hog pancreas enzyme	-
Robertson and VanSoest 1981 also Mongeau and Brassard 1979	ND boil 30m, remove add amylase, reboil 30m[b]	B
Giger et al 1981	boil in H_2O 30 min incubate 80C w. amylase 30m, ND boil 1 h.[b]	C
Wainman et al 1981	overnight w. amylase 38C ND boil 1h[b]	D
McQueen and Nicholson 1979	similar to D but w. less enzyme b[b]	E
Wainman et al Modified by Mascarenhas-Ferreira	incubate w. amylase overnight, room & ND boil 1 hr.[b]	F
ibid	incubate 30 m at Room T. ND boil 1hr[b]	G
Mascarenhas-Ferreira et al 1983	ND without EDTA + enzyme 15m room + boil 1hr add EDTA 5m before end.[b]	H
ibid	Same as H except ND + EDTA used throughout[b]	I
Sills et al 1983	Gel Starch, Amylase pH 4.6 overnight 38°C; ND boil 1h	-
Jeraci et al in press c	Add 0.25 ml termamyl 120 to ND boil 1 h.[b]	-

[a] uses sulfite and decalin. Approved by the AACC.
[b] Sulfite and decalin omitted.

Table 3. Comparison of Three Neutral-Detergent Methods Using Various Operating Conditions and Two Amylases[a].

Reference	Conditions	Acronym
Jeraci et al. in press c:	Single addition of heat-stable α-amylase to ND solution, boil 60 min.	ND-T
Robertson and Van Soest 1981:	Boil sample wtih 50 mL of ND solution for 30 min and then add 50 mL of ND solution and _Bacillus_ _subtilus_ α-amylase, boil 30 min. Filter and add _Bacillus_ _subtilus_ α-amylase to crucible, incubate for 10-15 min.	ND-S
Mertens 1988:	Boil sample wtih 45 mL of ND solution for 40 min and then add 55 mL of ND solution and _Bacillus_ _subtilus_ α-amylase, boil 20 min. Filter and add _Bacillus_ _subtilus_ α-amylase to crucible, incubate for 3 min.	ND-S2

[a] Adapted from Jeraci et al., in press c.

120 L (Table 3). The use of these three methods gave similar NDF values for corn starch, corn starch-casein mixture, and brewers grains.

Both the Termamyl amylase and the B. subtilis amylase were relatively effective in removing starch from the samples (Jeraci, in press c). Thus, the modified AACC starch assay did not detect starch in the NDF of six of the samples and only a negligible amount of starch was detected in the NDF residues from the other samples. However, Termamyl was somewhat more effective than the B. subtilis amylase in removing starch. Thus, although the feces and the commercial brown bread gave less NDF by the ND-S assay using the B. subtilis amylase, the starch content was large in these NDF residues (Table 4 and 5). Subsequently, the difference in NDF values for the feces (9.5%) and the commercial brown bread (6.9%) was accounted for by the difference in the hemicellulose values. It is apparent from these studies that the differences between the NDF values for the samples can not be attributed to starch and probably reflect a loss in hemicellulose. Studies are in progress to characterize the composition of these hemicelluloses that are sensitive to the B. subtilis amylase and to identify the contaminating enzymes. The nitrogen recovered in the ND fiber values for eight samples were not affected by the ND-T and the ND-S methods (Jeraci et al., 1989 in press c).

The composition of commercially available processed cereal products was determined by Jeraci et al, (in press c) using the ND-T and ND-S methods. These methods gave similar NDF, ADF, hemicellulose, cellulose, and lignin values for nine of the samples. Different NDF and hemicellulose values for the enriched self-rising wheat flour and barley were obtained by the NDF methods.

Descriptive Properties

The physico-chemical properties of fiber can greatly modify physiological and nutritional responses. For dietary fiber the most important physical properties are cation exchange, hydration capacity, fermentation rate and particle size. We have explored and developed feasible methods for these descriptive attributes.

Cation exchange - Originally we measured cation exchange capacity (CEC) using lithium (Van Soest and Robertson, 1976). The method seemed limited because it requires elution with acid that might alter fibers and it also showed a poor repeatability. Later we applied and modified the copper method of Keijbets and Pilnik (1974), (McBurney et al., 1983). This method is more repeatable but is limited to measurements at pH 4. To obtain exchange values at pH 7 the use of praseodymium acetate was introduced (Allen et al., 1986). Use of trivalent praseodymium offers a convenient method in which the endpoint is read spectrophotometrically. However, it is essential that the buffer system is phosphate free to avoid interference. This method at pH 4 gives equivalent CEC values compared to the copper procedure but at pH 7 the CEC values almost double (Figure 2).

The most important functional groups in fiber that contribute to cation exchange are pectins, phenolics, lignin and tannin (McBurney et al., 1986). Among human foods pectin is characteristic of many vegetables while lignin is in brans, particularly wheat bran. Mineral binding is not necessarily a negative attribute of fiber and indeed for some cations may promote positive balances (Table 6). Wood cellulose, very low in exchange due to delignification, when added to bread had negative effects on mineral balance of iron, copper, zinc and magnesium (Table 6). However, bran and cabbage balances are more positive except for iron and calcium. Cabbage balances are positive except in the case of calcium.

Table 4. Evaluation of Two Amylases in Two Neutral-Detergent (ND) Methods. Analysis of Neutral-Detergent Fiber (NDF) and the Content of Starch in NDF[a,b,c,d].

NDF Method:	ND-S			ND-T	
Sample	NDF%	SEM		NDF%	SEM
Timothy hay	65.5	1.31		66.3	0.98
Wheat straw	83.8	0.11		85.0	0.58
Artichoke hearts	11.6	0.68		12.3	0.34
Feces (human)	13.8	0.37		23.8	2.62
Alfalfa hay	46.8	0.92		47.1	1.31
Broccoli	18.3	1.06		18.7	0.50
Okra	11.9	0.03		12.6	0.28
Kale	15.9	1.01		16.0	0.52
Wheat bran, hard red	48.2	3.29		54.5	3.53
Corn silage	51.5	0.10		52.8	1.12
Green peas	15.7	1.12		16.5	0.21
Commercial brown bread	16.9	2.74		23.9	3.70
Baby lima beans	12.9	2.44		13.4	1.84
Ethanol-extracted corn[e]	8.6	1.72		9.8	0.72

[a] Adapted from Jeraci et al., in press c.

[b] Values corrected for water content.

[c] Bacillus subtilis amylase (Sigma Chemical Co.) is used in the ND-S method (Robertson and Van Soest,1981) and heat-stable α-amylase (Sigma Chemical Co.) is used in the ND-T method described in this paper.

[d] SEM: standard error of the mean for NDF from three replicated experiments.

[e] Sweet corn was extracted with 80% ethanol and the insoluble residue was analyzed.

Table 5. Starch Content of the NDF Residues Expressed as Percent of the Original Sample Dry Matter Using _Bacillus subtilis_ or Heat-stable α-Amylase[a,b,c].

	% Starch			
Amylase:	_B. subtilis_		Heat-stable amylase	
NDF Method:	ND-S		ND-T	
Sample	Mean	SD	Mean	SD
Artichoke hearts	0.4	0.05	0.0	0
Feces (human)	0.7	0.27	0.0	0
Alfalfa hay	0.4	0.10	0.8	0.20
Broccoli	0.3	0.01	0.1	0.05
Kale	0.7	0.10	0.0	0
Wheat bran, hard red	0.2	0.20	0.9	0
Corn silage	0.2	0.10	0.7	0.20
Green peas	2.1	0.60	1.1	0.10
Commercial brown bread	0.3	0.20	0.0	0
Baby lima beans	0.9	0.30	0.0	0

[a] Adapted from Jeraci et. al. in press c.

[b] Values corrected for water content.

[c] _Bacillus subtilis_ amylase (Sigma Chemical Co.) used in the ND-S method (Robertson and Van Soest, 1981) and heat-stable α-amylase (Sigma Chemical Co.) used in the ND-T method described in this paper.

[c] Sweet corn was extracted with 80% ethanol and the insoluble residue was analyzed.

Hydration capacity - Originally we applied a filtration-suction procedure for measuring water-holding capacity (Van Soest et al., 1983). This procedure can only be used for the insoluble sources. However, a later a modification of the techniques of Robertson and Eastwood 1981 was developed using an osmotic system with polyethylene glycol. This procedure allows the measurement of gums and other soluble fibers. Values obtained by the two procedures for insoluble fibers are related but not identical (Figure 3). For both methods it is necessary to prepare isolated fiber fractions for assay so that there is not interference from non fiber components in the sample.

It has been suggested that water-holding capacity promotes fecal response (McConnell et al., 1974) which Stephen and Cummings (1979) attacked by pointing out that a high degree of fermentation may ruin this capacity. However, hydration capacity and cation exchange are positive factors in promoting fermentability (McBurney et al., 1983; Van Soest et al., 1983) and that if one considers the replacement of fermented fiber with produced microbes, the original hypothesis may still stand (McBurney et al., 1985; Jeraci and Horvath, 1989). A comparison of the physical properties of some fiber sources is shown in Table 7. A correlation (R=+0.7) between cation exchange and hydration capacity has been observed (Van Soest and Robertson, 1976).

Stephen and Cummings (1979) calculated the incremental responses of microbial matter and water for cabbage and bran. They postulated that the Two fiber sources had different mechanisms for pulling fecal water: microbes dominating in the case of cabbage and undigested fiber in the case of bran. Their data are compared with those of the Cornell study in Figure 4 and are in essential agreement. Cabbage is very fermentable and yields a major proportion of bacteria. The Cornell experiment showed a smaller response in the case of cabbage than did Stephen and Cummings. The finely ground wood cellulose and fine bran produced less water and bacteria but relatively equal fiber residues. The bacterial response was the least in the case of wood cellulose and water response poorest in the case of fine bran.

Particle Size - Fibers with significant unfermentable fractions will contain water that is associated with the undigested fiber. This latter capacity can be greatly affected by reduction in particle size, both in the direct measurement (Table 7) and in human fecal responses (Table 8). This is particularly the case for fine wheat bran and wood celluloses (Heller et al., 1980 and Van Dokkum et al., 1982). Particle size of insoluble fibers is related to hydration capacity because of the sponge effect in the insoluble cellular matrix. Fine grinding results in a loss of interior space, resulting in an increase in density and a decrease in water-holding capacity.

We have applied and modified systems for particle size measurement used for animal feeds to human foods. A special apparatus was developed for dealing with small samples (Allen et al., 1984) has been used on fiber pills and other similar products. The separation of particle sizes within a compound pill is of great aid in identifying sources of fiber components in mixed preparations.

Fermentation rate - The rate of fermentation can be measured by means of a modification of the procedure of Tilley and Terry (1963) using rumen fluid. Human fecal inoculum can be substituted with usually similar results (Jeraci, 1984). If the whole food is fermented, the analysis of the residual fiber is required at certain times during the incubation.

An alternative procedure is the measurement of gas generated by fermentation (Menke et al., 1979; Jeraci and Horvath, 1989). This requires the

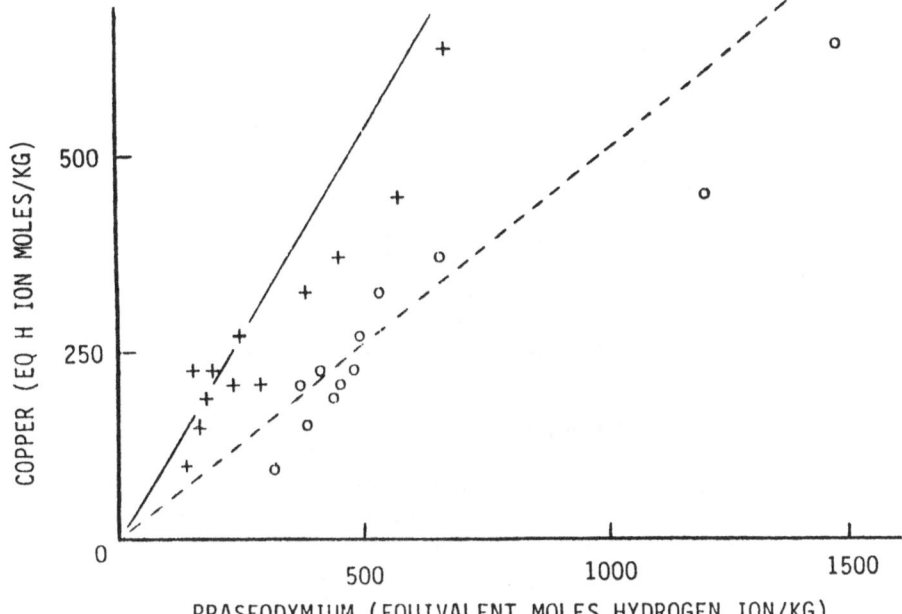

Figure 2. Comparison of three different measurements of cation exchange: Copper II at pH 3.5 and praseodymium III at pH 3.5 (+) and 7.0, (+) (Van Soest and Jones 1988).

Table 6. Effect of fiber sources upon nutrient balances in young men (mg/day). Differences relative to a low fiber control (Van Soest and Jones 1988).

Fiber Sources	Mineral					
	Mg	Zn	Cu	Mn	Ca	Fe
Cabbage	+ 70[*]	+2.4[*]	+.39[+]	+ .1	- 80[*]	-1.4[*]
Wheat bran[a]	+110[**]	-1.9	-.011	- .03	+ 15	-3.3[**]
Cellulose[b]	- 80[*]	-5.4[**]	-.29	-3.0[**]	+110[*]	-0.9

[a]Coarse and fine no significant difference averaged values

[b]Solka floc[R]

[+] P < .1; [*] P < .05; [**] P < .01

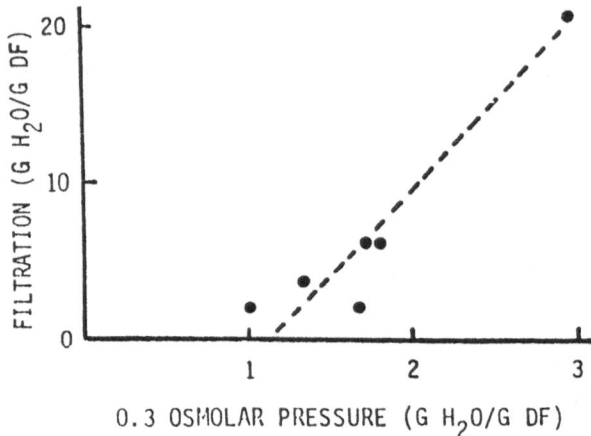

Figure 3. Relationship between filtration (suction) and osmolar
pressure measurements of hydration capacity for some
insoluble fiber sources.

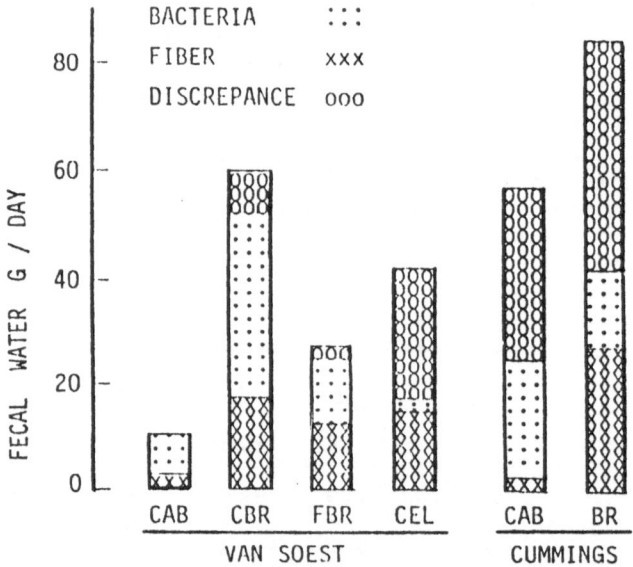

Figure 4. Histogram showing the calculated increments of water over
controls in feces based on the expected hydration
capacities of residual dietary fiber and bacteria.
Microbial mass was estimated in the Cornell Study by
metabolic fecal nitrogen, in the Cummings experiment
by different procedure (Cummings et al., 1979).

Table 7. Physical properties and fermentability of some sources of
dietary fiber (McBurney et al 1985, Horvath 1984, Van Soest
et al 1983).

Source	Cation exchange meq/kg	Hydration ml/g Dietary fiber	Dietary fiber	Unfermentable residue %
Cabbage	1920	2.8	2.3	9
Coarse bran	870	1.7	1.9	54
Fine bran	870	1.1	ND[a]	53
Wood cellulose	50	.6	.64	80
Psyllium	30	.9	3.1	72
Pectin	2270	11.0	---	0
Bacteria	---	4.0	---	---

[a]Not determined.

Table 8. Stool characteristics and transit (mean retention) time
in the Cornell Study (Heller et al 1981).

	Basal	Coarse bran	Fine bran
Mean size microns	---	744[b]	173[a]
Stool wt g/day	79[a]	144[c]	111[b]
Percent water	75.2[b]	77.6[c]	73.5[a]
Fecal nitrogen g/day	1.0[a]	1.9[b]	1.1[a]
Transit time hr	62[a]	41[d]	57[a]
Stools/week	5.2[a]	7.7[c]	7.1[b]

Numbers in the same row with different superscripts (a,b,c,d) are
significantly different.

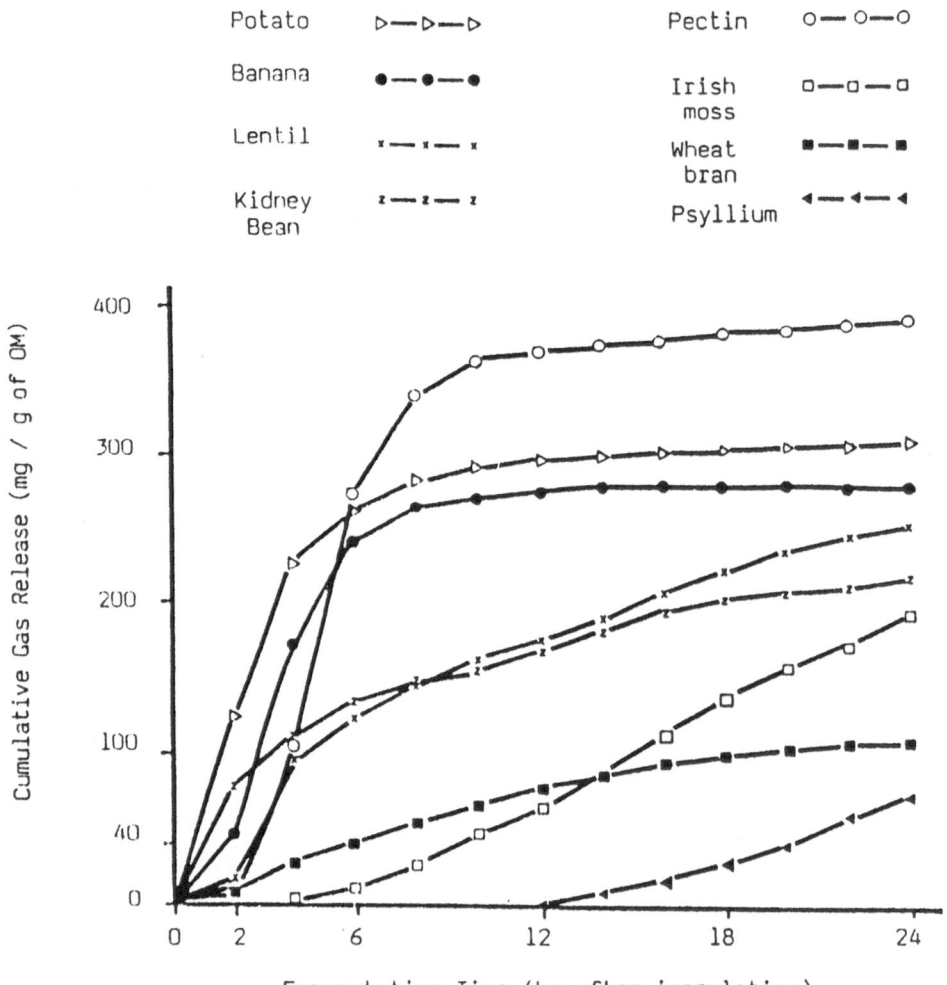

Figure 5. Cumulative gas release from eight dietary fibers by human
in vitro fermentation (banana, Irish moss, kidney bean,
lentil, pectin, potato, psyllium, and wheat bran) (Jeraci
and Horvath, 1989).

isolation of the dietary fiber free of non fiber components. Fibers vary greatly in their fermentation rates. Some kinds of pectin are the most fermentable being complete in about 12 hours, while other fiber sources are slower and often incomplete (Figure 5). Gas entrapment in cases of gel-like fibres like psyllium can lead to the underestimation of gas production (Jeraci and Horvath, 1989).

References

Allen, M.S., M.I. McBurney and P.J. Van Soest. 1986. Cation-exchange capacity of plant cell walls at neutral pH. J. Sci. Food Agric. 36:1065-1072.

Allen, M.S., J.B. Robertson, and P.J. Van Soest. 1984. A comparison of particle size methodologies and statistical treatments. In: Kennedy, P.M. ed. Techniques in Particle Size Analysis of Feed and Digesta in Ruminants. Occasional publication no. 1. Edmonton: Canadian Society of Animal Science; pp39-56.

Cummings, J.H., D.A.T. Southgate, W.J. Branch, H.S. Wiggins, H. Houston, D.J.A. Jenkins. 1979. Digestion of pectin in the human gut and its effect on calcium absorption and large bowel function. Br. J. Nutr. 41:(3):477-485.

Fonnesbeck, P.V. 1976. Estimating nutritive value from chemical analyses. P. 219-231 In P.V. Fonnesbeck, L.E. Harris and L.C. Kearl (Ed.) First International Symposium, Feed Composition, Animal Nutrient Requirements and Compterization of Diets. Utah State Univ., Logan.

Giger, S., M. Dorleans, and D. Sauvant. 1981. Adaptation of the Van Soest method to a routine determination of concnetrate feedstuffs. Commission of European Communities Workshop on Methodology of Analysis of Feedingstuffs for Ruminants. European Van Soest Ring Test. Report of meeting to discuss analytical results. MAFF, Slough Laboratory, England, 17 pp.

Glatter, S., J. Carr and B.A. Lewis. 1988. Comparative study of the (1/3 1/4) -ß-D-glucan isolated from cereal grains. (abstract). International Symposium on Cereal Carbohydrate. Edinburgh, Scotland.

Grace, A. and D.T. Gordon. 1986. Total dietary fiber analysis using enzymes of higher purity and specificity. Amer. Assoc. of Cereal Chem. (AACC), Toronto, Canada.

Heller, S.N., L.R. Hackler, J.M. Rivers, P.J. Van Soest, D.A. Roe, B.A. Lewis and J.B. Robertson. 1980. Dietary fiber: the effect of particle size of wheat bran on colonic function in young adult men. Am. J. Clin. Nutr. 33:1734.

Horvath, P.J. 1984. The measurement of dietary fiber and the effects of fermentation. PhD Thesis, Cornell Unviersity, Ithaca, NY.

Jeraci, J.L. 1984. Use of the high fiber chemostat to study interactions among gut microflora. PhD Thesis, Cornell University.

Jeraci, J.L., B.A. Lewis, P.J. Van Soest and J.B. Robertson. in press a. J. Assoc. Off. Anal. Chem.

Jeraci, J.L., B.A. Lewis, J.B. Robertson and P.J. Van Soest. in press b. Analysis of foodstuffs for dietary fiber by the urea enzymatic dialysis method. Amer. Chem. Soc. National Meeting, April 1989. Dallas, Texas.

Jeraci, J.L., T.H. Hernandez, J.B. Robertson and P.J. Van Soest. in press c. Heat-stable amylases and triethylene glycol in the neutral-detergent fiber method. J. Assoc. Off. Anal. Chem.

Jeraci, J.L. and P.J. Horvath. 1989. In vitro fermentation of dietary fiber by human fecal organisms. Animal Feed and Technology 23:121.

Keijbets, M.J.H. and W. Pilnik. 1974. Some problems in the analysis of pectin in potato tuber issue. Potato Res. 17:169-177.

Lee, S.C., E.L. Anderson and V.A. Hicks. 1988. Recommended modifications to AOAC total dietary fiber method. 102nd AOAC Annual International Meeting and Exposition. Palm Beach, FLorida, August 29-Sept. 1, pp.31.

Li, B.W. and K.W. Andrews. 1988. Simplified method for determination of total dietary fiber in foods. J. Assoc. Off. Anal. Chem. Vol. 71 p.1063.

Mascarenhas-Ferreira, A., J. Kerstens and C.H. Gast. 1983. The Study of several modifications of the neutral-detergent fibre procedure. Animal Feed Science and Technology, 9:19-26.

McBurney, M.I., M.S. Allen and P.J. Van Soest. 1986. Praseodymium and copper cation-exchange capacities of neutral-detergent fibres relative to composition and fermentation kinetics. J. Sci. Food Sgric. 37:666-672.

McBurney, M.I., P.J. Horvath, J.L. Jeraci and P.J. Van Soest. 1985. Effect in vitro fermentation using human fecal inoculum on the water-holding capacity of dietary fibre. Brit. J. Nutr. 53:17-24.

McBurney, M.I., P.J. Van Soest and L.E. Chase. 1983. Cation exchange capacity and buffering capacity of neutral-detergent fibres. J. Sci. Food Agric. 34:910-916.

McConnell, A.A., M.A. Eastwood and W.D. Mitchell. 1974. Physical characteristics of vegetable foodstuffs that could influence bowel function. J. Sci. Fd. Agric. 25:1457-1464.

McQueen, R.E. and J.W.G. Nicholson. 1979. Modification of the neutral-detergent fiber procedure for cereals and vegetables by using alpha-amylase. J. Assoc. Off. Anal. Chem. 62:676- 680.

Menke, K.H., L. Raab, A. Salewski, H. Steinglass, D. Fritz and W. Schneider. 1979. The estimation of the digestibility and metabolizable energy content of ruminant feedingstuffs from the gas production when they are incubated with rumen liquor in vitro. J. Agric. Sci., Camb. 93:217-22.

Mertens, D.R. 1988. Sources of variation in neutral detergent fiber analysis. 102nd AOAC Annual International Meeting and Exposition. Palm Beach, FLorida, August 29-Sept. 1, pp.31.

Mongeau, R. and R. Brassard. 1982. Determination of neutral detergent fiber in breakfast cereals: Pentose, hemicellulose, cellulose and lignin content. J. Food Sci. 47:550-555.

Prosky, L., N.G. Asp, I. Furda, J.W. DeVries, I.F. Schweitzer and B.F. Harland. 1984. Determination of total dietary fiber in foods and food products and total diets: Interlaboratory study. J. Assoc. Off. Anal. Chem. 67, 1044-1052.

Prosky, L., A.S.P. Nils-Georg, T.F. Schweizer, I. Furda and J.W. Devries. 1988. Determination of insoluble, soluble, and total dietary fiber in foods and food products: interlaboratory study. J. Assoc. Off. Anal. Chem. 71:1017.

Robertson, J.B. and P.J. Van Soest. 1977. Dietary fiber estimation in concentrate feedstuffs. J. Anim. Sci. 45:254.

Robertson, J.B. and P.J. Van Soest. 1981. In: The Analysis of Dietary Fiber in Foods, W.P.T. James & O. Theander (Eds.), Marcel Dekker, New York, NY. pp123-158.

Schaller, D. 1976. Analysis of cereal products and ingrediants. Cereal Foods World, 21(8):426.

Sills, V.E. and G.M. Wallace. 1982. A semi-micro neutral-detergent fibre method for cereal products. Fibre in Human and Animal Nutr., Massey Univ., New Zealand.

Stephen, A.M. and J.H. Cummings. 1979. Water-holding by dietary fibre in vitro and its relationship to faecal output in man. Gut 20, 722-729.

Taverner, M.R. and R.H. King. 1975. Prediction of the digestible energy in pig diets from analyses of fibre contents, Animal Production, 21:275.

Theander, O. & E.A. Westerlund. 1986. J. Agric. Food Chem. 34:330-338.

Tilley, J.M.A. and R.A. Terry. 1963. A two-stage technique for the in vitro digestion of forage crops. J. Br. Grassld. Soc. 18:104-111.

Trowell, H. 1976. Definition of dietary fiber and hypothesis that is a protective factor in certain diseases. Am. J. Clin. Nutr. 29:417.

Van Dokkum, W., N.A. Pikaar and J.T.N.M. Thissen. 1982. Physiological effects of fibre-rich types of bread. 2. Dietary fibre from bread: digestibility by the intestinal microflora and water-holding capacity in the colon of human subjects. Brit. J. Nutr. 50:61-74.

Van Soest, P.J. 1982. Nutritional Ecology and the Ruminant. Cornell University Press.

Van Soest, P.J., P.J. Horvath, M.I. McBurney, J.L. Jeraci and M.S. Allen. 1983. Some in vitro and in vivo properties of dietary fibers from noncereal sources. Ed. I. Furda. Amer. Chem. Soc. Washington, D.C. p.135-141.

Van Soest, P.J. and L.H.P. Jones. 1988. Analysis and classification of dietary fibre. In: Trace Element Analytical Chemistry in Medicine and Biology Volume 5. Peter Bratter and Peter Schramel Eds. Walter de Gruyter & Co., Berlin, New York. p.351.

Van Soest, P.J. and R. W. McQueen. 1973. Proc. Nutr. Soc. 32:123-130.

Van Soest, P.J. and J.B. Robertson. 1976. Chemical and physical properties of dietary fibre. Proc. of the Miles Symposium. Nutr. Society of Canada. Halifax, N.S. pp.13-25.

Van Soest, P.J. and R.H. Wine. 1967. Use of detergents in the analysis of fibrous feeds. IV. Determinations of plant cell wall constituents. J. Assoc. Off. Anal. Chem. 50:50.

Wainman, F.W., P.J.S. Dewey, and A.W. Boyne. 1981. Compound feedingstuffs for ruminants. Feedingstuffs Evaluation Unit, Rowett Research Insititute, Bucksburn, Aberdeen, Third report, 49 pp.

Key words: Total dietary fiber, neutral-detergent fiber, pectin, physical properties and fermentation

DIETARY FIBER ANALYSIS AND NUTRITION LABELLING

Thomas F. Schweizer

Nestlé Research Centre
Nestec Ltd., Vers-chez-les-Blanc
CH - 1000 Lausanne 26, Switzerland

INTRODUCTION

Ever since the rediscovery of dietary fiber in the early Seventies there has been a lively discussion about the most valid way of defining and analyzing this nutritionally important entity. In recent years, the increasing interest from consumers and the food industry in dietary fiber has resulted in more and more food labels carrying a declaration of fiber content. While, at first sight, such an extension of consumer information could appear as trivial it has attracted much more interest among nutritionists, analysts and legislative bodies than other food components, probably because of the simultaneous increase in interest in health claims and because of the complex nature of dietary fiber.

This interest has been strongly focused on unresolved conceptual and analytical questions but less on whether these questions are relevant for food labelling. In addition, the discussions led in a controversial manner and the failure to reach a consensus gave the impression that data good enough to appear on food label data could not be obtained. Indeed, the recently published Surgeon General's Report on Nutrition and Health stated that total fiber might be labelled "to the extent permitted by analytical methods", whereas calories, protein, carbohydrate, fats, cholesterol, sodium, vitamins and minerals were mentioned without such a query (1).

This chapter therefore attempts to relate a few selected conceptual and analytical fiber issues to the needs of an informative and equilibrated nutrition labelling.

DIETARY FIBER DEFINED BY AN OPERATIONAL CONVENTION

The minimum information usually provided by nutrition labels relates to the contents of the macronutrients protein, fat and carbohydrate, the latter corresponding most often to a calculated "by difference" value. When introducing dietary fiber as a separate entity on the label, it should obviously be distinguished from the remaining carbohydrates in the nutritionally most relevant manner.

In Trowell's original concept (2) dietary fiber was described as being resistant to hydrolysis by the enzymes of man. This definition respects the fundamental difference between the "available" carbohydrates which are digested and absorbed in the small intestine, and the "unavailable" carbohydrates which escape small intestinal digestion, but are partially fermented by the colonic microflora. The most widely accepted working definition of dietary fiber, as the sum of lignin and polysaccharides that are not hydrolysed by the endogenous secretions of the human digestive tract (3,4), is also based on the difference between digestible and non-digestible polysaccharides.

This physiologically meaningful definition cannot be met perfectly by current analytical methods. For example, the extent of starch malabsorption cannot be precisely mimicked in vitro. Therefore , a purely chemical definition of dietary fiber as non-starch polysaccharide has been proposed (5). However, this would exclude lignin and any form of undigestible starch. In addition, these compounds would unnecessarily appear as "available" carbohydrates on the food labels. Finally, because of the enormous complexity of food carbohydrates, even such a narrow and apparently simple definition could not be met by any known analytical method. Indeed, a number of delimitation problems exist for any definition, as discussed in this book by Asp.

It can therefore be concluded that an operational definition of dietary fiber must include conventions and that for the purpose of food labelling dietary fiber should be defined as what is measured by the method agreed upon. Similar operational definitions are in use for other food components (protein is defined as total nitrogen times a factor) without major application problems. Consequently, it is important to know the possibilities and limitations of the various analytical methods proposed.

METHODS FOR LABELLING OF TOTAL DIETARY FIBER

The botanical, physiological and chemical aspects of the dietary fiber concept find their correspondence in three main analytical principles and a considerable number of methods which have been repeatedly reviewed and updated (6-9). Therefore, only the perspectives for labelling will be considered for each method type.

Plant Cell Wall Methods

These methods are certainly suitable tools for providing more knowledge about cell wall architecture, especially when combined with polysaccharide structural analysis and microscopy. In general, however, they are not well adapted for routine use, pose major problems with mixed and processed foods, and have difficulties in coping with soluble non-wall polysaccharides.

Recently, the well-known and established procedure for measuring neutral detergent fiber (NDF) has been combined with an entirely separate determination of hot-water soluble (SOL) fiber (10). This combination of methods is claimed to give values for total dietary fiber (TDF), as the sum of NDF and SOL, which are in good agreement with TDF measured according to the official AOAC method. The method has of course the practical disadvantage that it needs two separate runs for only one TDF value. A conceptual disadvantage is the risk of analyzing some fiber components twice or not at all. It is unlikely that a buffered detergent

266

solution at pH 7 solubilises as much fiber as hot water at an uncontrolled pH (11).

Chemical Methods

The chemical methods aim at measuring dietary fiber as the sum of its constituents, usually after acid hydrolysis of the fiber polysaccharides. Two methods widely used in fiber research (12, 13) analyse the derivatized monosaccharides by gas-liquid chromatography (GLC). Because the methods for labelling the basic constituents of foods should be suitable for routine use, these quite difficult and demanding GLC methods are generally considered as unsuitable.

Recently, therefore, there has been renewed interest in colorimetric procedures (14, 15). However, colorimetric methods for measuring total reducing sugars, e.g. by the dinitrosalicylic acid assay (15), exhibit different colour yields for various sugars and are therefore not well adapted to the complex mixtures resulting from acid hydrolysis of fiber polysaccharides. Alternatively, different colorimetric assays can be combined to account separately for different sugar classes, but this complicates the analysis and has other limitations (14, 16).

Finally, it appears that the most critical step in chemical methods, namely the acid hydrolysis, may not be as yet optimal in some of the methods described (9, 10). Nevertheless, the GLC and colorimetric methods of Englyst have been recommended as an official method in the UK (12).

Enzymatic-Gravimetric Methods

These methods go back to the determination of indigestible residues, already performed in the last century (17). They have been improved and refined ever since, and the major developments have been reviewed (6, 7, 9, 18). Recently, the joined efforts of several groups have resulted in an official AOAC-method for total dietary fiber which was accepted after two international collaborative studies (19, 20). Subsequently, this method has been approved as the legal or recommended procedure for food analysis in at least ten countries including the USA, the Nordic countries, West Germany and Switzerland. Further validation studies have been published recently (21-24).

The AOAC-method is at present probably the most widely used fiber method for food-labelling and quality control. It is admittedly less rapid than previous gravimetric routine methods such as the crude fiber or NDF methods, but gives much more meaningful values for TDF because it includes soluble fiber. The similarly rapid method of Asp uses different enzymes (25), but gives results in agreement with the AOAC-method (19).

The AOAC-method and other enzymatic-gravimetric methods have been criticized for measuring a chemically undefined entity without any predictive value for physiological effects (5, 12). These criticisms do not appear founded when remembering what can be reasonably expected from a routine analytical method. First, none of the currently accepted methods for establishing the analytical composition of foods in terms of total protein, fat, ash, starch, etc. predict physiological effects. Second, gravimetric residues can be characterized in terms of monosaccharide and uronic acid components of the fiber polysaccharides and of Klason lignin, to obtain the same detailed information as in the chemical

GLC methods discussed above. This has been repeatedly demonstrated for different gravimetric methods including the official AOAC method (26-28). It has also been shown that the agreement between enzymatic-gravimetric and advanced GLC methods is good (29).

SEPARATE LABELLING OF INSOLUBLE AND SOLUBLE FIBERS

There is an increasing tendency to subdivide broadly dietary fibers into insoluble (IDF) and soluble dietary fibers (SDF), and several of the more recent methods for dietary fiber measurement include the possibility of determining IDF and SDF separately (12, 13, 22, 25, 26). This is usually done either by analyzing IDF and SDF separately or by calculating SDF from the analysis of TDF and IDF in two separate runs. The latter approach could indeed avoid some technical difficulties encountered with SDF determinations (22).

The main question, however, is whether there is a sufficient rationale for making such a subdivision on a food label. It is true that numerous particulate fibers which are highly insoluble, such as cellulose or corn bran or seed hulls, have different physiological effects compared to those of typical soluble fibers such as pectin or guar. However, most fibers found in a mixed diet are intermediate types and are organized in a cell-wall structure. Their physiological effects certainly do not fit into such a simplified classification. Furthermore, for none of the methods providing separate measurement of IDF and SDF, be it directly or by difference, has the physiological relevance of the separation been convincingly demonstrated. In addition, the solubility of dietary fibers is very much method-dependent and several methods bear little resemblance to physiological conditions. It has been shown (11, 26, 30) that "physiological" methods including both an acidic and a neutral protein digestion step (25, 26) would lead to higher fiber solubility than other methods.

Reservations concerning the separate labelling of insoluble and soluble dietary fibers are also supported by recent in vitro and in vivo experiments pertaining to colonic fermentation of dietary fiber. Cellulose, a typical insoluble dietary fiber, is highly fermentable when present in apple cell-walls, but is much less fermentable when present in wheat bran (31). Similarly, in a rat experiment, it has been shown that pea and carrot fibers were equally fermentable (32) despite different proportions of insoluble and soluble dietary fiber (28).

At the present time, therefore, the evidence for labelling insoluble and soluble fibers separately is not sufficiently convincing. Indeed, such an indication could be irrelevant or even misleading.

COLLABORATIVE STUDIES

For obvious reasons, only reproducible methods can be accepted for nutritional labelling, and it is therefore interesting to compare the results of collaborative studies conducted with the AOAC method internationally and with the Englyst method in Great Britain.

For the AOAC-method, the average reproducibilities (R) for TDF, calculated according to ISO-norm 5725, decreased from a high 5.4 in the interlaboratory study (19) to an acceptable 2.0 and 2.7 in two AOAC collaborative studies (20,

22). A further improvement of R, namely to 1.1, was reached in a Swiss collaborative trial (21), although the 15 participating laboratories had little experience in fiber analysis. The fact that a pre-trial sample with a known target fiber content was provided may have contributed to such an excellent result.

In comparison, the average R-values of 2.6 and 3.2 reached with the GLC and colorimetric methods of Englyst in a study with a pre-trial sample appear somewhat high (33). In addition, only cereals with fiber contents below 11% were included in the study. Interestingly, in the same study, the colorimetric method gave higher fiber contents than the GLC method in all seven samples analyzed. A possible reason for this could be incomplete acid hydrolysis which affects dietary fiber values from GLC methods more than from colorimetric methods, since the latter measure in part the di- and oligosaccharides.

In summary, only the AOAC method for TDF and perhaps the Englyst GLC method have given acceptable results in collaborative studies, but a GLC method appears to be unsuitable for the purpose of labelling TDF.

PRESENT AND FUTURE CHOICES

When balancing the various analytical principles against global nutritional considerations and the practical needs of food and nutrition labelling, it appears that the enzymatic-gravimetric methods can offer more advantages and versatility than other approaches, for both present and future needs.

1. The official AOAC method (20, 22) provides a value for total dietary fiber (TDF) which can be measured and reproduced in food control laboratories, without special equipment or training. After suitable collaborative validation, the method of Asp (25) with its cheap and generally available enzymes could provide a useful alternative. Such TDF values, taken together with the ingredients listed on a food label in descending order of predominance according to weight, even give some information on expected physiological effects. Labelling of TDF is relevant, clear and more informative than total carbohydrate.

2. The same enzymatic-gravimetric methods can provide separate values for insoluble and soluble dietary fiber, but the physiologically most relevant separation remains to be found. Until then, solubility alone should not be singled out and overemphasized on food labels. However, the proteolytic step(s) included in enzymatic-gravimetric methods are likely in the future to confer one more advantage over other methods.

3. Gravimetric residues are a suitable starting material for the detailed characterization of all fiber components. In contrast to chemical methods, however, the detail is an option and not a must.

   - The sugar composition of the non-starch polysaccharides can be determined as with GLC methods. This information is unlikely to play a direct role on food labels in the nearer future, but it could become useful for establishing ingredient identities. For example, "soy fiber" could designate fiber from either hulls or cotyledons, which have very different compositions (34). Similarly, both oat hulls and oat bran could hide behind the loose designation "oat fiber" (35).

- The analysis of <u>resistant starch</u> (RS) in fiber residues is convenient and easier than with other methods. At present, however, there is little reason to indicate RS separately or even to exclude it from the definition of dietary fiber.

- In principle, the measurement of <u>lignin</u> should be possible, but it is generally known that the material remaining after acid hydrolysis, i.e. Klason lignin, does not only represent true lignin. This fact calls for improvements in lignin measurement rather than for excluding it from the fiber concept.

Formally, the component analysis of fiber can be compared with the analysis of fatty acids or amino acids. However, in contrast to these, the component sugars of fiber are not amenable to absorption and have no direct metabolic effects. Their value on a label would therefore be less, especially when considering the importance of the botanical and physical appearance of fiber in the food.

The above considerations about the advantages and shortcomings of various analytical approaches for food labelling have not touched upon an issue of increasing importance not only for food labelling. It is well recognized that certain fiber polysaccharides escape determination in fiber methods because they are soluble in aqueous 78% ethanol and/or are dialyzable (chapter of Asp, 7), but these are generally considered of minor importance quantitatively, similarly as the indigestible sugars in legumes. With the increasing interest in industrially available special indigestible poly- and oligosaccharides (polydextrose, partially hydrolysed polysaccharides, fructooligosaccharides, sugar alcohols, etc.) new aspects of both analysis and labelling call for new choices which need to be prepared for the future.

REFERENCES

1   Anonymus. In: The Surgeon General's Report on Nutrition and Health. *U.S. Department of Health and Human Services* . Washington, p. 18 (1988)

2   Trowell, H. Crude fibre, dietary fibre and atherosclerosis. *Atherosclerosis* 16, 138-140 (1972)

3   Trowell, H., Southgate, D.A.T., Wolever, T.M.S., Gassull, A.R. and Jenkins, D.J.A. Dietary fibre redefined. *Lancet* I, 967 (1976)

4   Southgate, D.A.T., Hudson, G.J. and Englyst, H. The analysis of dietary fibre. The choices for the analyst. *Journal of the Science of Food and Agriculture* 29, 979-988 (1978)

5   Englyst, H.N., Trowell, H., Southgate, D.A.T. and Cummings, J.H. Dietary fiber and resistant starch. *American Journal of Clinical Nutrition* 46, 873-874 (1987)

6   Asp, N.-G. and Johansson, C.-G. Dietary fiber analysis. *Nutrition Abstracts and Reviews* 54, 735-752 (1984)

7   Schweizer, T.F. Progress in dietary fibre analysis. *Mitteilungen aus dem Gebiete der Lebensmitteluntersuchung und Hygiene* 75, 469-483 (1984)

8   Selvendran, R.R., Stevens, B.J.H and DuPont, M.S. Dietary fiber: chemistry, analysis and properties. *Advances in Food Research* <u>31</u>, 117-209 (1987)

9   Schweizer, T.F. Dietary Fibre Analysis. *Lebensmittel Wissenschaft und Technologie* <u>22</u>, 54-59 (1989)

10   Mongeau, R. and Brassard, R. A rapid method for the determination of soluble and insoluble dietary fiber: Comparison with AOAC total dietary fiber procedure and Englyst's method. *Journal of Food Science* <u>51</u>, 1333-1336 (1986)

11   Graham, H., Rydberg, M.G. and Aman, P. Extraction of soluble dietary fiber. *Journal of Agricultural and Food Chemistry* <u>36</u>, 494-497 (1988)

12   Englyst, H.N. and Cummings, J.H. Improved method for measurement of dietary fiber as non-starch polysaccharides in plant foods. *Journal of the Association* of Official Analytical Chemists <u>71</u>, 808-814 (1988)

13   Theander, O. and Westerlund, E.A. Studies of dietary fiber. 3. Improved procedures for analysis of dietary fiber. *Journal of Agricultural and Food Chemistry* <u>34</u>, 330-336 (1986)

14   Faulks, R.M. and Timms, S.B. A rapid method for determining the carbohydrate component of dietary fibre. *Food Chemistry* <u>17</u>, 273-287 (1985)

15   Englyst, H.N. and Hudson, G. Colorimetric method for routine mesurement of dietary fibre as non-starch polysaccharides. A comparison with gas-liquid chromatography. *Food Chemistry* <u>24</u>, 63-76 (1987)

16   Theander, O. and Westerlund, E. Determination of individual components of dietary fiber. In: Handbook of Dietary Fiber in Human Nutrition (G.A. Spiller, ed.), CRC Press, Boca Raton, 57-75 (1986)

17   Stutzer, A. and Isbert, A. Untersuchungen über das Verhalten der in Nahrungs- und Futtermitteln enthaltenen Kohlenhydrate zu den Verdaungsfermenten. *Zeitschrift für Physiologische Chemie* <u>12</u>, 72-130 (1888)

18   Schweizer, T.F. Methoden zur Bestimmung von Nahrungsfasern. In: Nahrungsfasern - Dietary fibres (R. Amadò and T.F. Schweizer, eds.), Academic Press, London, 53-73 (1986)

19   Prosky, L., Asp, N.-G., Furda, I., DeVries, J.W., Schweizer, T.F. and Harland, B. Determination of total dietary fiber in foods, food products, and total diets: Interlaboratory study. *Journal of the Association of Official Analytical Chemists* <u>67</u>, 1044-1052 (1984)

20   Prosky, L., Asp, N.-G., Furda, I., DeVries, J.W., Schweizer, T.F. and Harland, B. Determination of total dietary fiber in foods and food products: Collaborative study. *Journal of the Association of Official Analytical Chemists* <u>68</u>, 677-679 (1985)

21   Schweizer, T.F., Walter, E. and Venetz, P. Collaborative study for the enzymatic, gravimetric determination of total dietary fibre in foods. *Mitteilungen aus dem Gebiete der Lebensmitteluntersuchung und Hygiene* <u>79</u>, 57-68 (1988)

22   Prosky, L., Asp, N.-G., Schweizer, T.F., DeVries, J.W. and Furda, I. Determination of insoluble, soluble and total dietary fiber in foods and food products: Interlaboratory study. *Journal of the Association of Official Analytical Chemists* 71, 1017-1023 (1988)

23   Rabe, E. Bestimmung der unlöslichen und löslichen Ballaststoffe. *Getreide, Mehl und Brot* 41, 302-305 (1987)

24   Rabe, E., Seibel, W., Suckow, P. and Meuser, F. Vergleichende Bestimmung von unlöslichen, löslichen und Gesamtballaststoffen in Getreideerzeugnissen. *Getreide, Mehl und Brot* 42, 297-305 (1988)

25   Asp, N.-G., Johansson, C.-G., Hallmer, H. and Siljeström, M. Rapid enzymatic assay of insoluble and soluble dietary fiber. *Journal of Agricultural and Food Chemistry* 31, 476-482 (1983)

26   Schweizer, T.F. and Würsch, P. Analysis of dietary fibre. *Journal of the Science of Food and Agriculture* 30, 613-619 (1979)

27   Marlett, J.A. and Navis, D. Comparison of gravimetric and chemical analyses of total dietary fiber in human foods. *Journal of Agricultural and Food Chemistry* 36, 311-315 (1988)

28   Nyman, M., Pålsson, K.-E. and Asp, N.-G. Effects of processing on dietary fibre in vegetables. *Lebensmittel-Wissenschaft und Technologie* 20, 29-36 (1987)

29   Nyman, M., Siljeström, M., Pedersen, B., Bach Knudsen, K.E., Asp, N.-G., Johansson, C.-G. and Eggum, O. Dietary fiber content and composition in six cereals at different extraction rates. *Cereal Chemistry* 61, 14-19 (1984)

30   Marlett, J.A., Chesters, J.G., Longacre, M.J. and Bogdanske, J.J. Recovery of soluble dietary fiber is dependent on the method of analysis. *American Journal of Clinical Nutrition* 50, 479-485 (1989)

31   Stevens, B.J.H. and Selvendran, R.R., Bayliss, C.E. and Turner, R. Degradation of cell wall material of apple and wheat bran by human faecal bacteria in vitro. *Journal of the Science of Food and Agriculture* 44, 151-166 (1988)

32   Nyman, M., Schweizer, T.F., Tyrèn, S., Reimann, S. and Asp, N.-G. Fermentation of vegetable fiber in the intestinal tract of rats and effects on fecal bulking and bile acid excretion. *Journal of Nutrition* (in press)

33   Englyst, H.N. and Cummings, J.H. Analyse der Polysaccharide pflanzlicher Lebensmittel. In: 22nd Congress Proceedings. DGQ (Deutsche Gesellschaft für Qualitätsforschung), Kiel, 21-50 (1987)

34   Schweizer, T.F., Bekhechi, A.R., Koellreutter, B., Reimann, S., Pometta, D. and Bron, B.A. Metabolic effects of dietary fiber from dehulled soybeans in humans. *American Journal of Clinical Nutrition* 38, 1-11 (1983)

35   Nyman, M. and Asp, N.-G. Fermentation of oat fiber in the rat intestinal tract: a study of different cellular areas. *American Journal of Clinical Nutrition* 48, 274-278 (1988)

THE UPPSALA METHOD FOR RAPID ANALYSIS OF TOTAL DIETARY FIBER

Olof Theander[1], Per Åman[1,2], Eric Westerlund[1] and
Hadden Graham[2]

[1]Department of Chemistry and
[2]Department of Animal Nutrition and Management
Swedish University of Agricultural Sciences
S-750 07 Uppsala

INTRODUCTION

An understanding of the nutritional effects of dietary
fiber has been considerably hampered by the lack of an appropriate definition, and consequently of adequate analysis
methods, for this component. Based on physiological criteria,
Trowell (1972) defined dietary fiber as "the remnants of the
plant cell-wall that are not hydrolysed by the alimentary
enzymes of man", and this was later simplified and expanded to
include "the plant polysaccharides and lignin which are resistant to hydrolysis by the enzymes of man" (Trowell et al.,
1976). In 1979 we proposed that dietary fiber could be defined
as the sum of non-starchy polysaccharides and Klason lignin,
and, in conjunction with this chemical definition, published
the first method (the Uppsala method) for the analysis and
characterization of fibers (Theander and Åman, 1979). As this
analytical procedure includes 'starch' resistant to α-amylases
and, as part of the Klason lignin complex, other indigestible
components such as tannins, cutins and Maillard products, this
definition conforms well with the original definitions of
Trowell and co-workers.

The Uppsala method is based on the prior removal of free
sugars and starch, including treatment with thermostable α-
amylases, followed by the determination of neutral non-starch
polysaccharide residues as alditol acetates by GLC. Klason
lignin, the non-carbohydrate fraction of dietary fiber, is determined gravimetrically as the acid-insoluble material and
uronic acid residues by decarboxylation. Recently the method
was improved to essentially a one-tube procedure with an increased speed of analysis (Theander and Westerlund, 1986a).

Besides the Uppsala method, two other methods, the 'UK'
method and the AOAC method, are widely employed in the analysis
of dietary fiber. The former, which is based on the work of
Southgate (1969), is today relatively similar to the Uppsala
procedure, and, for example, recently included correction fac-

tors in the GLC determination of neutral polysaccharide residues (Englyst and Cummings, 1988). However, the 'UK' method does
not include any measure of Klason lignin, and has an extraction
step for the removal starch not hydrolyzed by α-amylases but
soluble in aqueous alkali or dimethylsulfoxide, the so-called
'resistant' starch (Englyst and Cummings, 1988). The 'Lund'
procedure, on which the AOAC method (Prosky et al., 1985) is
mainly based, is also similar to the Uppsala method in that it
includes the degradation of starch by the thermostable α-
amylases (Asp et al., 1983). However this step is followed by
the enzymatic degradation of protein and the gravimetric
determination of the resultant dietary fiber residue. Although
this method was originally designed to be fast and simple, the
need to correct the fiber residue for contamination by protein,
ash and, in some cases, other components has rendered this
procedure rather laborious.

In this paper we describe the improved Uppsala method and
compare this method with the 'UK' and 'Lund' methods for the
determination of dietary fiber in divergent food samples. We
also briefly discuss some pertinent problems with fiber analysis, including the recovery of soluble fibers during ethanol
precipitation, the occurrence of minor fiber constituents such
as acetyl and phenolic groups and the effects of heat treatment
on the fiber content in foods.

THE UPPSALA METHOD FOR ANALYSIS OF TOTAL DIETARY FIBER

A brief scheme for the analysis of total dietary fiber and
individual dietary fiber components by the Uppsala method is
presented in Fig. 1 (Theander et al., 1989). Representative
samples of dry materials are ground to pass a 0.5 mm screen.
Samples with higher water contents are freeze-dried and ground
or homogenized directly. If samples contain more than 6% fat
pre-extraction with petroleum ether is recommended. Starch and
sugars are removed by incubation with a potent thermostable α-
amylase (Termamyl), minimizing the risk for retrogradation, and
thereafter with amyloglucosidase. Soluble fibers are precipitated with 80% ethanol and soluble and insoluble fibers recovered from low-molecular weight sugars and degraded starch by
centrifugation. Total dietary fibers are hydrolyzed, using a
Klason (sulfuric acid) procedure including swelling in 12.0 M
sulfuric acid at 30°C and hydrolysis in 0.136 M sulfuric acid
in an autoclave at 121°C after addition of myo-inositol as
internal standard. The hot solution is filtered through a glass
filter and the ash-free residue weighed as Klason lignin. The
neutral polysaccharide residues in the filtrate are reduced and
acetylated using 1-methylimidazole as a catalyst (Blakeney et
al., 1983) and the formed alditol acetates quantified by GLC on
a DB-225 capillary column using correction factors for the
individual sugars. Uronic acids are determined on a separate
sample of the original material using a decarboxylation procedure. Released CO_2 is trapped in aqueous sodium hydroxide and
quantified conductivimetrically (Theander and Åman, 1979).

The method has been applied to a number of divergent samples, including foods, feeds, digesta and feces, and has proven
to be robust, reproducable, adaptable and accurate. For example, eight totally independant analyses of a purified cotton
linter gave an average cellulose content of 99.0% (SD 1.3) and

xylan content of 0.25%. Using this improved procedure one skilled analyst can now run over forty samples per week. This method can easily be adapted to separately analyze soluble and insoluble fiber components. In this case the soluble fibers are separated after the starch degradation and recovered by dialysis and freeze-drying (Theander and Åman, 1979). It is, however, important to remember that the yield of soluble fibers is very much dependant on the experimental conditions used (Graham et al., 1988).

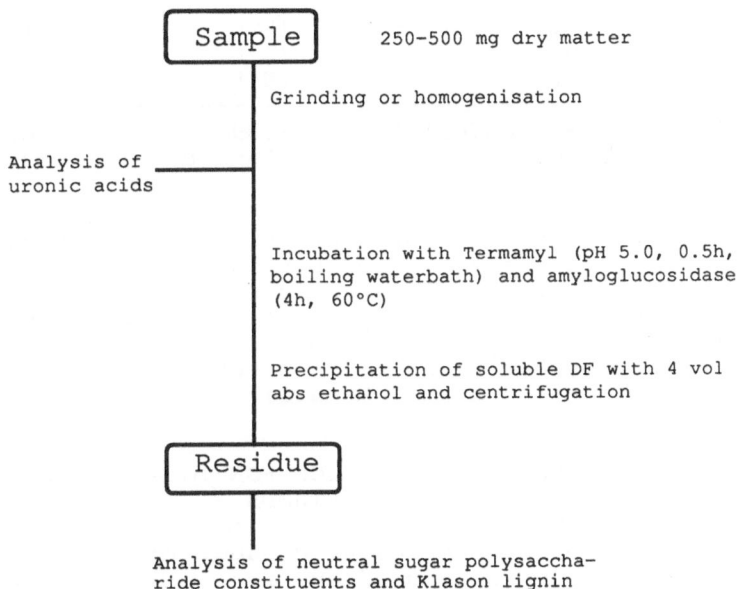

Fig. 1. Rapid analysis of dietary fiber by the Uppsala method.

If desired, 'resistant' starch can be determined on the original sample after initial removal of starch by two incubations with Termamyl and one with amyloglucosidase. 'Resistant' starch is then solubilized in 2 M aqueous potassium hydroxide, hydrolyzed with amyloglucosidase and released glucose determined by the glucose oxidase method (Westerlund et al., 1989).

It should be emphasised that almost all fiber analysis methods depend on the enzymatic removal of starch, and thus it is imperative that these enzymes are free from fiber-degrading activity. Unfortunately it would appear that many of the amyloglucosidase preparations at present in use have, for example, β-glucanase activity, and will thus give low values for the fiber contents in cereals. Therefore all new enzyme preparations and batches should be checked for fiber-degrading activity prior to use for fiber analysis.

Table 1. Content of dietary fiber (DF) as determined by the
Uppsala[a] and 'Lund'[b] methods, non-starch polysaccharides (NSP) as determined by the 'UK'[c] method,
and of resistant starch[d] (RS) and Klason lignin[a] (KL)
in some food products (% of dry matter).

Product	Uppsala method			Lund method	UK method	NSP + KL + RS
	DF	KL	RS	DF	NSP	
Corn flakes	2.7	0.3	1.5	4.5	0.8	2.6
Bread crust	3.7	0.3	0.5	4.6	2.7	3.5
Bread crumb	3.8	0.3	0.8	4.7	2.7	3.8
Rye crisp	23.2	2.1	0.2	27.0	20.0	22.3
Green peas	14.5	0.3	0.3	15.6	12.8	13.4
Soybean	16.8	0.8	nd	18.6	15.1	15.9
Deskinned onion	20.1	1.0	nd	22.6	19.2	20.2
Sugar-beet fiber	64.7	1.3	nd	76.5	67.0	68.3

nd not detected.   [a]Theander et al., 1989.   [b]Asp et al., 1983.
[c]Englyst and Cummings, 1988.   [d]Westerlund et al.,1989.

In the determination of neutral non-starch polysaccharide
residues by GLC it is possible to omit the reduction step and
directly acetylate the hydrolysate. Quantification of the
multiple peaks of the individual sugars can be based either on
simple isometric peaks by conventional multi-points calibration
or, with somewhat better results, on several peaks by partial
least square (PLS) calibration (Hämäläinen et al., 1989).

COMPARISON BETWEEN THE UPPSALA METHOD FOR TOTAL DIETARY FIBER
ANALYSIS AND THE 'UK' AND 'LUND' METHODS

In this study eight samples were used. These included corn
flakes, dry green peas, deskinned and freeze-dried onions, rye
crisp bread containing whole-rye flour and wheat bran (Wasabröd
AB), whole-fat soybeans, sugar-beet fiber (Swedish Sugar AB)
and freeze-dried crumb and crust from rolls made from high
quality wheat flour. All materials were ground to pass a 0.5 mm
screen. The rye crisp bread and whole-fat soybeans contained
more than 6% fat and were therefore pre-extracted with petroleum ether.

Total dietary fibers, including Klason lignin and 'resistant' starch, were analyzed according to the Uppsala method and
the results compared to the 'UK' method (Englyst and Cummings,
1988) and to the enzymatic-gravimetric 'Lund' method after
correction for contaminating crude protein and ash (Asp et al.,
1983). The last two analyses were kindly carried out by The
Rowett Research Institute, Aberdeen ('UK') and AnalyCen AB,
Lidköping ('Lund'), in laboratories which routinely use these
methods. 'Resistant'starch was analysed according to Westerlund
et al. (1989).

The content of total dietary fibers as determined by the
Uppsala method varied from 2.7% in corn flakes to 64.7% in
sugar-beet fiber (Table 1). Only small amounts of Klason lignin
were present in most samples but the rye crisp bread with wheat

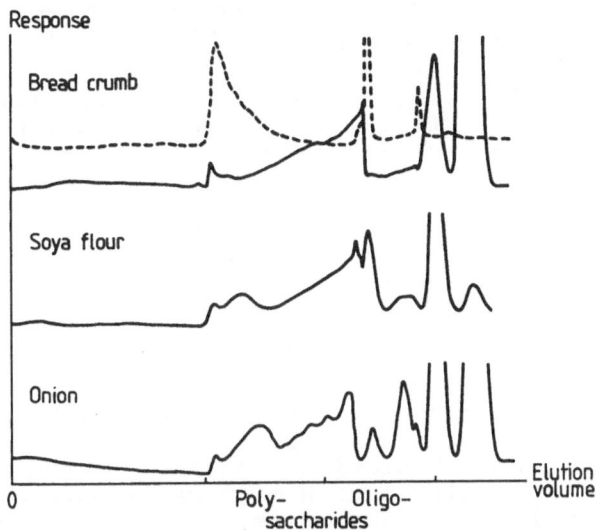

Fig. 2. Gel filtration on Biogel P-2 of carbohydrates in supernatants obtained during the 80% ethanol precipitation
of fibers in the Uppsala method. Refractometer response is shown by the solid lines and UV detector (280
nm) response by the broken line. Oligosaccharides
(DP<10) and polysaccharides (DP≥10) were separated in
comparison with a maltodextrin standard.

bran contained 2.1%. The starch-containing samples were analysed for 'resistant' starch, and corn flakes and the crumb and
crust were found to contain significant amounts. Results from
the 'UK' method were, as expected, generally lower than those
from the Uppsala method. After adding the contents of Klason
lignin and 'resistant' starch, not incuded in the 'UK' method,
to the content of non-starch polysaccharides analysed results
similar to the Uppsala method were obtained. On average, however, the results from the 'UK' method still were 2.5% lower
but varied from -7.6% for green peas to +5.6% for sugar-beet
fiber. The enzymatic-gravimetric 'Lund' method gave, on
average, 22% higher figures for dietary fibers in analyzed
samples compared to the Uppsala method. The largest difference
was obtained for corn flakes (+67%) and the smallest for green
peas (+8%). The residues from the 'Lund' method evidently contained substantial amounts of non-fiber components. An overestimation of dietary fibers by enzymatic-gravimetric methods
has previously been reported (Marlett and Navis, 1988; Marlett,
1989).

POLYSACCHARIDE CONTENTS IN SUPERNATANTS AFTER PRECIPITATION OF
SOLUBLE FIBERS WITH 80% ETHANOL

Water-soluble polysaccharides are commonly precipitated
with 80% ethanol in gravimetric as well as GLC methods used for
determination of dietary fiber. In order to investigate the

Table 2. Content of ester substituents in some high-fiber products (% of dry matter)[a]

Product	Dietary fiber	Coumaroyl groups	Feruloyl groups	Acetyl groups
Maize bran	76.4	0.2	2.4	1.5
Wheat bran	37.2	trace	0.2	0.3
Potato fiber	71.9	trace	trace	0.7
Sugar beet fiber	70.8	trace	0.5	1.6

[a]Torneport and Theander, unpublished data.

effectiveness of polysaccharide precipitation, the 80% ethanolic solutions (supernatants) obtained during the dietary fiber analysis in the present study were fractionated by gel-filtration. Based on the elution profiles observed, in comparison with that of a maltodextrin mixture, separate fractions containing poly- or oligosaccharides were collected as indicated in Fig. 2. Sugar analysis of the polysaccharide fractions established that only 1-6% of the total fiber content in the original sample was lost in the supernatants, with greater losses in the heat-treated samples. These findings are in agreement with previous observations that marginal losses of soluble fibers may occur during the ethanol precipitation step (Theander and Westerlund, 1986b).

ACETIC AND CINNAMIC SUBSTITUENTS

Acetic and cinnamic acids are often esterified to dietary fiber polymers. Acetyl substituents may be quantified as 1-acetylpyrrolidine by GLC (Månsson and Samuelsson, 1981) and phenolic acids, after release by sodium hydroxide treatment, by HPLC (Ternrud et al., 1987). In some cases these components may contribute significantly to the dietary fiber content (Table 2). Acetyl and cinnamoyl residues constituted 4.1 % of maize bran and 2.1 % of sugar beet fiber, accounting for significant amounts of the fiber polymers. Highest amounts of cinnamoyl residues were found in maize bran and high amounts of acetyl groups in sugar beet fiber and again maize bran.

EFFECTS OF HEAT-TREATMENT ON DIETARY FIBER VALUES

A number of reactions which may affect the nutritional and functional properties and the dietary fiber content of the product can take place during heat-treatment of foods. In starch-rich food items, for instance bread, formation of 'resistant' starch (Englyst et al., 1982) leads to an increase in the dietary fiber value (Siljeström and Asp, 1985). A recent study dealing with the effects of baking on polysaccharides in white bread fractions (Westerlund et al., 1989) showed that the formation of 'resistant' starch decreased in going from crumb to the outer crust (Table 3). This progressive effect was probably due to water and temperature gradients formed during the baking procedure. The total content of dietary fiber glucans (not including 'resistant' starch), on the other hand, increased in the order crumb to outer crust, mainly as a result

Table 3. Contents of dietary fiber (DF) polysaccharides and resistant starch in dough and white bread fractions following baking at 210°C for 35 minutes (% of dry matter)[a]

Component	Dough	Crumb	Inner crust	Outer crust
Resistant starch	0.09	1.02	0.62	0.30
Soluble DF-glucans	0.11	0.08	0.10	0.21
Insoluble DF-glucans[b]	0.49	0.60	0.65	0.65
Total DF-glucans[b]	0.60	0.68	0.75	0.86
Soluble arabinoxylans	0.72	0.58	0.64	0.86
Insoluble arabinoxylans	1.20	1.09	1.03	0.70
Total arabinoxylans	1.92	1.67	1.67	1.56
Total dietary fiber	2.61	3.37	3.04	2.72

[a]Westerlund et al., 1989.   [b]Resistant starch not included.

of increasing contents of water-soluble glucans. This observation strongly suggested that formation of non-starch glucans had, at least in part, occurred via fragmentation of starch and subsequent reactions of the fragments. These fragments often have 1,6-anhydroglucopyranose as end units (Theander and Westerlund, 1987;). The content of insoluble arabinoxylans decreased during baking and this was particularly apparent for the outer crust. This decrease was partly accompanied by an increase in the content of water-soluble arabinoxylans. The decrease in total arabinoxylans during baking is due to various degradation reactions. On the whole, dietary fiber values were higher in the bread fractions than in the dough, mainly due to the formation of 'resistant' starch.

Polymeric products of the Maillard reaction are also of importance for the assessment of dietary fiber in heat-treated foods. This is because such products are often insoluble in 72% sulfuric acid, giving a higher recovery of Klason lignin and

Table 4. Effect of extrusion cooking on Klason lignin content (% of dry matter) and the percent of nitrogen recovered in the Klason lignin fraction[a]

Sample	Klason lignin content	% Nitrogen recovered in Klason lignin
Wheat flour		
untreated	0.2	0.3
extruded at 168°C	0.4	2.3
Whole wheat flour		
untreated	1.4	4.7
extruded at 180°C	2.8	11.0

[a]Theander, 1983.

thus dietary fiber (Van Soest, 1965; Theander, 1983). It has been observed that the Klason lignin content in wheat flours can be increased by a comparatively short technical heat-treatment such as extrusion-cooking (Table 4). This increase partly resulted from an increase in the amount of nitrogenous compounds recovered in the Klason lignin residue due to the occurrence of Maillard reactions. The nitrogen level in the Klason lignin fraction may therefore be considered as a marker of the extent of Maillard reactions. The Klason lignin fraction has been designated as the 'non-carbohydrate' part of the dietary fiber (Theander and Westerlund, 1986a).

ACKNOWLEDGEMENTS

Tommy Ericson, AnalyCen AB, Lidköping, Sweden, and Peter Dewey, Rowett Research Institute, Aberdeen, Scotland, are gratefully acknowledged for carrying out some of the dietary fiber analysis.

REFERENCES

Asp, N.-G., Johansson, C.-G., Hallmer, H., and Siljeström, M., 1983, Rapid enzymatic assay of insoluble and soluble dietary fiber, **J. Agric. Food Chem.**, 31:476.

Blakeney, A. B., Harris, P. J., Henry, R. J., and Stone, B. A., 1983, A simple and rapid preparation of alditol acetates for monosaccharide analysis, **Carbohydr. Res.**, 113:291.

Englyst, H. N., Cummings, J. H., 1988, Improved method for measurement of dietary fiber as non-starch polysaccharides in plant foods, **J. Assoc. Off. Anal. Chem.**, 71:808.

Englyst, H., Wiggins, H. S., and Cummings, J. H., 1982, Determination of the non-starch polysaccharides in plant foods by gas-liquid chromatography of constituent sugars as alditol acetates, **Analyst,** 107:307.

Graham, H., Grön Rydberg, M.-B., and Åman, P., 1988, Extraction of soluble dietary fiber, **J. Agric. Food Chem.**, 36:494.

Hämäläinen, M., Theander, O., Nordkvist, E., and Ternrud, I. E., 1989, Multivariate calibration in the determination of aldoses as acetates by gas chromatography, to be published.

Månsson, P., and Samuelsson, B., 1981, Quantitative determination of O-acetyl and other O-acyl groups in cellulosic material, **Sven. Papperstidn.**, 84:R15.

Marlett, J., 1989, Measuring dietary fiber, **Anim. Feed Sci. Technol.**, 23:1.

Marlett, J., and Navis, D., 1988, Comparison of gravimetric and chemical analyses of total dietary fiber in human foods, **J. Agric. Food Chem.**, 36:311.

Prosky, L., Asp, N.-G., Furda, I., DeVries, J. W., Schweizer, T. F., and Harland, B. F., 1985, Determination of total dietary fiber in foods and food products: Collaborative study. **J. Assoc. Off. Anal. Chem.**, 68:677.

Siljeström, M., and Asp, N.-G., 1985, Resistant starch formation during baking. Effect of baking time and temperature and variations in the recipe, **Z. Lebensm. Unters. Forsch.**, 181:4.

Southgate, D. A. T., 1969, Determination of carbohydrates in foods. II. Unavailable carbohydrates, **J. Sci. Food Agric.**, 20:331.

Ternrud, I. E., Lindberg, J. E., and Theander, O., 1987, Continuous changes in straw carbohydrate digestion and composition along the gastro-intestinal tract in ruminants, **J. Sci. Food Agric.**, 41:315.

Theander, O., 1983, Advances in the chemical characterization and analytical determination of dietary fibre components, **in** "Dietary Fibre," G. G. Birch and K. J. Parker, eds., Applied Scientific Publishers, London.

Theander, O., and Åman, P., 1979, Studies on dietary fibres. 1. Analysis and chemical characterization of water-soluble and water-insoluble dietary fibres, **Swedish J. Agric. Res.**, 9:97.

Theander, O., Åman, P., Westerlund, E., and Graham, H., 1989, The Uppsala method for rapid analysis of total dietary fiber and its individual components, to be published.

Theander, O., and Westerlund, E., 1986a, Studies on dietary fiber. 3. Improved procedures for analysis of dietary fiber, **J. Agric. Food Chem.**, 34:330.

Theander, O., and Westerlund, E., 1986b, Determination of individual components of dietary fiber, **in** "Handbook of Dietary Fiber in Human Nutrition," G. Spiller, ed., CRC Press, Boca Raton, FL.

Theander, O., and Westerlund, E., 1987, Studies on chemical modification in heat-processed starch and wheat flour, **Stärke,** 39:88.

Trowell, H., 1972, Ischemic heart disease and dietary fiber, **Amer. J. Clin. Nutr.**, 25:926.

Trowell, H., Southgate, D. A. T., Wolever, T. M. S., Leeds, A. R., Gassull, M. A., and Jenkins, D. J., 1976, Dietary fibre redefined. **Lancet,** 1:967.

Van Soest, P. J., 1965, Use of detergents in analysis of fibrous feeds. III. Study of effects of heating and drying on yield of fiber and lignin in forages, **J. Assoc. Off. Anal. Chem.**, 48:785.

Westerlund, E., Theander, O., Andersson, R., and Åman, P., 1989, Effects of baking on polysaccharides in white bread fractions, **J. Cereal Sci.**, 10, in press.

SIMPLIFIED METHOD FOR THE DETERMINATION OF TOTAL DIETARY FIBER AND ITS

SOLUBLE AND INSOLUBLE FRACTIONS IN FOODS

Betty W. Li and Maria S. Cardozo

U. S. Department of Agriculture, Agricultural Research
Service, Beltsville Human Nutrition Research Center
Nutrient Composition Laboratory, Beltsville, MD 20705

INTRODUCTION

Since its adoption by the Association of Official Analytical Chemists
in 1985 (1, 2), an enzymatic–gravimetric procedure has been widely used
for the determination of total dietary fiber (TDF), particularly, in the
United States. Compared to other published methods (3, 4) the AOAC/TDF
method is relatively rapid, not so tedious and does not require highly
skilled analysts nor sophisticated instruments. However, its procedure
still calls for three separate enzyme hydrolysis steps and three periodic
pH adjustments.

Earlier (5) we described a simplified method based on the same
principle but somewhat different approach as the AOAC/TDF method.
Results on the TDF determinations of 12 food samples showed good
agreement between the two methods. Recently, we analyzed 25 foods in
blind duplicates as part of a collaborative study contracted by the
Nutrient Data Research Branch, Human Nutrition Information Service
(HNIS), USDA. Total dietary fiber content of these foods (based on dry
weight) was determined using the official method and the simplified
method. Soluble and insoluble fractions of dietary fiber for other
selected foods were also determined using the procedures described by
Prosky, et al. in 1988 (6) and the modifications of our simplified method
as presented in this paper.

MATERIALS

HNIS Collaborative Study samples were procured and processed by E.
Augustin of Washington State University. Samples with high water content
were initially diced in a Cuisinart food processor, then freeze–dried and
ground in an U.D. Cyclone Sample Mill fitted with a 40 mesh screen.   All
food items listed in Table I are self descriptive with the following
exceptions: apples were raw with skin; baked beans were in tomato sauce;
pie crust was baked; popcorn was air popped; Reduced calorie white bread
was made with added corn bran and soy bran. Samples were shipped in
brown plastic bottles. Portions from each of the 50 bottles were dried in
a vacuum oven at 60 °C for 3.5 h. After weighing a specific amount
(approximately 1 g for AOAC and 0.5 g for Simplified procedure), those
designated for defatting were extracted with n–hexane and dried in teflon

tubes prior to dietary fiber analysis. For AOAC determination, samples were transferred quantitatively into beakers by rinsing tubes with portions of 50 mL phosphate buffer. For Simplified procedure, 25 mL of deionized water was added to the tubes.

TOTAL DIETARY FIBER ANALYSIS

For AOAC/TDF determination, samples were analyzed by incorporating the latest revisions described by Prosky, et al. (6).

For Simplified TDF determination, samples were analyzed according to the procedures described in Reference 5. A general outline of the two methods is given in Figure 1.

SOLUBLE AND INSOLUBLE DIETARY FIBER FRACTIONATION

After enzyme hydrolysis, sample reaction mixture was filtered to separate the soluble and insoluble dietary fiber fractions.

For the AOAC procedure, the insoluble residue was washed with two 10 mL portions of deionized water. To the combined filtrate and washing was added 360 mL of 95% ethanol to precipitate the soluble dietary fiber. Subsequent treatments of the residues from both the insoluble and soluble fractions are the same as described in Reference 6 or as outlined in Figure 2.

For the Simplified procedure, the insoluble residue was washed with 35 mL of 95% ethanol and later rinsed with 10 mL of acetone. To the combined filtrate and washing was added another 100 mL of 95% ethanol. The mixture was allowed to stand at room temperature for 1 h and the precipitates were filtered; washed with two portions of 20 mL of 78% ethanol, two portions 10 mL 95% ethanol and 10 mL of acetone. Acetone-washed residues were dried at 105 °C in an air oven for at least 2 h or overnight. The dried residues were transferred to a desiccator and weighed after at least 2 h. Ash content of the residues was determined after heating at 525 °C for 5 h. Protein content of the residues was determined as Kjeldahl nitrogen (N x 6.25) on a Kel-Foss autoanalyzer. An outline of the above procedure is given in Figure 3.

RESULTS AND DISCUSSION

Table I gives the mean TDF values of 4 replicates (2 for each blind duplicates), the standard deviations and the coefficients of variation for 25 foods which were analyzed by two methods within our laboratory. The data from other participants in the collaborative study will be published elsewhere. Omitting the values for corn bran, which has the highest TDF content on the list, a regression equation was obtained indicating good agreement between the methods even for this group of diverse food samples: $y = 0.053 + 1.05x$; $r=0.99$. Statistical analysis of the variance between the duplicate runs on the same day indicated that the AOAC/TDF method has variability 3 times greater than that of the Simplified method. This is not surprising given the fact that the Simplified method has fewer operating steps, uses only one enzyme and a single buffer which does not precipitate in dilute alcohol; thus, resulting in a smaller ashless blank residue. Typically, in the AOAC/TDF method, the blank residue is between 17.5 and 21.5 mg, but in the Simplified method, it is between 4.9 and 6.1 mg. Table II demonstrates the feasibility of correcting for a relatively large protein content in

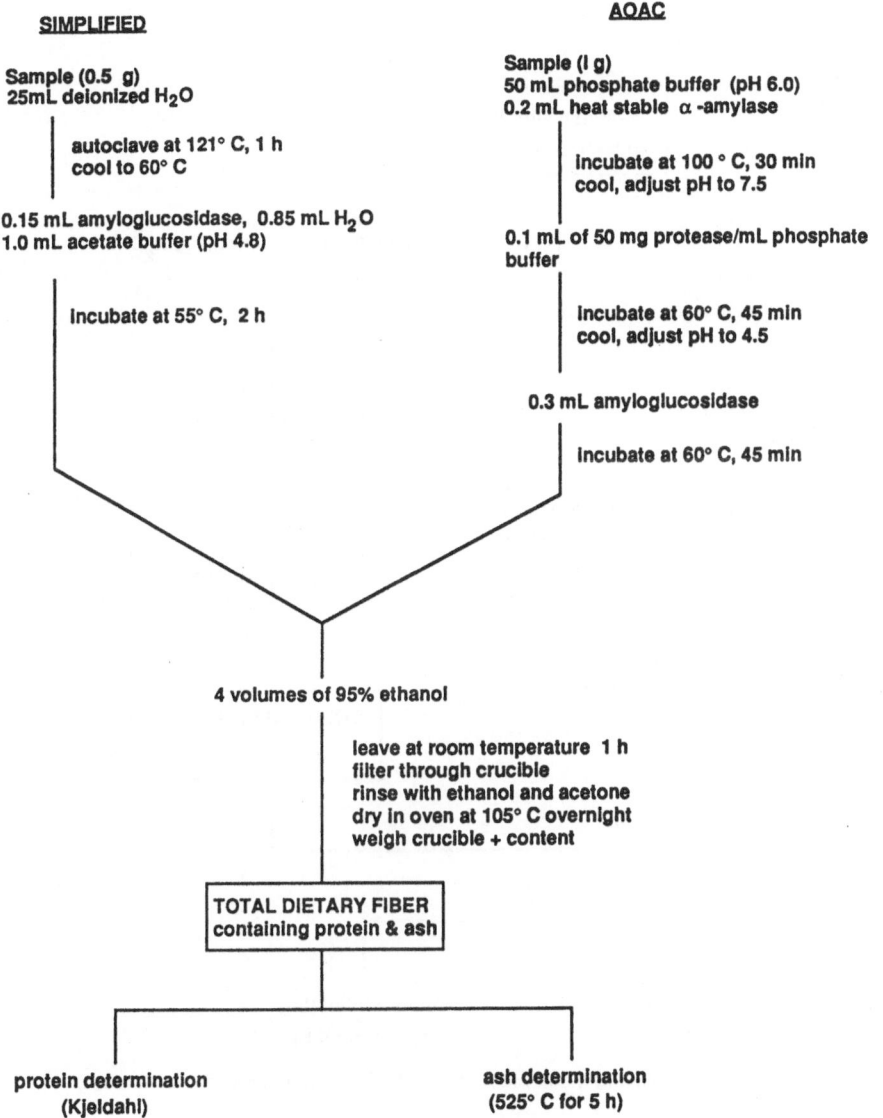

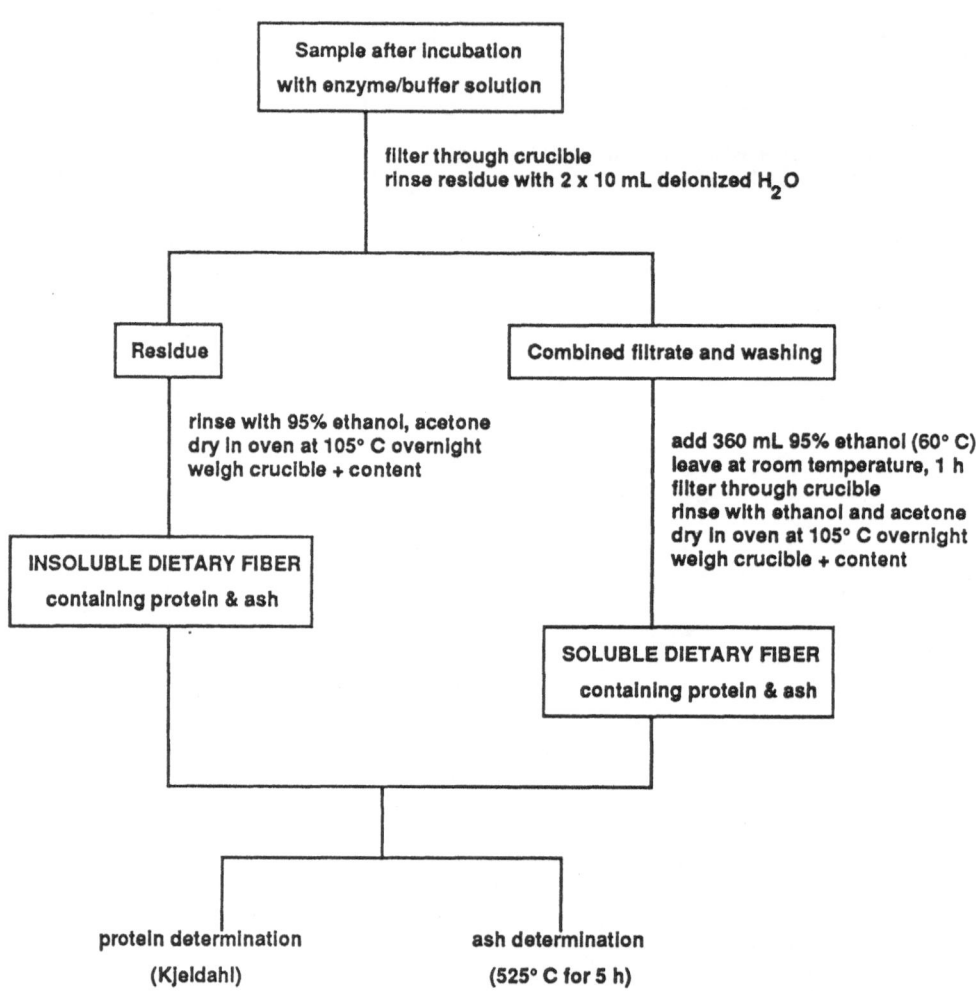

| Sample after incubation with enzyme/buffer solution |

filter through crucible
rinse residue with 2 x 10 mL deionized H_2O

| Residue | | Combined filtrate and washing |

rinse with 95% ethanol, acetone
dry in oven at 105° C overnight
weigh crucible + content

add 360 mL 95% ethanol (60° C)
leave at room temperature, 1 h
filter through crucible
rinse with ethanol and acetone
dry in oven at 105° C overnight
weigh crucible + content

| INSOLUBLE DIETARY FIBER containing protein & ash |

| SOLUBLE DIETARY FIBER containing protein & ash |

protein determination
(Kjeldahl)

ash determination
(525° C for 5 h)

Figure 2 Soluble and Insoluble Dietary Fiber Determination - AOAC Method

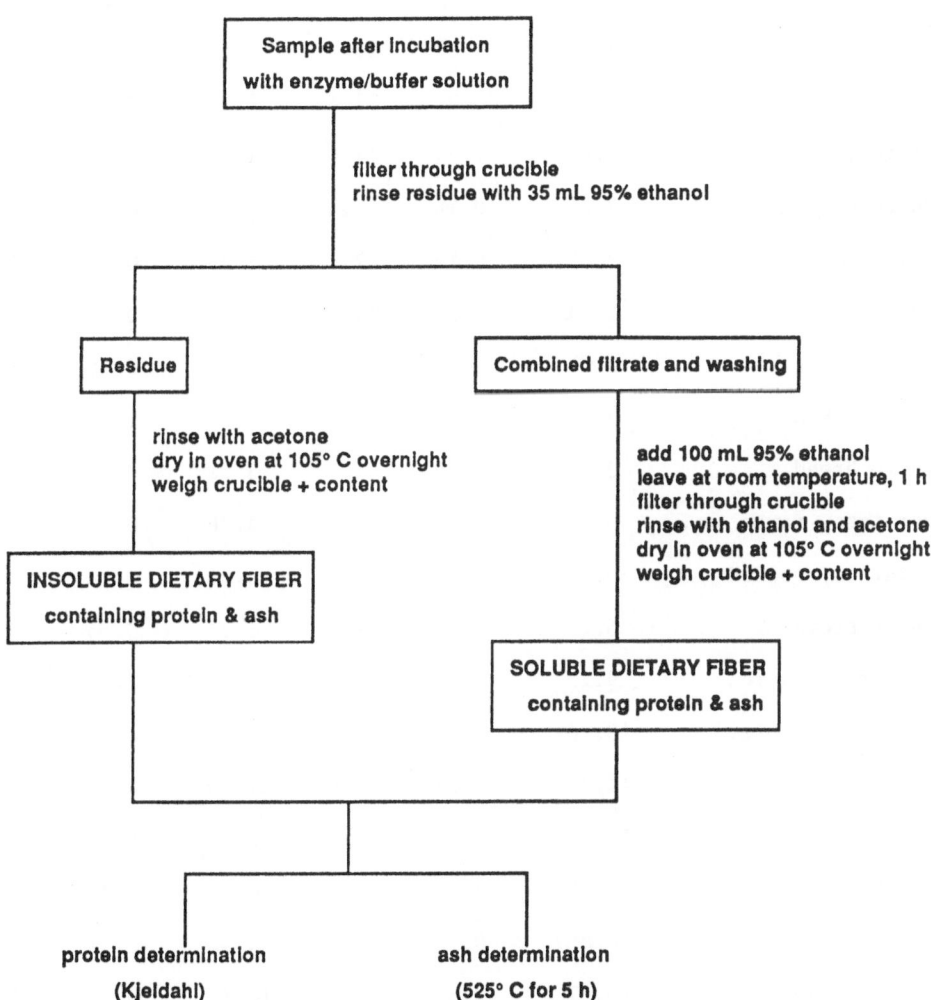

Figure 3 Soluble and Insoluble Dietary Fiber Determination - Simplified Method

Table I

Total Dietary Fiber Content of Foods (Simplified vs. AOAC Method)

Food	SIMPLIFIED TDF g/100g dry wt	S.D.	C.V.	AOAC TDF g/100g dry wt	S.D.	C.V.
White rice, raw	0.54	0.03	5.6	0.68	0.20	29.7
Pie crust	1.92	0.35	18.2	1.58	0.44	27.9
Cake donuts	1.97	0.34	17.4	1.98	0.14	6.9
White rice, cooked	2.09	0.31	14.6	2.21	0.47	21.4
Chocolate chip cookies	2.19	0.16	7.3	2.67	0.22	8.2
Raised donuts	2.38	0.17	7.2	2.38	0.36	15.2
Oatmeal cookies	2.71	0.50	18.4	2.94	0.54	18.5
Brown rice, raw	3.07	0.24	7.9	2.89	0.46	16.0
Corn flakes	3.08	0.09	2.8	2.17	0.57	26.4
White bread	3.08	0.45	14.7	3.52	0.13	3.8
Potatoes, raw	5.16	0.18	3.6	5.78	0.14	2.4
Potatoes, cooked	6.12	0.15	2.5	6.84	0.10	1.4
Wheat bread	6.23	0.15	2.5	6.21	0.56	9.1
Peanuts, roasted	9.25	0.25	2.7	11.35	1.45	12.7
Whole wheat bread	10.76	0.13	1.2	11.02	0.44	4.0
Apples	11.29	0.29	2.6	11.50	0.22	1.9
Popcorn	12.20	1.70	13.9	12.55	1.45	11.6
Oat bran	13.00	2.46	18.9	16.01	1.98	12.4
Soy flour	14.85	0.35	2.3	14.98	0.67	4.5
White bread (reduced calories)	15.57	0.51	3.3	16.11	0.27	1.7
Baked beans	19.16	0.40	2.1	20.01	2.35	11.7
Cabbage, raw	20.03	0.23	1.2	21.46	0.46	2.1
Carrots, raw	20.79	0.50	2.4	23.48	0.89	3.8
Cocoa	27.91	1.35	4.8	27.98	0.43	1.5
Corn bran	85.45	0.84	1.0	89.95	0.24	0.3

the TFA residue from the AOAC method. Using the equations in Figure 4, the final values for the TDF content of a sample of whole wheat bread proved to be very similar for these two methods. A simpler equation has been proposed by Prosky et al. (6) whereby, the protein and ash corrections are no longer expressed as % of residue, but as absolute amounts in milligrams. This is valid if and only when the residue weights of the replicates are identical.

For lack of standard reference materials and in order to check the accuracy of the two methods, a recovery study was carried out by analyzing two pairs of mixtures containing samples (listed in Table I) whose dietary fiber values had the least variance between blind duplicates and between methods. The analyzed values for the mixture of known proportions were compared with those expected based on the TDF values as determined earlier. The Results given in Table III for mixtures (containing cabbage and cooked potatoes, corn bran and whole wheat bread) showed good agreement for each of the two methods; the linear regression equations are as follows:

Simplified: TDF(found) = 1.34 + 0.97xTDF(expected); r=0.998
AOAC: TDF(found) = 0.85 + 1.02xTDF(expected); r=0.998

From these data, we can be assured that the methods are not only comparable but are equally reliable for TDF determinations of certain foods.

With increasing interest in and demand for the amount of soluble dietary fiber in foods, attempts have been made by various analysts to provide such information by modifying existing TDF methods. For example, Prosky, et al. (6) reported on a collaborative study which was conducted for the determination of insoluble and soluble dietary fiber in 10 foods and food products. As an extension of our Simplified method for TDF, we, likewise, made a few minor changes and proceeded to separate the enzyme hydrolyzate into soluble and insoluble fractions. Again, for comparison purposes, we analyzed a number of foods using both AOAC and Simplified methods, and the results on 5 foods are presented in Table IV. Overall, the sum of soluble and insoluble dietary fiber agrees well with the TDF value obtained as such for each of these foods. In the case of soy polysaccharide, the proportion of soluble to insoluble fractions are different for the two methods; in fact the two fractions vary even within the Simplified method depending on the autoclave temperature. At 121 °C (the temperature we recommend at present) the mean of duplicate values for soluble fraction is 18.6 and for insoluble 55.1 g/100g as is; at 130 °C (recommended for starch gelatinization by the American Association of Cereal Chemists) the values are 47.6 for soluble and 29.8 for insoluble fraction. Then, there are other foods, such as cabbage and potatoes whose soluble, insoluble and total dietary fiber values are not affected by different autoclave temperatures. Nevertheless, there is a definite need for more collaboration between those who conduct physiological studies and those who develop or improve analytical methods involving dietary fiber.

Among other foods which we analyzed for soluble and insoluble dietary fiber by the Simplified method, there were several whose sums of soluble and insoluble fractions were much larger than their corresponding TDF values (% dry weight). For example, soy isolate and soy flour: 16.3 vs. 2.31 and 19.5 vs 14.8 respectively. Similar discrepancies were observed with the AOAC/TDF procedure. Further studies are underway in assessing the possibility of interference or inhibition in the Kjeldahl nitrogen analysis due to the presence of substances, such as condensed tannins (7).

In conclusion, we wish to point out that the Simplified method for total dietary fiber determination is as rapid and less labor intensive

$$\text{Blank} = \text{mg blank residue} - \left(\frac{\% \text{ protein in blank residue} + \% \text{ ash in blank residue}}{100} \right) \times \text{mg blank residue}$$

$$\text{TDF (\%)} = \frac{\left[\text{mg residue} - \left(\frac{\% \text{ protein in residue} + \% \text{ ash in residue}}{100} \right) \times \text{mg residue} - \text{blank} \right] \times 100}{\text{mg sample}}$$

Figure 4 Equations for calculations

Table II

TDF Content of Whole Wheat Bread: AOAC vs. Simplified Method

	Sample g	Residue g	Protein mg(%)	Ash mg(%)	Blank residue, mg	Blank corr. mg	TDF %
AOAC	1.0002	0.1607	22.2(13.8)	17.7(11.0)	20.0	8.9	11.19
Simplified	0.6124	0.1702	99.3(58.4)	2.8(1.62)	5.2	2.6	10.69

TABLE III

TDF CONTENT OF MIXED REFERENCE SAMPLES – RECOVERY STUDIES

Mixture	SIMPLIFIED			AOAC		
	Ratio %	TDF (found) g/100g dry wt	TDF (expected) g/100g dry wt	Ratio %	TDF (found) g/100g dry wt	TDF (expected) g/100g dry wt
Cabbage/cooked potatoes	44.0/55.0	13.3	12.4	50.0/50.0	15.4	14.4
	44.0/55.0	13.4	12.4	51.7/48.3	14.9	14.7
Corn bran/whole wheat bread	47.6/52.4	40.3	38.4	55.0/45.0	53.3	54.2
	49.3/50.6	40.7	39.4	50.1/49.9	50.1	50.4
	76.6/23.4	30.0	28.0	33.0/67.0	37.7	36.9
	77.4/22.6	29.7	27.4	31.4/68.6	36.7	35.6

Table IV

Soluble, Insoluble and Total Dietary Fiber Contents of Foods*
(Simplified vs. AOAC Method)

		Sol		Insol		Sol + Insol		TDF	
Apples	(S)	4.22;	4.27	6.60;	6.87	10.8;	11.1	10.9;	11.6
	(A)	4.03;	3.55	8.81;	8.63	12.8;	12.3	11.3	11.5
Baked Beans	(S)	5.40;	5.57	13.8;	13.7	19.2;	19.3	19.6;	19.9
	(A)	6.03;	6.05	12.3;	11.6	18.3;	17.6	17.3;	17.2
Carrots	(S)	7.12;	7.23	11.4;	10.9	18.3;	18.1	18.8;	18.6
	(A)	10.9;	11.4	12.0;	11.8	22.9;	23.2	21.5;	22.5
Oats	(S)	5.79;	5.68	4.85;	5.42	10.6;	11.0	11.8;	12.0
	(A)	6.64;	6.80	4.59;	4.68	11.2;	11.5	12.6;	11.6
Soy Polysacch.	(S)	18.4;	18.7	56.6;	53.6	75.0;	72.3	75.7;	76.2
	(A)	6.37;	4.24	68.1;	69.6	74.5;	73.8	71.2;	70.9

*   = g/100g dry wt
(S) = Simplified
(A) = AOAC

than the AOAC/TDF method. It has better repeatability and less filtration problems. For quality control and labelling purposes, the simplified method could be a viable alternative to the official method as it stands now.

ACKNOWLEDGMENT

This work was partially supported by the Nutrient Data Research Branch/HNIS/USDA. The authors thank Ms. Rhoda Barnes for the Kel-Foss nitrogen analyses.

LITERATURE CITED

1. Prosky, L., Asp, N-G., Furda, I., DeVries, J.W., Schweizer, T.F., Harland, B. J. Assoc. Off. Anal. Chem. 1985, 68, 677–679.

2. "Changes in Methods" J. Assoc. Off. Anal. Chem. 1985, 68, 399; 1986, 69, 370.

3. Englyst, H.N., Cummings, J.H. J. Assoc. Off. Anal. Chem. 1988, 71, 808–814.

4. Theander, O., Westerlund, E.A. J. Agric. Food Chem. 1986, 34, 330–336.

5. Li, B.W., Andrews, K.W. J. Assoc. Off. Anal. Chem. 1988, 71, 1063–1064.

6. Prosky, L., Asp, N G., Schweizer, T.F., DeVries, J.W., Furda, I. J. Assoc. Off. Anal. Chem. 1988, 71, 1017–1023.

7. Saura-Calixto, F. J. Food Sci. 1988, 53, 1769–1771.

DIFFERENTIATION OF DIETARY FIBER SOURCES BY CHEMICAL CHARACTERIZATION

Michael A. McLaughlin and Martha L. Gay

U.S. Food and Drug Administration
Division of Food Chemistry and Technology, HFF-413
200 C St. S.W., Washington, DC, 20204

INTRODUCTION

The term "dietary fiber" is generally defined as the components of plant cell walls which are indigestible by humans.[1,2] Interest in fiber has risen recently as various physiological effects, such as the lowering of serum cholesterol[3] and a decrease in the incidence of colon cancer,[4] have become widely reported. From a chemical point of view, dietary fiber consists of the nonstarch polysaccharides (NSP) and lignin, which are not metabolized by human small intestinal enzymes. The NSP are represented by such chemically diverse compounds as hemicellulose, cellulose, pectin, carrageenan, guar gum and agar. Lignin is a highly cross-linked polymer of phenylpropane units derived from coniferyl, sinapyl and coumaryl alcohols. Information about various types of polysaccharides found in plants is used in the present study to differentiate fiber sources by characterizing their polysaccharide components. From this type of chemical characterization we can then "fingerprint" or analyze for chemical differences among the many fiber sources that are claimed to produce different physiological effects.

The variability in the nonstarch components of the various fiber sources is evident from an examination of the literature. Surveys and comparisons such as those by Englyst et al.,[5] MacArthur and D'Appolonia[6] and Henry[7] clearly show that cereals, as well as some processed foods, contain different types of NSP. Studies of the literature on the composition of individual grains enhance the picture greatly. Investigations of oat bran by Aspinall and Carpenter[8] and Wada and Ray[9] provide much information on the proportion of β-glucan, a hemicellulose of considerable interest because of its physiological effects.[10] Research into the nonstarch components of barley,[11] wheat[12-14] and soy[15,16] indicates the presence of a broad range of derivatives of glucose, galactose, xylose, arabinose and uronic acids in the component polysaccharides. Chemical approaches to dietary fiber analysis, such as those reported by Theander[17] and Englyst et al.,[5] provide a guide by which a procedure based on standard carbohydrate chemistry may be used to analyze and differentiate dietary fiber materials.

New Developments in Dietary Fiber
Edited by I. Furda and C. J. Brine
Plenum Press, New York, 1990

We describe here a procedure in which we use the previously mentioned variations in NSP to differentiate fiber sources by physical (molecular weight determinations) and chemical characterizations (methylation analysis). The procedure consists of a combination extraction/enzyme hydrolysis[18,19] scheme which first removes starch and protein, with further fractionation of the materials into one of three solvents. The extraction procedures produce a water-soluble fraction and two separate water-insoluble fractions. The water-insoluble material is fractionated by the sequential use of dimethylsulfoxide (DMSO) and 5% NaOH solvent systems. The three fractions are analyzed to ascertain dry weight, molecular weight composition and methylated alditol acetate profile.

EXPERIMENTAL

Fiber Materials

The fiber materials studied in these experiments were obtained from the following sources:

Soy Fiber - Ralston Purina Company, St. Louis, MO;

Soft White Wheat and Hard Red Wheat - American Association of Cereal Chemists, St. Paul, MN;

Oat Bran - National Oat Company, Cedar Rapids, IA.

Solvent Extractions

All organic solvents used were spectrophotometric grade (Burdick & Jackson Laboratories, Inc., Muskegon, MI). Aqueous solutions were prepared with Milli-Q grade deionized water (Millipore Corp., Bedford, MA). A Soxhlet extractor was used for the initial chloroform and 95% ethanol extractions; 500 ml of the appropriate solvent was needed to extract ~20 g of fiber. The material left over from the enzyme hydrolysis sequence (<1.0 g) was sequentially extracted with 200 ml of DMSO and 200 ml of 5% NaOH aqueous solution (at room temperature). The 5% NaOH extract was neutralized and both extracts were dialyzed to remove DMSO or ionic salts. The extracts were then lyophilized for quantitation and storage. The overall extraction procedure is shown in Fig. 1.

Enzyme Hydrolysis Procedure

A modified Association of Official Analytical Chemists (AOAC) method[19] was used for enzymatic removal of starch and protein. Enzymes used in this procedure were obtained from Sigma Chemical Corp. (St. Louis, MO). For the enzymatic hydrolysis, three aqueous solutions were prepared: 0.05 M phosphate buffer (pH = 6.0), 0.171 N NaOH and 0.205 M H_3PO_4. The procedure was as follows. Three 1.00 g test portions of each chloroform/95% ethanol-extracted fiber fraction were weighed to ±0.1 mg into 125 ml Erlenmeyer flasks. Blank flasks containing no fiber material were used as controls. To each flask was added 50 ml of 0.05 M phosphate buffer (pH = 6.0 ± 0.1). Sodium hydroxide or phosphoric acid solution was used to adjust the pH to 6.0. After adjustment of pH, 0.2 ml of α-amylase (Sigma Chemical No. A-0164) was added to each flask. Aluminum foil was used to cover each flask before incubation. The flasks were placed in an oil bath and heated to 95°C. Initial incubations were 30 min, beginning when the temperature of the solution reached 95°C. The flasks were shaken at 5 min intervals to ensure thorough mixing. After incubation the solutions were allowed to cool to room temperature and the pH was

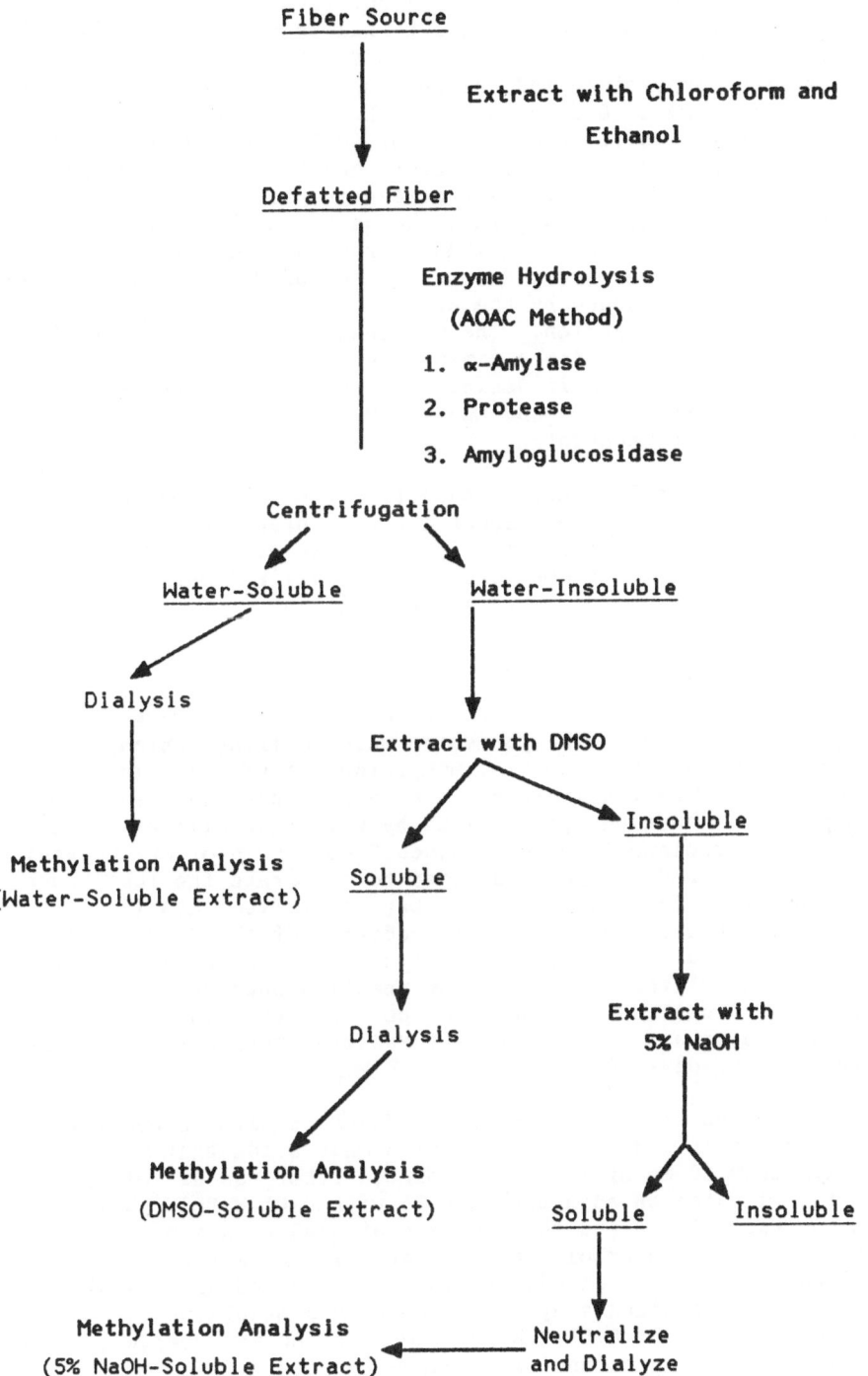

Fig. 1. Enzyme hydrolysis and extraction scheme for the fractionation
of dietary fiber.

adjusted to 7.5 ± 0.1 by the addition of 10 ml of 0.171 N NaOH. With the pH at 7.5, 5 mg of protease (Sigma Chemical No. P-3910) was added as a 5 mg/ml solution. The protease solutions were incubated 30 min in a shaking water bath once the internal temperature of the solutions reached 60°C. After protease hydrolysis, the flasks were allowed to cool to room temperature and the pH was adjusted to 4.5 ± 0.1 by the addition of 10 ml of 0.205 M H_3PO_4. The final incubation was initiated by the addition of 0.3 ml of amyloglucosidase (Sigma Chemical No. A-9913). The temperature and time of this incubation were identical to those of the protease incubation. The solutions were then centrifuged to separate the water-soluble and water-insoluble fractions. Aliquots from the water-soluble fractions and controls were removed and four volumes of 95% ethanol were added. The precipitate produced by the addition of the 95% ethanol was isolated by centrifugation. The supernatant was removed and the precipitate was washed three times with absolute ethanol. The precipitates were dried in desiccators over P_2O_5. The dry precipitates were weighed and control weights were subtracted in order to determine fiber content.

The buffer salts were removed from the remaining water-soluble fraction by dialysis with a Spectra-Por membrane (Spectrum Medical Industries, Inc., Los Angeles, CA; molecular weight cutoff = 3,500). The water-soluble fractions were then lyophilized before size-exclusion chromatography and methylation.

Methylation of Fiber Extracts

A modified version[20] of the Hakomori procedure[21] was used for methylation (Fig. 2). The potassium methylsulfinyl anion was generated by the use of potassium hydride and DMSO.[22] The dimsyl anion was produced routinely at room temperature under an argon atmosphere. Potassium hydride (35% by weight in oil) was washed three times with petroleum ether to produce 0.45-0.50 g of dry potassium hydride. A stream of argon was used to evaporate the last traces of petroleum ether from the hydride. Dry, distilled DMSO (6.8 ml) was added to the hydride powder. Slow addition of the DMSO was necessary to control the vigorous hydrogen evolution and resultant foaming of the solution. After 10-15 min, the reaction subsided, producing a yellow-green solution and some grey precipitate. The reaction mixture was centrifuged and the supernatant was transferred to 1.0 ml capped vials for storage at 0°C.

Reduction of the water-soluble fraction with carbodiimide/$NaBD_4$ was necessary in order to reduce any uronic acids that were present.[23] Approximately 50 mg of the water-soluble fiber was dissolved in 25 ml of water with the pH adjusted to 4.75 with 0.01 M HCl. While the solution was stirred and the pH was monitored with a meter, 0.5 g of 1-cyclohexyl-3-(2-morpholinoethyl)-carbodiimide-metho-p-toluenesulfonate was added. As the pH increased as a result of the formation of the mixed anhydride, the pH was maintained at ~7.0 by the dropwise addition of 0.05 M HCl. The reaction was complete after 2 hr. Two drops of n-octanol were added to reduce foaming and dropwise addition of 10 ml of 2 M $NaBD_4$ was begun. As the first milliliter of the $NaBD_4$ solution was added, the pH was maintained at 7.0 with 0.1 M HCl. After this addition, sufficient borate was present in the mixture to buffer the solution during the addition of the remaining $NaBD_4$. The pH was eventually allowed to rise to 8.5-9.0, and the solution was stirred for an additional hour. The reaction mixture was

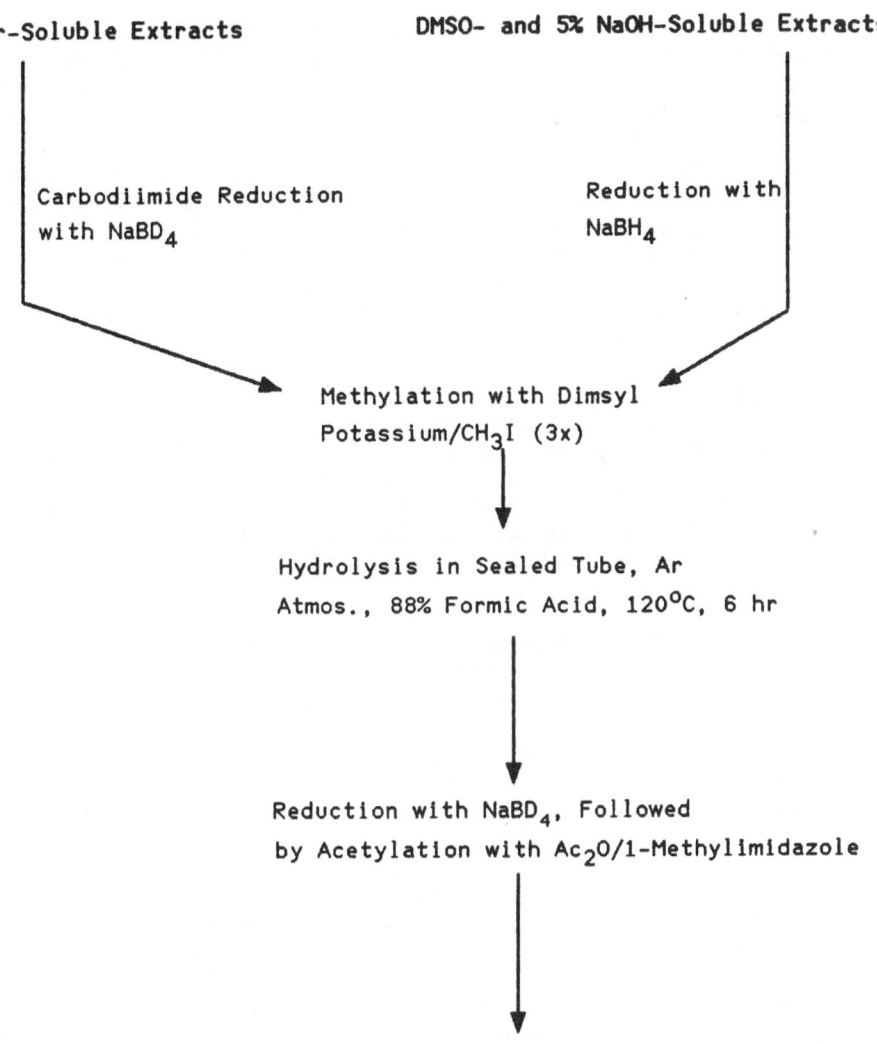

Water-Soluble Extracts   DMSO- and 5% NaOH-Soluble Extracts

Carbodiimide Reduction   Reduction with
with $NaBD_4$ $NaBH_4$

Methylation with Dimsyl
Potassium/CH_3I (3x)

Hydrolysis in Sealed Tube, Ar
Atmos., 88% Formic Acid, 120°C, 6 hr

Reduction with $NaBD_4$, Followed
by Acetylation with Ac_2O/1-Methylimidazole

GC/MS Analysis

Fig. 2. Methylation scheme for dietary fiber.

then titrated to pH 5.5 with glacial acetic acid. The solution was dialyzed against 10% ethylene glycol solution to remove complexed borate and then against deionized water. The solutions were lyophilized to recover the reduced polysaccharides.

The preliminary reduction of the DMSO extract was a simple matter of adding a DMSO solution containing $NaBH_4$ at 1.0 mg/ml to the DMSO-soluble extract and incubating it for 90 min at 40°C.[23] The solution was then neutralized with glacial acetic acid and dialyzed to remove the salts.

The three fiber extracts (water-soluble, DMSO-soluble and 5% NaOH-soluble) were methylated by the following procedure. A 10 mg portion of the material was dissolved in 200 μl of dry, distilled DMSO under an argon atmosphere. The dimsyl potassium solution (200 μl) was added and the mixture was stirred under argon. After 30 min the solution was cooled with ice and 150 μl of methyl iodide was added. The solution was allowed to warm to room temperature and was mixed during thawing. After 15 min, 3.0 ml of 2:1 (v/v) chloroform/methanol was added, followed by 2.0 ml of water. The layers were thoroughly mixed and the upper phase was removed. This washing procedure was repeated with four 2.0 ml portions of water. After the final wash, 2 ml of 2,2-dimethoxypropane and 20 μl of glacial acetic acid were added. This mixture was lightly heated (60°C) while a stream of argon was used to evaporate and dry the methylated polysaccharides.

In the cases where the fiber extracts were not totally soluble in DMSO, a stepwise methylation procedure was used. The extract was again placed in 200 μl of DMSO containing 20 μl of dimsyl potassium. After mixing, the solution was allowed to cool and 5 μl of ice-cold methyl iodide was added. An additional 60 μl of dimsyl potassium was added and the solution was mixed and cooled with ice; then 15 μl of cold methyl iodide was added. At this point the material was in solution and could be methylated as described in the preceding paragraph.

The methylated polysaccharides were hydrolyzed by dissolution in 1.0 ml of 88% formic acid solution and heating at 90°C for 6 hr. The tubes were evacuated and flushed with argon three times before heating in order to remove oxygen. Removal of oxygen from the hydrolysis mixture was necessary because oxidative degradation under these acidic conditions would result in degradation of much of the methylated material.

The hydrolysis products were reduced and acetylated by a simplified process predicated on the use of 1-methylimidazole to catalyze the acetylation.[24] The sugar residues were dissolved in 100 μl of 1 M NH_4OH. After the addition of 0.5 ml of a 20 mg/ml solution of $NaBD_4$ in DMSO, the solution was incubated at 40°C for 1.5 hr. The excess deuteride was then destroyed by addition of 100 μl of glacial acetic acid. The acetylation was carried out by the stepwise addition of 100 μl of anhydrous 1-methylimidazole and 0.5 ml of acetic anhydride. After 10 min at room temperature, the reaction was complete. The alditol acetates were then stored in a freezer at 0°C until gas chromatography/mass spectrometry (GC/MS) analysis.

Size-Exclusion Chromatography of Water-Soluble Fiber

Weight-average molecular weights (M_w) and polydispersity indices were determined under the following conditions. The columns used were a Shodex S-804 (500 x 8 mm, exclusion limit = 5 x 10^5 daltons; Showa Denko, Tokyo, Japan) and a Shodex S-806 (500 x 8 mm, exclusion limit = 5 x 10^7 daltons) held at a temperature of 40°C. The mobile phase was 0.1 M $NaNO_3$ at a flow rate of 1.0 ml/min. Test solutions and standards consisted of 0.1% (w/v) concentrations with an injection volume of 20 μl. The column was standardized by use of eight molecular weight standards (pullulans from Polymer Laboratories, Inc., Amherst, MA), ranging in size from 5,800 to 853,000 daltons. Test solutions were de-proteinated by cation-exchange chromatography. A refractive index detector was used in this analysis. Calculations were performed with the assistance of a Spectra-Physics GPC+ software package. The best fit for the resulting calibration curve was found to be a cubic equation.

GC/MS Determination of Partially Methylated Alditol Acetates

In order to identify the monosaccharides present in the extracts, as well as the substitution pattern of the polysaccharides, it was necessary to determine them by GC/MS. The gas chromatograph used was a Finnigan Model 9610 equipped with a split/splitless injector. The column was a fused silica SP-2330 (30 m x 0.25 mm, 0.20 μm film; Supelco, Inc., Bellefonte, PA), using a linear flow of helium at 40 cm/sec. Because of the polarity of the column, it was necessary to use a temperature program[25] which started with a hold at 160°C for 5 min and then ramped the temperature to 210°C at 2°/min. At that point the program was changed to increase the temperature to 240°C at 5°/min. The column was then held at 240°C for 25 min to ensure that all of the compounds of interest had eluted. A direct insertion interface of the fused silica column into the mass spectrometer source was used.

A Finnigan 4000 mass spectrometer was used in these investigations. The operating conditions were as follows: electron potential 70 eV, collector current 0.35 mA, multiplier set at 700 V, mass range analyzed 40-400 amu.

RESULTS AND DISCUSSION

Weight Percent Composition of Fiber Fractions

The weight percentages of the fractions from the extraction procedure are shown in Table 1. The extraction procedure was applied to separate the NSP on the basis of their solubility in certain solvents. The quantities found in the water-soluble fractions are consistent with trends found in previous experiments by other investigators.[19] The wheat sources contained a relatively small amount of water-soluble fiber when contrasted with the soy and oat. The two solvent systems for the water-insoluble material were chosen to provide complementary solvating abilities. The DMSO treatment

Table 1. Weight Percentage Composition of Solvent Extracts from Fiber Sources[a]

Fiber	Extract		
	Water	DMSO	5% NaOH
Oat Bran	6.0 ± 0.3	9.4 ± 0.8	7.1 ± 0.1
Soy Fiber	4.0 ± 2.1	1.6 ± 0.4	25.4 ± 0.4
Soft White Wheat	1.2 ± 0.1	4.6 ± 1.2	22.1 ± 1.0
Hard Red Wheat	1.6 ± 0.1	5.0 ± 0.9	15.8 ± 1.0

[a]All fibers were pretreated by enzyme hydrolysis as described in the experimental section. Data represent three analyses of each fraction ± 1 standard deviation.

resulted in extraction of a fraction of the noncellulosic polysaccharides, which are not structurally confined in or covalently attached to the cell wall matrix. This solvent is also known to be nondegradative and to exhibit the ability to solvate a wide variety of carbohydrates. The 5% NaOH solution was chosen for its capacity to cause swelling of cellulose fibers and to release some polysaccharides which are covalently linked to lignin. The sodium hydroxide solution contained a low concentration of $NaBH_4$ (10 mM) to negate polysaccharide peeling reactions.[26] The sodium hydroxide extracts from soy, soft white wheat and hard red wheat comprised a surprisingly large proportion of the extracted material when compared to the other extracts. As previously reported by other investigators,[19] oat bran was found to have a low proportion of water-insoluble fiber.

Size-Exclusion Chromatography Experiments

Analysis of the water-soluble fraction by the use of high-performance size-exclusion chromatography indicated that each of the extracts from the various fiber sources had a unique elution pattern consisting of two or three components of various molecular weights. These results are listed in Table 2. The chromatogram of the oat bran extract contained a component (peak 1) which consistently exhibited a molecular weight in excess of one million daltons. This component also was found to have the lowest polydispersity index in this study. These two properties of this component, although not diagnostic, might with additional study be used in the confirmation of identity of a source. Some of the other fiber extracts (soy and soft

Table 2. Size-Exclusion Chromatographic Analysis of Water-Soluble Fiber Fractions

Peak	Retention Time (min)	M_w[a] (daltons)	Polydispersity Index
		Soy Fiber	
1	10.45	7.02×10^5	1.15
2	12.35	1.64×10^5	1.41
3	14.72	2.6×10^4	1.17
		Hard Red Wheat	
1	13.77	4.66×10^4	1.31
2	16.86	9.1×10^3	1.19
		Soft White Wheat	
1	10.45	6.29×10^5	1.61
2	16.86	9.12×10^3	1.28
		Oat Bran	
1	10.21	1.01×10^6	1.02
2	14.72	4.35×10^4	1.32
3	16.86	1.01×10^4	1.09

[a]Weight-average molecular weight.

white wheat) contained components of rather large molecular weight
(>10^5 daltons), but these components were so close in M_w and the peaks
were so broad that they were not useful as indicators. The
water-soluble fractions from hard red wheat contained two components
which were at least an order of magnitude smaller in M_w than the
components of highest molecular weight from the other fiber extracts.
These components are sufficiently similar in M_w to other components
(i.e., peak 2 in the chromatogram for oat bran and peak 2 in the
chromatogram for soft white wheat) that they can be used only to help
in characterizing fibers from different sources.

All of the extracts examined by this process exhibited a
characteristic pattern in the size-exclusion chromatographic analysis.
However, we found that if the extracts from the various fiber sources
were mixed, the resulting chromatogram appeared totally useless for
analysis as the peaks tended to blend together producing one large,
undifferentiated peak.

Interpretation of Mass Spectral Data

Interpretation of the mass spectra obtained from the GC/MS
analysis of the partially methylated alditol acetates from the four
fiber sources was based upon the theoretical rules of fragmentation
derived from the work of Bjorndal et al.,[27,28] Golovkina et al.[29] and
others.[30-35] These authors made extensive use of stable
isotope-labeling studies to determine fragmentation mechanisms.
Results in the present study are discussed below in terms of these
mechanisms. Fragmentation patterns from the partially methylated
alditol acetates proceed by initial cleavages of bonds adjacent to
methoxy ether groups followed by secondary fragmentations often
involving rearrangement of a hydrogen atom to the leaving group and
loss of a fragment corresponding to either methanol or acetic acid.
These rearrangements produce carbocations which are stabilized by the
additional charge delocalization. In the initial cleavage, the charge
is carried by the methoxy group. The carbocation then initiates
secondary cleavages at positions β to the charge site. The
incorporation of a deuterium atom during the reduction of the carbonyl
group simplifies the elucidation of the structures by labeling the
acetal carbon. This labeling distinguishes symmetrical structures
(such as the 2-O-Me and 4-O-Me pentitol tetraacetates) that produce
identical spectra by the derivatization process used in this study.
The following two examples are provided to illustrate the
fragmentaton principles.

The mass spectrum produced by the derivative 1,4-di-O-acetyl-
(1-deuterio)-2,3,5-tri-O-methyl pentitol (Fig. 3) is an excellent
example of the fragmentation pattern of an alditol acetate generated
from a nonreducing terminal sugar. The base peak for this compound
and most of the other alditol acetates is the m/z 43 ion produced by
the ready cleavage of an acetate ester. This fragment is not
structurally significant but its presence is characteristic of almost
all of these alditol acetate derivatives. The alditol acetate
represented in Fig. 3 is preferentially cleaved between the adjacent
methoxylated carbons (C-2 and C-3). The fission yields two primary
fragments at m/z 118 and 161. Other primary fragments, some of which
are not readily apparent in the spectrum, result from other cleavages
of bonds adjacent to methoxylated carbons. In this case these are the
fragments at m/z 45, 162 and 205. The m/z 45 ion is characteristic of

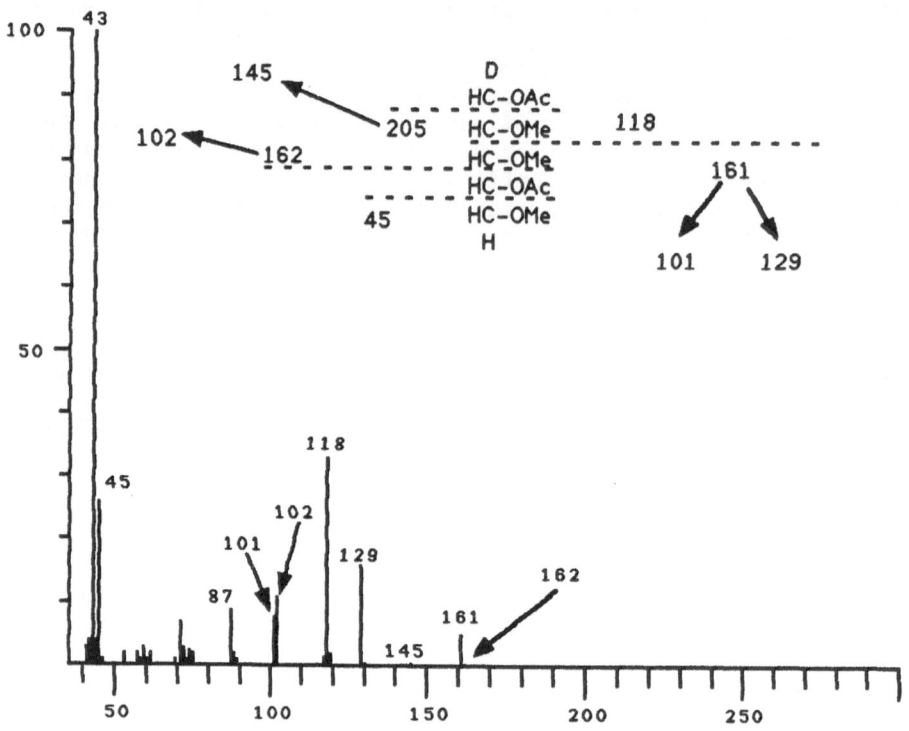

Fig. 3. Mass Spectrum of 1,4-di-O-acetyl-(1-deuterio)-2,3,5-tri-O-methyl pentitol.

these types of alditols as it is always found when the terminal C-5 (in pentitols) or C-6 (in hexitols) is methylated. The m/z 162 and 205 ions are found in very low abundance and their presence is usually inferred from smaller secondary fragments (m/z 102 and 145, respectively). The secondary fragments are quite easy to explain. The m/z 145 ion is produced after initial cleavage between C-1 and C-2, which results in the m/z 205 ion. The m/z 205 ion produces the m/z 145 ion by elimination of acetic acid, through a McLafferty rearrangement β to the C-2 charge site. The m/z 101 and 129 ions are both derived from the m/z 161 primary ion through eliminations α and β to the charge site. The m/z 129 ion is produced by elimination of methanol. The ion at m/z 101 is generated by elimination of acetic acid, one of the rare eliminations α to the charge site found with these types of compounds. The ion at m/z 102 results from elimination of acetic acid from the m/z 162 ion. The more favorable elimination of acetic acid β to the charge site results in a very low abundance of the parent ion. The final ion, m/z 87, is derived from the elimination of ketene from the m/z 129 fragment.

The mass spectra produced by the hexopyranosyl derivatives reveal many similarities to those found for the pentitols, although they can be readily distinguished by the presence of the larger four, and in some cases five, carbon fragments. The spectrum from 1,4,5-tri-O-acetyl-(1-deuterio)-2,3,6-tri-O-methyl hexitol (Fig. 4) illustrates these differences. This spectrum has several fragments that are generated by the same processes found with the pentitol, i.e., the fragments at m/z 43, 45, 87, 102, 118, 129 and 162. The hexitol is

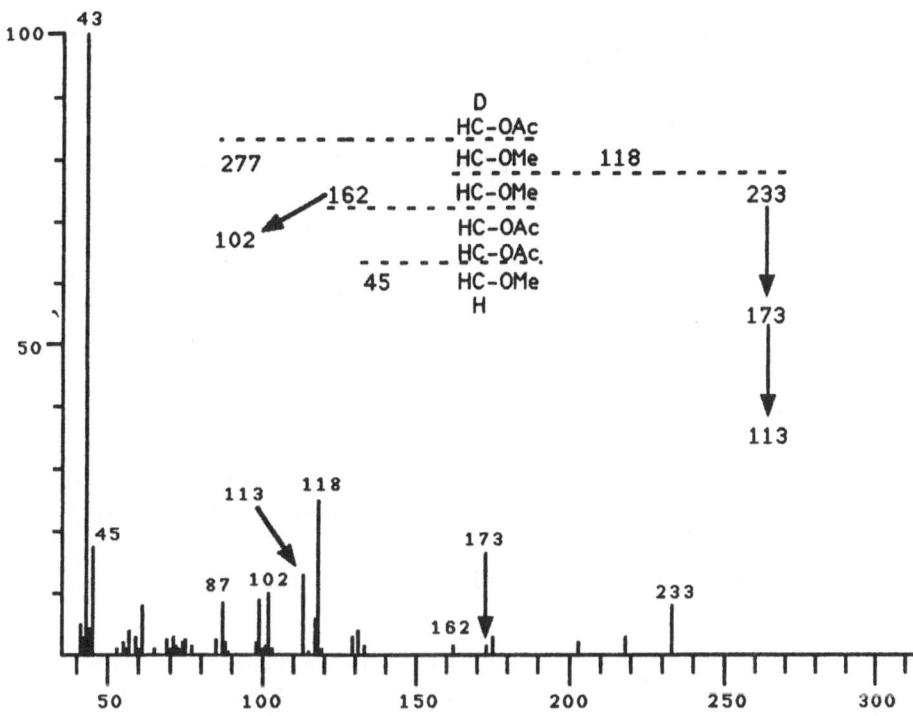

Fig. 4. Mass spectrum of 1,4,5-tri-O-acetyl-(1-deuterio)-2,3,6-tri-O-methyl hexitol.

distinguished from the pentitol by an ion at m/z 233 (one of the two major fragments generated by fragmentation of the molecular ion between C-2 and C-3) and, in some cases, by an ion at m/z 277. The m/z 233 parent ion sequentially eliminates two molecules of acetic acid α and β to the carbocation, thus generating the daughter ions m/z 173 and 113. Further fragmentation of the m/z 113 ion occurs through nonspecific fragmentation pathways which provide no structural information. The overall mass spectrum from this derivative is one that is frequently seen in methylation analysis in the examination of extracts containing starch and/or cellulose, two of the most abundant polysaccharides found in nature.

The mass spectra obtained from our fiber extracts were interpreted by the rules described in the two examples. The results from analysis of the 5% NaOH extract from soy fiber are representative of the alditol acetates found in the other fiber extracts. Table 3 summarizes our findings for this fraction.

It is apparent from Table 3 that this fraction consists of polysaccharide(s) derived from arabinose (Ara) and/or galactose (Gal). The arabinose clearly dominates the mixture (75.3% of the partially methylated alditol acetates identified) and provides four of the alditol acetates found in the analysis. The presence of branch points, such as the 3-OMe Ara derivative, and terminal groups (2,3,5-OMe Ara and 2,3,4,6-OMe Gal) signify the existence of a branched polysaccharide or polysaccharides in this extract. It is the variation in monosaccharide composition and branching of the NSP as

305

Table 3. Partially Methylated Alditol Acetates from the 5% NaOH
Extract from Soy Fiber

Methylated Alditol	Relative Retention Time[a,b]	% of Total
2,3,5-OMe Ara	0.62	20.7
2,3,4,6-OMe Gal	1.09	4.3
2,3-OMe Ara	1.15	25.8
3,4-OMe Ara	1.14	1.2
2,3,6-OMe Gal	1.37	20.4
3-OMe Ara	1.38	27.6

[a]Relative to 2,3,4,6-OMe glucitol diacetate.
[b]Peaks of compounds with similar relative retention times (i.e.,
1.15, 1.14 and 1.37, 1.38) were distinguishable in the chromatogram.

illustrated in this example that allow us to "fingerprint" the
different fiber sources.

The partially methylated alditol acetates present in the DMSO and
5% NaOH extracts demonstrate the compositional variations that
differentiate the fiber sources. The data from the DMSO extracts
(Table 4) indicate that the polysaccharides in this fraction are
relatively simple due to the almost total lack of derivatives from
branch points. The DMSO extracts for soy, oat bran and soft white
wheat consist entirely of alditols derived exclusively from straight
chain arabinans, galactans and glucans. The DMSO extract for hard red
wheat is the exception as the presence of 3-OMe Ara indicates some
type of branched arabinan. Although there is not a great deal of
information available from this particular solvent fraction, it does
provide a simple contrast of the various fiber types.

Examination of the 5% NaOH extracts (Table 5) reveals a greater
diversity of derivatives compared to those found in the DMSO extracts.
The results range from a simple straight chain xylan found in the soft
white wheat extract to the six derivatives isolated in the soy

Table 4. Partially Methylated Alditol Acetates from the DMSO Extract

Methylated Alditol	Fiber			
	Soy	Oat Bran	Soft White Wheat	Hard Red Wheat
2,3,5-OMe Ara	-	-	-	+
2,3-OMe Ara	+	-	+	+
2,3,6-OMe Glu[a]	-	-	+	-
3-OMe Ara	-	-	-	+
2,3,6-OMe Gal	+	+	-	-

[a]Glu = Glucitol.

306

Table 5. Partially Methylated Alditol Acetates from the 5% NaOH Extract

| Methylated Alditol | Fiber | | | |
	Soy	Oat Bran	Soft White Wheat	Hard Red Wheat
2,3,5-OMe Ara	+	+	-	-
2,3,5-OMe Xyl[a]	-	-	-	+
2,3,4,6-OMe Gal	+	-	-	-
2,3-OMe Ara	+	-	-	-
3,4-OMe Ara	+	-	-	-
2,3-OMe Xyl	-	+	+	+
2,3,6-OMe Glu	-	+	-	-
3-OMe Ara	+	-	-	-
2,3,6-OMe Gal	+	-	-	-
3-OMe Xyl	-	+	-	+

[a]Xyl = Xylitol.

extract. The results for the soy extract were discussed in a previous paragraph. The oat bran was found to produce four different derivatives, i.e., one branched alditol (3-OMe Xyl), two internal chain types (2,3-OMe Xyl and 2,3,6-OMe Glu) and a terminal residue (2,3,5-OMe Ara). The hard red wheat produced three alditols, two also found in the oat bran extract (2,3-OMe Xyl and 3-OMe Xyl) and the terminal residue 2,3,5-OMe Xyl. Both of these extracts apparently contain a branched xylan but differ in that the oat bran extract is heterogeneous due to the presence of the arabinose and glucose derivatives. The hard red wheat data point to a simple, branched xylan as the sole component of this extract. The 3-OMe Xyl residue in the oat bran and hard red wheat extracts indicates a branch point at the C-2 position.

CONCLUSION

As the claims of the beneficial effects of specific types of dietary fiber are expanded and new claims are made, characterization of the chemical composition of fiber will become increasingly important to our understanding of various health effects. In this preliminary study we have applied the well-defined procedures of size-exclusion chromatography and methylation analysis to characterize and distinguish the various types of dietary fiber. By using size-exclusion chromatography for preliminary screening, coupled with methylation analysis of the fiber extracts, we can obtain a "fingerprint" of the fiber. Continuing experiments with water-soluble fractions and expansion of our database of fiber sources are in progress and will provide further data for this particular analytical scheme.

REFERENCES

1. G.A. Spiller and R.M. Kay, Recommendations and conclusions of the dietary fiber workshop of the XI international congress of nutrition, Rio de Janeiro, Am. J. Clin. Nutr. 32:2102 (1979).

2. H. Trowell, D.A. Southgate, T.M. Wolever, A.R. Leeds, M.A. Gassell, and D.J.A. Jenkins, Dietary fiber redefined, Lancet 1:967 (1976).

3. J.W. Anderson, L. Story, B. Sieling, W.-J.L. Chen, M.S. Petro, and J. Story, Hypocholesterolemic effects of oat bran and bean intake for hypercholesterolemic men, Am. J. Clin. Nutr. 40:1146 (1984).

4. M.A. Howell, Diet as an etiological factor in the development of cancers of the colon and rectum, J. Chronic Dis. 27:67 (1975).

5. H.N. Englyst, V. Anderson, and J.H. Cummings, Starch and non-starch polysaccharides in some cereal foods, J. Sci. Food Agric. 32:1434 (1983).

6. L.A. MacArthur and B.L. D'Appolonia, Comparison of nonstarchy polysaccharides in oats and wheat, Cereal Chem. 57:39 (1980).

7. R.J. Henry, A comparison of the non-starch carbohydrates in cereal grains, J. Sci. Food. Agric. 36:1243 (1985).

8. G.O. Aspinall and R.C. Carpenter, Structural investigations on the non-starchy polysaccharides in oat bran, Carbohydr. Polym. 4:271 (1984).

9. S. Wada and P.M. Ray, Matrix polysaccharides of oat coleoptile cell walls, Phytochemistry 17:923 (1978).

10. J.B. Wyman, K.W. Heaton, A.P. Manning, and A.C.B. Wicks, The effect on intestinal transit and the feces of raw and cooked bran in different doses, Am. J. Clin. Nutr. 29:1474 (1976).

11. A.G.J. Voragen, H.A. Schols, F.M. Marijs, and F.M. Rombouts, Non-starch polysaccharides from barley: Structural features and breakdown during malting, J. Inst. Brew. 93:202 (1987).

12. J.M. Brillouet and C. Mercier, Fractionation of wheat bran carbohydrates, J. Sci. Food. Agric. 32:243 (1981).

13. S.G. Ring and R.R. Selvandren, Isolation and analysis of cell wall material from Beeswing wheat bran (Triticum aestivum), Phytochemistry 19:1723 (1980).

14. P.B. Schwarz, W.H. Kunerth, and V.L. Youngs, The distribution of lignin and other fiber components within hard red spring wheat bran, Cereal Chem. 65:59 (1988).

15. G.O. Aspinall, I.W. Cottrell, S.V. Egan, I.M. Morrison, and J.N.C. Whyte, Polysaccharides of soy-beans. Part IV. Partial hydrolysis of the acidic polysaccharide complex from cotyledon meal, J. Chem. Soc. C, 1071 (1967).

16. G.O. Aspinall, K. Hunt, and I.M. Morrison, **Polysaccharides of soy-beans. Part V. Acidic polysaccharides from the hulls**, J. Chem. Soc. C, 1080 (1967).

17. O. Theander, **Advances in the characterization and analytical determination of dietary fibre components**, in: "Dietary Fibre," C.C. Birch and K.J. Parker, eds., Applied Science Publishers, London (1983).

18. L. Prosky , N.-G. Asp, I. Furda, J.W. DeVries, T.F. Schweizer, and B.F. Harland, **Determination of total dietary fiber in foods, food products, and total diets: Interlaboratory study**, J. Assoc. Off. Anal. Chem. 67:1044 (1984).

19. AOAC, "Changes in Official Methods of Analysis," 14th Ed., 1st Suppl., AOAC, Arlington, VA (1985) secs 43.A14-43.A20.

20. P.J. Harris, R.J. Henry, A.B. Blakeney, and B.A. Stone, **An improved procedure for the methylation analysis of oligosaccharides and polysaccharides**, Carbohydr. Res. 127:59 (1984).

21. S.I. Hakomori, **A rapid permethylation of glycolipid and polysaccharide catalyzed by methylsulfinyl carbanion in dimethyl sulfoxide**, J. Biochem. (Tokyo) 55:205 (1964).

22. L.R. Phillips and B.A. Fraser, **Methylation of carbohydrates with dimsyl potassium in dimethyl sulfoxide**, Carbohydr. Res. 90:149 (1981).

23. N.C. Carpita and E.M. Shea, **Linkage structure of carbohydrates by gas chromatography-mass spectrometry (GC-MS) of partially methylated alditol acetates**, in: "Analysis of Carbohydrates by GLC and MS," C.J. Biermann and G.D. McGinnis, eds., CRC Press, Boca Raton (1988).

24. A.B. Blakeney, P.J. Harris, R.J. Henry, and B.A. Stone, **A simple and rapid preparation of alditol acetates for monosaccharide analysis**, Carbohydr. Res. 113:291 (1983).

25. E.M. Shea and N.C. Carpita, **Separation of partially methylated alditol acetates on SP-2330 and HP-1 vitreous silica capillary columns**, J. Chromatogr. 445:424 (1988).

26. C.E. Ballou, **Alkali sensitive glycosides**, Adv. Carbohydr. Chem. 9:59 (1954).

27. H. Bjorndal, B. Lindberg, and S. Svensson, **Mass spectrometry of partially methylated alditol acetates**, Carbohydr.Res. 5:433 (1967).

28. H. Bjorndal, B. Lindberg, A. Pilotti and S. Svensson, **Mass spectra of partially methylated alditol acetates II. Deuterium labeling experiments**, Carbohydr. Res. 15:339 (1970).

29. L.S. Golovkina, O.S. Chizhov, and N.S. Wulfson, **Mass-spektrometritcheskoe issledowanie uglewodow soobshenie 9. Acetaty polilow**, Izv. Akad. Nauk SSSR, Ser. Khim. 1915 (1966).

30. P.E. Jansson, L. Kenne, H. Liedgren, B. Lindberg, and J.

Lonngren, A practical guide to the methylation analysis of carbohydrates, Chem. Commun. Univ. Stockholm 8 (1976).

31. B. Lindberg, Methylation analysis of polysaccharides, Methods Enzymol. 28:178 (1972).

32. B. Lindberg and J. Lonngren, Methylation analysis of complex carbohydrates: General procedure and application for sequence analysis, Methods Enzymol. 50:3 (1978).

33. H. Rauvala, J. Finne, T. Krusius, J. Karkainen, and J. Jarnefelt, Methylation techniques in the structural analysis of glycoproteins and glycolipids, Adv. Carbohyr. Chem. Biochem. 38:389 (1981).

34. H. Bjorndal, C.G. Hellerquist, B. Lindberg, and S. Svensson, Gas-liquid chromatography and mass spectrometry in methylation analysis of polysaccharides, Angew. Chem. Int. Ed. Engl. 9:610 (1970).

35. J. Lonngren and S. Svensson, Mass spectrometry in structural analysis of natural carbohydrates, Adv. Carbohydr. Chem. Biochem. 29:41 (1974).

ANALYSIS OF FOODSTUFFS FOR DIETARY FIBER BY THE UREA ENZYMATIC DIALYSIS METHOD

Joseph L. Jeraci, Betty A. Lewis, James B. Robertson and Peter J. Van Soest

Division of Nutritional Sciences
Savage Hall, Cornell University, Ithaca NY 14853

ABSTRACT

Total dietary fiber (TDF) values for cereal grains, fruits, vegetables, processed foods, and purified or semi-purified dietary fiber products were determined by a new method using 8M urea and enzymes (urea enzymatic dialysis, UED, method). The results are compared with the official AOAC procedure. Soluble and insoluble dietary fiber were determined for several of these foodstuffs and compared with the NDF values. Crude protein and ash contamination was usually lower with the UED method compared with the AOAC method, particulary for samples that formed gels during ethanol precipitation. Urea and the heat stable amylase were effective in removing starch even at relatively low temperatures of the assay (50°C). The new assay is relatively economical in use of equipment, enzymes, and reagents. Studies are currently in progress to minimize the assay time for the UED method while further improving its flexibility and robustness. The results of the studies will be discussed.

INTRODUCTION

An improved method for total, insoluble and soluble dietary fiber is described in this paper. The urea enzymatic dialysis (UED) method employs 8M urea to hydrate and extract starch and other water soluble polysaccharides, a heat-stable amylase and dialysis to remove starch, and protease digestion of plant proteins. The content of dietary fiber for foodstuffs by the UED is usually determined using temperatures of 20°C and 50°C, but the step using the 50°C can be omitted. Thus the UED method can be used to evaluate the effect of cooking on the dietary fiber of various feedstuffs. The method described by Prosky et al (7) exposes the foodstuffs to 95°C-100°C for up to 30 min.

MATERIALS AND METHOD

(1) Urea enzymatic dialysis method for total, soluble, and insoluble dietary fiber

The urea enzymatic dialysis method as described by Jeraci et al, (5) was modified as follows. Termamyl 120L (Novo Laboratories, Inc. Wilton, CT 06897) was used to prepared a 0.33% (v/v) heat- stable αamylase - 8M urea solution. To determine soluble and insoluble dietary fiber the method was further modified (Figure 1). The dialysis tube was removed from the water bath and the contents of the tube was filtered under vacuum through pre-weighed Whatman #54 filter paper (12.5 cm) on a conical funnel. The tube, filter paper, and sample were washed with distilled water. The vacuum flask containing the filtrate was removed and the paper and residue were washed with two 10 mL portions of 80% ethanol and two 10 mL portions of acetone. This residue was used to determine insoluble dietary fiber. The filtrate was transferred from the vacuum flask into a 600 mL beaker (calibrated for volume and weight) and distilled water was used to rinse the inside of flask into the beaker. Four volumes (or use weight instead of volume) of absolute ethanol was added to the beaker and held for at least 4 h.

The contents of the beaker was filtered under vacuum through pre-weighed Whatman #54 filter paper (12.5 cm) on a conical funnel and processed as previously described. In this study, the hot weighing procedure described by Goering and Van Soest (4) was used in the urea enzymatic dialysis method.

(2) AOAC total, soluble, and insoluble dietary fiber method

The AOAC total dietary fiber method as described by Prosky et al. (7) was modified as follows. Each sample was analyzed in quadruplicate, the sample size was reduced from 1.0 g to 0.75 g and all of the reagents except the enzymes were proportionally scaled down. Preliminary studies indicated that these modifications improved the filtration of the ethanol precipitate and improved reproducibility.

(3) Pectin method

The meta-hydroxydiphenyl assay described by Blumencrantz and Asboe-Hansen (1) was modified by Bucher (2). This modified assay for uronic acids maximizes the response of galacturonic acid and minimizes the response of other uronides and non-uronide carbohydrates. The modified assay omits the borate in the sulfuric acid reagent, and generates the chromogen at 80°C for 5 min. The samples are cooled to 4°C before adding the reagents.

(4) Starch method

The AACC method 76-11 was modified as follows. In the gelatinization step, flasks containing sample and buffer solution were placed in an ultrasonic bath at 85°C for 30 min. The 4-aminoantipyrene chromogen was used.

(5) Neutral detergent method

The neutral detergent method described by Robertson and Van Soest (8) was used to determine the neutral detergent fiber (NDF) in some of the foodstuffs, while the neutral detergent method described by Jeraci and Van Soest (6) was used to determine the neutral detergent fiber in others.

Sample Preparation

The dietary fiber was determined on freeze-dried samples that were cooled in a desiccator before weighing. These dried samples were milled to pass 0.5- 1mm mesh sieve. Fat content of the samples in this study was < 10%.

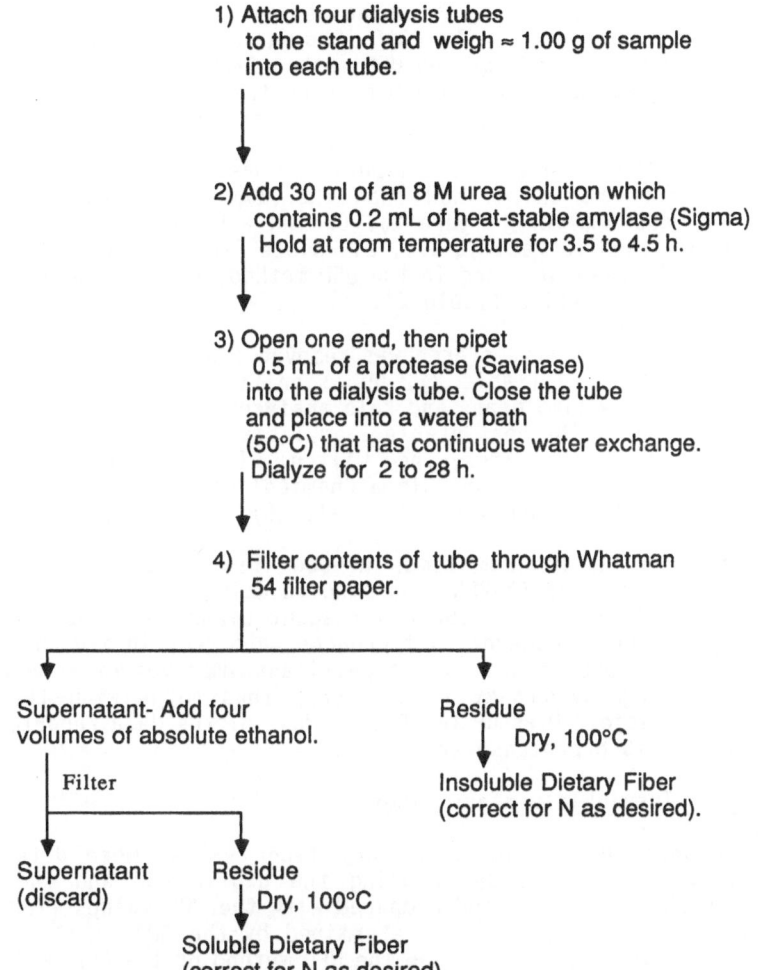

1) Attach four dialysis tubes
   to the stand and weigh ≈ 1.00 g of sample
   into each tube.

2) Add 30 ml of an 8 M urea solution which
   contains 0.2 mL of heat-stable amylase (Sigma)
   Hold at room temperature for 3.5 to 4.5 h.

3) Open one end, then pipet
   0.5 mL of a protease (Savinase)
   into the dialysis tube. Close the tube
   and place into a water bath
   (50°C) that has continuous water exchange.
   Dialyze for 2 to 28 h.

4) Filter contents of tube through Whatman
   54 filter paper.

Supernatant- Add four
volumes of absolute ethanol.

Residue
  Dry, 100°C

Insoluble Dietary Fiber
(correct for N as desired).

Filter

Supernatant
(discard)

Residue
  Dry, 100°C

Soluble Dietary Fiber
(correct for N as desired).

Figure 1.   Flowchart of the Urea Enzymatic Dialysis Method
             to Determine Insoluble and Soluble Dietary Fibers.

313

RESULTS AND DISCUSSION

Total Dietary Fiber

The total dietary fiber (TDF), crude protein (CP), and ash have been determined for 90 foodstuffs and several sources of purified dietary fiber using the AOAC-1988 method described by Prosky et al. (7) and the urea enzymatic dialysis (UED) method described by Jeraci et al. (5). Several of these foodstuffs are presented in the study, while others have already been presented (Carr et al., 3; Jeraci et al., 5; Jeraci and Van Soest., 6). TDF for the samples by the AOAC-1988 was usually higher (or the same) than the values obtained for the UED method (Table 1). Smaller correction factors for CP and ash were obtained using the UED method compared to the AOAC method (Jeraci et al., 5; Jeraci and Van Soest 6).

For two of the purified or semi-purified dietary fiber products the TDF was higher for the UED method than the AOAC-1988 method and standard deviations were smaller (Figure 2). CP and ash contamination were lower for the UED method (Jeraci et al., 5). Since ß-glucanase activity was not detected in the reagents used in the UED method, recovery of water soluble ß-glucan is quantitative (Table 2).

Essentially all of the starch was removed from the total dietary fiber using the UED method (Table 3), but significant levels of starch were observed with baby lima beans and kale analysed by the AOAC method (Jeraci et al., 5). Both of the heat-stable α-amylases (Termamyl 240L and Termanyl 120L) from Novo Laboratories, Inc. Wilton, CT 06897, and the heat-stable α-amylases (No. A-0164) from Sigma Chemical Co. St. Louis, MO 63178 can be used in the UED method (Jeraci et al., 5).

The minimum and maximum times needed for the Savinase 8.0L in the protease-dialysis step (2-28h) was studied using corn meal and casein. No significant differences in TDF and reagent blanks were detected between the 2 and 8 h times; however, differences were seen in ash and CP (Jeraci et al., 5). The UED method is a significant improvement compared to the method described by Prosky et al. (7), thus minimizing the error in treating all residue N as CP (Nx 6.25). Most of the data reported here are based on 5 to 16 h protease treatment.

Soluble and Insoluble Dietary Fiber

The soluble and insoluble dietary fiber values were determined for three oat-based food products using the UED method and the AOAC-1988 method (Prosky et al., 7) and compared with the NDF values (Table 4). The insoluble dietary fiber values determined by the AOAC-1988 method were larger than values obtained using the UED method or the NDF method (Jeraci and Van Soest 6). More soluble dietary fiber was determined with the UED method compared to the AOAC-1988 method for the oat-based food products. The ß-glucan content of these food products (Carr et al., 3) was larger than the soluble fiber content determined using the AOAC-1988 method.

The soluble and insoluble dietary fiber (Table 5) for several other foodstuffs were determined and compared with the NDF values (Robertson and Van Soest, 8). In other studies by Jeraci et al., (unpublished) and B. Li (personal commucation) 80 % ethanol may not be sufficient to precipitate all of the soluble dietary fiber from some foodstuffs. Both of these research groups are currently evaluating the recovery of soluble dietary fiber.

Table 1. Composition of Various Food Samples using the Urea Enzymatic Dialysis Method (UED) and the AOAC Total Dietary Fiber method[a,b,c].

	% Total Dietary Fiber	
Sample	UED	Prosky
Bran cereal	37.1	37.8
Corn cereal	3.1	3.6
Asparagus spears	23.5	23.3
Green beans	31.8	34.7
Beet root	19.6	25.3
Corn	9.4	7.9
Kidney beans	22.7	23.9
Lima beans	19.5	18.9
Onions	17.9	16.4
Peas	29.7	26.5
Pork & beans	19.8	22.2
Vegetable soup	11.9	12.3
White bread	4.4	5.5
Egg noodles	3.9	3.4
White flour	4.5	3.7
Macaroni	5.9	4.4
Rice	2.7	1.5
Spaghetti	5.1	4.4
Apple, cored & peeled	14.1	14.7
Blueberries	22.0	22.2
Cantelope	7.6	9.2
Orange, FL, peeled, seedless	17.2	18.0
Pear, Bartlett, cored	26.9	26.3
Plum, Friar	12.3	15.6
Tangerines, peeled, seedless	15.5	15.2
Watermelon	5.0	4.3

[a] Corrected for water.
[b] The AOAC Total Dietary Fiber Method (Prosky et. al., 7).
[c] Samples were prepared by J. Marlett.

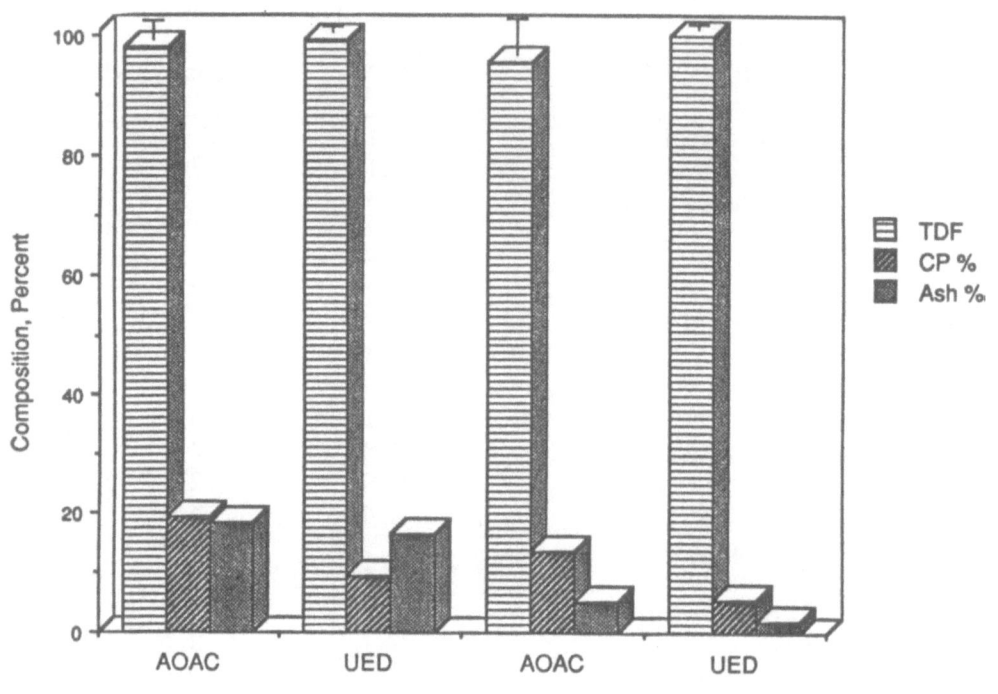

Figure 2.   Recovery of Total Dietary Fiber (TDF) from Carboxymethylcellulose
(Medium Viscosity, Sigma Co.) and Locust Bean Gum (Nutritional
Biochemical Co., Cleveland, Ohio) Expressed as Percent of Protein-
free and Ash-free Organic Matter in Original Sample by the AOAC-
1985 and the UED Methods.   Crude Protein (CP) and Ash in the
Dietary Fiber Residues are EXpressed as Percent of Original Sample.
TDF has bee Corrected for Crude Protein and Ash.

Table 2. Recovery of Water Soluble ß-Glucan and 80% Ethanol Precipitate Expressed as % of the Original Oats[a][b].

Treatment	ß-Glucan[c]		Ethanol Precipitate[d]	
		SD		SD
8 M Urea	3.8	0.1	29.3	0.9
+ Amylase	3.8	0.1	18.9	0.8
+ Protease	3.7	0.2	13.4	0.7

[a] Adapted from Jeraci et al. (5)

[b] Values corrected for moisture (ß-Glucan: vacuum oven 55°C; 80 % Ethanol precipitate: 105°C air oven), n=3.

[c] Contents of dialysis tubes were centrifuged and the supernatant was analysed for water soluble ß-glucan.

[d] 80 % Ethanol precipitate not corrected for crude protein, ash and the reagent blank residue.

Table 3. Composition (Starch and Pectin) of Various Foodstuffs and Resistant Starch in the Total Dietary Fiber [a,b,c,d,e].

Sample	Original Sample		% Starch in TDF	
	%Starch	%Pectin	AOAC	UED
Kale	8.1	4.2	3.5	0.2
Green Peas	15.8	1.2	0.4	0.4
Broccoli	2.2	3.4	0.0	0.1
Baby Lima Beans	44.9	1.1	0.9	0.2

[a] Adapted from Jeraci et al. (5).

[b] Corrected for water.

[c] Starch analysis is by a modified AACC assay.

[d] Pectin analysis is by a modification of the Blumencrantz and Asboe-Hansen method by Bucher (2).

[e] Total dietary fiber of the foodstuffs assayed by AOAC-1985, AOAC-1988 or UED methods. The resistant starch is expressed as % of the original sample.

Table 4. Composition (Insoluble Fiber, and Soluble Fiber) for Three Ready to Eat Cereals Using the Urea Enzymatic Dialysis (UED) Method, the AOAC Total Dietary Fiber Method, and the Neutral-Detergent fiber Method[a,b,c].

| | % Dietary Fiber ± SD | | | | |
| | UED | | AOAC | | NDF |
Method:	Insoluble	Soluble	Insoluble	Soluble	Insoluble
Cereals					
Oat bran hot cereal	9.6 ± 0.0	10.2 ± 0.7	23.8 ± 0.2	1.6 ± 0.9	14.0 ± 0.3
Oat bran	16.8 ± 1.1	2.9 ± 0.7	18.2 ± 1.5	4.2 ± 2.1	14.9 ± 0.2
Rolled oats	3.5 ± 0.4	6.8 ± 1.0	12.8 ± 2.1	1.0 ± 1.0	4.9 ± 0.8

[a]Corrected for water.
[b]The AOAC Total Dietary Fiber Method (Prosky et. al., 7).
[c]The Neutral-Detergent fiber described by Jeraci and Van Soest (6) modified to calculate insoluble fiber corrected for ash and crude protein.

318

Table 5. Composition of Various Food Samples using the Urea Enzymatic Dialysis Method (UED) and the Neutral Detergent Fiber Method (NDF) a,b,c,d.

	% Dietary Fiber		
Method:	UED[e]		NDF[f]
Sample	Insoluble	Soluble	Insoluble
Bran cereal	32.9	4.2	35.4
Asparagus spears	18.0	2.8	17.5
Green beans	23.3	3.4	22.7
Kidney beans	15.6	5.7	17.2
Macaroni	2.4	1.9	4.4

[a] Adapted from Jeraci et al. (5).

[b] Corrected for water.

[c] Neutral detergent fiber method described by Robertson and Van Soest (8).

[d] Samples were prepared by J. Marlett.

[e] In the UED method, dietary fiber values were corrected for ash and crude protein.

[f] In the NDF method, insoluble dietary fiber values were not corrected for crude protein.

SUMMARY

It is apparent from the 90 foodstuffs investigated thus far that the UED method affords several advantages for TDF. Thus, the UED method (1) removes essentially all starch, (2) generates smaller correction factors for CP and ash, (3) gives better quantitative recovery of purified and semi-purified dietary fiber products, (4) can handle more samples for a similar investment of labor and equipment, (5) does not expose the foodstuff to temperatures above 50C and (6) requires only one reagent solution and two enzymes.

REFERENCES

1. Blumencrantz, N. and G. Asboe-Hansen. 1973. Analyt. Biochem. 54:484-489

2. Bucher, A.C. 1984. MS Thesis, Cornell University, Ithaca, NY.

3. Carr, J., S. Glatter, J.L. Jeraci and B.A. Lewis. in press. Cereal Chemistry.

4. Goering, H.K. and P.J. Van Soest. 1970. Forage Fiber Analysis, Agriculture Handbook No. 379. U.S. Department of Agriculture, Washington, D.C.

5. Jeraci, J.L, B.A. Lewis, J.B. Robertson and P.J. Van Soest. 1989.
    J.A.O.A.C. 72:677-680.

6. Jeraci, J.L. and P.J. Van Soest. 1989. 197th ACS National Meeting.
    Dallas, TX, April 9-14, 1989.

7. Prosky, L., A.S.P. Nils-Georg, T.F. Schweizer, I. Furda and J.W. Devries.
    1988. Determination of insoluble, soluble, and total dietary fiber in
    foods and food products: interlaboratory study. J. Assoc. Off. Anal.
    Chem. 71:1017.

8. Robertson, J.B. and P.J. Van Soest. 1981. In: The Analysis of Dietary
    Fiber in Foods, W.P.T. James & O. Theander (Eds.), Marcel Dekker, New
    York, NY. pp123-158.

INDEX